The use of fractal concepts in understanding various growth phenomena, such as molecular beam epitaxy (MBE) or fluid flow in porous media, is increasingly important these days. This book introduces the basic models and concepts that are necessary to understand in a pedagogical way the various growth processes leading to rough interfaces. The text will be accessible to readers not familiar with the field.

Nature provides a large number of rough surfaces and interfaces. Similarly, rough surfaces are regularly observed in the laboratory during various technologically important growth technologies, such as MBE. In an attempt to understand the origin of the roughening phenomena, several computer models and theoretical approaches have recently been developed. The principal goal of this book is to describe the basic models and theories as well the principles one uses to develop a model for a particular growth process. Furthermore, having described a particular growth model, the authors show how one can address and answer questions such as whether the surface will be rough, how rough it will be, and how to characterize this roughness. Having introduced the basic methods and tools needed to study a growth model, the authors discuss in detail two classes of phenomena: fluid flow in a porous medium and molecular beam epitaxy. In both cases, in addition to the models and analytical approaches, the authors describe the relevant experimental results as well.

This text contains extensive homework problems at the ends of the chapters, and will be invaluable for advanced undergraduates, graduate students and researchers in physics, materials science, chemistry and engineering, and especially those interested in condensed matter physics and surface growth.

Fractal Concepts in Surface Growth

Cover design: Joyce Hempsted.

Sergey V. Buldyrev is responsible for both fractal images on the cover, which concern surface growth in a disordered three-dimensional medium. The computer simulations were generated using the directed percolation depinning model of interface growth, described in Chapter 10, which is characterized by a depinning threshold that occurs only in the presence of a directed surface of pinning sites.

The coloring of the inset is due to Krisztina A. Barabási.

Albert-László Barabási obtained his Ph.D. at Boston University in 1994. Currently he is with the Physical Sciences Department of IBM T. J. Watson Research Center.

H. Eugene Stanley is University Professor and Professor of Physics at Boston University. He was chosen 1992 Massachusetts Professor of the Year, and was awarded a Guggenheim Fellowship in 1979–1981. Professor Stanley is a Fellow of the American Physical Society and the American Association for the Advancement of Science, and was elected chair of the 1986 IUPAP International Conference on Statistical Physics. He is the author of *Introduction to Phase Transitions and Critical Phenomena*, which won the Choice Award for Outstanding Academic Book of 1971, and is co-author with D. Stauffer of *From Newton to Mandelbrot: A Primer in Theoretical Physics.*

FRACTAL CONCEPTS IN SURFACE GROWTH

Albert-László Barabási

IBM Thomas J. Watson Research Center, Yorktown Heights, New York

H. Eugene Stanley

Physics Department, Boston University Boston, Massachusetts

Published by the Press Syndicate of the University of Cambridge
The Pitt Building, Trumpington Street, Cambridge CB2 1RP
40 West 20th Street, New York, NY 10011-4211, USA
10 Stamford Road, Oakleigh, Melbourne 3166, Australia

First published 1995

Printed in Great Britain at the University Press, Cambridge

A catalogue record of this book is available from the British Library

Library of Congress cataloguing in publication data

Barabási, Albert-László
Fractal concepts in surface growth / Albert-László Barabási,
H. Eugene Stanley.
p. cm.
Includes bibliographical references and index.
ISBN 0 521 48308 5 – ISBN 0 521 48318 2 (pbk.)
1. Surfaces (Physics). 2. Interfaces (Physical sciences)
3. Crystal growth. 4. Molecular beam epitaxy. 5. Fractals.
I. Stanley, H. Eugene (Harry Eugene), 1941– . II. Title.
QC173.4.S94B37 1995
530.4′17–dc20 93-37111 CIP

ISBN 0 521 48308 5 hardback
ISBN 0 521 48318 2 paperback

TAG

To Krisztina and Idahlia

Contents

Preface

This book is intended to serve as an introduction to the multidisciplinary field of disorderly surface growth. It is a reasonably short book, and is not designed to review all of the recent work in this rapidly developing area. Rather, we have attempted to provide an introduction that is sufficiently thorough that much of the current literature can be profitably read. This literature spans many disciplines, ranging from applied mathematics, physics, chemistry and biology on the one hand to materials science and petroleum engineering on the other.

It is envisaged that this book may be of use in courses in many different departments, so no specific background on the part of the reader is assumed. Part 1 of the book is an introduction that should bring readers from a variety of disciplines to a common place of discourse. Thus the first chapter illustrates the range of natural examples of disorderly surface growth, and mentions without any use of mathematics a few of the key new ideas that serve to provide some insights. The second chapter introduces the scaling approach to describing surface growth, by focusing on a single tractable model system – ballistic deposition. No prior exposure to scaling concepts is assumed. The third and last chapter of Part 1 introduces the key fractal concepts of self-similarity and self-affinity.

Part 2 comprises five chapters devoted to the general topic of nonequilibrium roughening. It begins with a brief chapter on an exactly solvable and well-understood limiting case in which particles are deposited on a substrate in an uncorrelated fashion. This limit of random deposition is utilized to introduce the concept of a stochastic growth equation – i.e., an equation that describes surface growth in the presence of noise. Subsequent chapters develop the stochastic growth equation approach to treat systems of increasing complexity. Topics not ordinarily presented, such as the renormalization group approach, are treated with both care and rigor.

Part 3 concerns a different category of surface growth, one in which the noise arises from the disorder in the medium in which the surface is growing. This form of growth arises naturally in a range of systems, such as flow in porous media, but also including more exotic systems such as rupture lines, bacterial colony growth, and the propagation of fire fronts. Quite recent work is emphasized, such as the recognition that one can regard the pinning threshold for surface growth as a critical point, and that dynamic motion near this pinning threshold occurs as a succession of 'avalanches', each successive avalanche stopping when it is pinned by a connected structure. These structures have the same statistics as those of the directed percolation problem in statistical mechanics.

By far the most industrially-relevant process treated in this book is that of molecular beam epitaxy (MBE), and the nine chapters comprising Part 4 are devoted to this important topic. After a careful introduction to the basic phenomena, both the linear and the nonlinear theories of MBE are developed in detail and their predictions compared with recent experiments. Submonolayer deposition, resulting in the formation of 'nanostructures', is described in Chapter 17, and the relevant experiments are compared in detail with the microscopic model encompassing deposition, diffusion and aggregation (DDA). Other topics such as the roughening transition and nonlocal growth models used to describe sputter deposition and roughening by ion bombardment are also discussed.

Part 5 comprises three chapters devoted to the basic topic of noise, including considerations related to situations in which the noise is not uncorrelated random events but rather is correlated. The physics of extreme events is described in Chapter 23 and exemplified by the Zhang model.

Part 6 concerns advanced topics suitable for the more advanced student, while Part 7 is a 'finale' summarizing systematically the zoo of different universality classes into which disorderly surface growth phenomena may be partitioned and ending in a brief 'outlook' on what the future may hold.

Three appendices – one on numerical methods, one on the dynamic renormalization group, and one on Hamiltonian descriptions – are included in order to make the book usable by the student who may desire to work in this field.

No introductory book can be complete, and the interested student will surely benefit from additional reading from among review articles and original sources. To aid in this endeavor, each chapter culminates not only in a set of exercises but also in a list of suggested further reading. An extensive and up-to-date bibliography – with full titles – is designed to facilitate the student's transition from studying textbooks to reading research material.

Since dynamic surface growth is a remarkably multidisciplinary field, there are almost as many notations as there are workers in the field. To help in this respect, we provide a list of symbols and acronyms, together with their definitions, in the *Notation Guide* that follows the list of contents.

The material treated herein is more than adequate for a one-semester 'stand-alone' course in surface science or engineering; with this use in mind, the number of chapters in the text coincides with the typical number of 90-minute lectures in a term. Alternatively, the instructor may wish to consider using selected topics from this book to supplement and bring up-to-date standard introductory courses in physics, chemistry, biology, applied mathematics, engineering or materials science. A suitable subset of the book that provides a reasonably coherent and relatively nonmathematical introduction to modern ideas in surface science is formed by Chapters 1-6 and 9-16. The student interested primarily in MBE should learn something from Chapters 12-17 and 19, which can be used as a supplement with occasional references to Chapters 2-6. Throughout, we have endeavored to keep the chapters reasonably independent of one another so that the reader who wishes may skip about in the text. To encourage such an adventurous spirit we have indicated by a horizontal arrow those equations that are referred to frequently.

A large number of individuals have been remarkably generous is assisting with the preparation of the manuscript. The earliest drafts were critiqued in detail by Boston University colleagues and collaborators L. A. N. Amaral, S. V. Buldyrev, R. Cuerno, M. F. Gyure, B. Jovanović, P. Jensen, K. B. Lauritsen, P. Maass, H. Makse, R. Mantegna, R. Sadr-Lahijany, and S. Zapperi. The first coherent text that could be exposed to anyone outside our research group elicited many helpful remarks from J. G. Amar, M. Cieplak, S. Havlin, R. Jullien, M. Kardar, B. D. Kay, M. Kotrla, J. Krim, P. Meakin, L. M. Sander, M. Siegert, H. Spohn, D. Wolf, P.-Z. Wong, and Y. Tu and especially S. Das Sarma, J. Kertész, M. O. Robbins, and T. Vicsek.

The final submitted version was scrutinized in considerable detail by most of the above-named colleagues, as well as by F. Family, G. Grinstein, H. G. E. Hentschel, T. Nattermann, R. Pandey, P. Šmilauer, J. D. Weeks, R. S. Williams, A. Zangwill and many talented young people – including A. Czirók, S. V. Ghaisas, R. Kotlyar, K. E. Khor, C. J. Lanczycki, M. Schroeder, E. Somfai, and S. Tomassone and, most especially, H. Leschhorn, who worked through the mathematics of the entire book with considerable care.

S. V. Buldyrev is responsible for the cover images of surface growth in a disordered three-dimensional medium, which he obtained using the directed percolation depinning model of interface growth. Countless colleagues generously provided us with selections of their most valuable (and, in some cases, irreplaceable) artwork, and we wish to thank all of them – especially M. G. Lagally.

J. D. Morrow demonstrated his considerable TEX skills in carrying out the countless improvements suggested by the individuals acknowledged above. D. Carr's innumerable trips to area libraries made possible the inclusion of titles and end page numbers for all 508 references in the bibliography.

Last but certainly not least, we wish to express our genuine appreciation to all the staff of Cambridge University Press – S. Capelin (*Editorial Manager in the Physical Sciences*), A. Tomlinson (*Production Manager*), A. J. Woollatt (TEX *"Genius"*), P.-J. Leone (*Marketing Manager, Science and Mathematics*) and most especially the entire *Technical Applications Group* – for their remarkable efficiency and good cheer throughout this entire project. Authors could imagine no finer group of co-workers than is exemplified by this team.

As we study the final page proof, we must resist the strong urge to re-write the treatment of several topics that we now realize can be explained more clearly and precisely. We do hope that readers who notice these and other imperfections will communicate their thoughts to us.

Yorktown Heights, New York — *Albert-László Barabási*

Boston, Massachusetts — *H. Eugene Stanley*

January 1995

Notation guide

Symbol	*Definition*	*Page*
AFM	atomic force microscope	168
AKPZ	anisotropic KPZ equation	272
a_0	lattice constant	193
BD	ballistic deposition model	19
C_a	capillary number	117
$C(\ell)$	height-height correlation function	303
$c_q(\ell)$	q-th order correlation function. Note: $c_2(\ell) \equiv [C(\ell)]^2$	263
d	*surface* dimension (in $(d+1)$-dimensional space)	36
d_{E}	embedding dimension	30
d_f	fractal dimension	31
d_T	topological dimension	31
D_n	diffusion constant for an n-particle cluster	184
D_s	surface diffusion constant	151
DDA	deposition-diffusion-aggregation model	176
DLA	diffusion-limited aggregation model	209
DP	directed polymer	277
DPD	directed percolation depinning model	100
E_0	activation energy for diffusion	134
E_D	activation energy for desorption	131
E_N	bonding energy	134
EW	Edwards-Wilkinson	50
F	particle flux; also: driving force	41
F_c	critical driving force	92
f	reduced force $(F - F_c)/F_c$	92
$f(u)$	scaling function of scaling variable u	24
g	coupling constant	71
$G_0(\mathbf{k}, \omega)$	propagator	317
$\mathscr{H}$	Hamiltonian	66
HRLEED	high-resolution low-energy electron diffraction	172
$\bar{h}(t)$	average height	20
$\mathbf{j}(\mathbf{x}, t)$	surface current	140
KPZ	Kardar-Parisi-Zhang	58
L	system size	20

Symbol	*Definition*	*Page*
L_1, L_2	crossover length scales	143, 241
ℓ_d, ℓ_2	characteristic length scales in submonolayer epitaxy	179, 182
m	local slope, or average slope	234
MBE	molecular beam epitaxy	128
N	average island size	178
p_c	critical probability in directed percolation	106
P_{cap}	capillary pressure	117
RD	random deposition model	38
RFIM	random field Ising model	109
RG	renormalization group	65
RHEED	reflection high energy electron diffraction	129
RSOS	restricted solid-on-solid model	86
SEM	scanning electron microscope	168
SOC	self-organized criticality	292
SOD	self-organized depinning model	108
SOS	solid-on-solid model	81
STM	scanning tunneling microscope	129
$t_\times$	crossover time	22
T_R	roughening temperature	192
TEM	transmission electron microscope	168
v	average growth velocity of surface	50
$w_L(\ell, t)$	'local' width on length scale ℓ	302
$w(L, t)$	total interface width, or 'roughness'	22
$w_{\text{sat}}(L)$	interface width after saturation	302
w_i	intrinsic width	309
z	dynamic exponent: $t_\times \sim L^z$	23
α	roughness exponent: $w_{\text{sat}} \sim L^\alpha$	23
α_q	multiaffine exponent	263
β	growth exponent: $w \sim t^\beta$	22
$\eta(\mathbf{x}, t)$	time-dependent stochastic (thermal) noise	41
$\eta(\mathbf{x}, h)$	time-independent 'quenched' noise	93
$\eta_d(\mathbf{x}, t)$	diffusive or conserved noise	240
θ	velocity exponent	92
Θ	coverage	176
λ	coefficient of nonlinear term in KPZ equation	58
μ	surface chemical potential	140
μ_w	dynamic viscosity of the wetting fluid	117
ν	'surface tension' parameter	50
ν	correlation length exponent	95
$\nu_\parallel, \nu_\perp$	parallel and perpendicular correlation length exponents	105
ξ	correlation length	26
$\xi_\parallel$	parallel correlation length	26
$\xi_\perp$	perpendicular correlation length	27
ρ	total island density	176
ρ_1	monomer density	176
ρ_n	density of islands that contain n particles	176
ϕ	exponent for long-range temporal noise correlations	246
ψ	exponent for long-range spatial noise correlations	246
ψ_d, ψ_2	scaling exponents for ℓ_d, ℓ_2 respectively	179, 177
ω_D	Debye frequency	133
$\equiv$	*defined* to be equal	20
$\sim$	*asymptotically* equal (in scaling sense)	22
$\approx$	*approximately* equal (in numerical value)	81

PART I Introduction

1 Interfaces in nature

Most of our life takes place on the surface of something. Sitting on a rock means contact with its surface. We all walk on the surface of the Earth and most of us don't care that the center of the Earth is molten. Even when we care about the interior, we cannot reach it without first crossing a surface. For a biological cell, the surface membrane acts not only as a highly selective barrier, but many important processes take place directly on the surface itself.

We become accustomed to the shapes of the interfaces we encounter, so it can be surprising that their morphologies can appear to be quite different depending on the scale with which we observe them. For example, an astronaut in space sees Earth as a smooth ball. However Earth appears to be anything but smooth when climbing a mountain, as we encounter a seemingly endless hierarchy of ups and downs along our way.

We can already draw one conclusion: surfaces can be smooth, such as the Himalayas viewed from space, but the same surface can also be rough, such as the same mountains viewed from earth. In general the *morphology depends on the length scale of observation*!

How can we describe the morphology of something that is smooth to the eye, but rough under a microscope? This is one question we shall try to answer in this book. To this end, we will develop methods to characterize quantitatively the morphology of an arbitrary interface. In fact, we shall see that concepts like roughness are replaced by exponents that refer not to the roughness itself, but to the fashion in which the roughness *changes* when the observation scale itself changes.

Fractal objects in nature are the same on different observation scales. However, the above examples look rather different when the scale is changed. Does this fact imply that the title of this book is a misnomer? Yes and no. In fact, many interfaces and surfaces are examples of *self-affine* objects, which are 'intermediate' between fractal objects and non-fractal objects in the following sense. When we make a scale change that is the *same in all directions*, self-affine objects change morphology. On the other hand, when we make a scale change that is *different for each direction*, then interfaces do not change morphology. Rather, they behave like fractal objects in that they appear the same before and after the transformation (see Fig. 1.1).

This book will explore in some detail the nature of such self-affine objects. We shall see that this feature is analogous to a 'symmetry principle.' Symmetry principles codified in group theory enable one

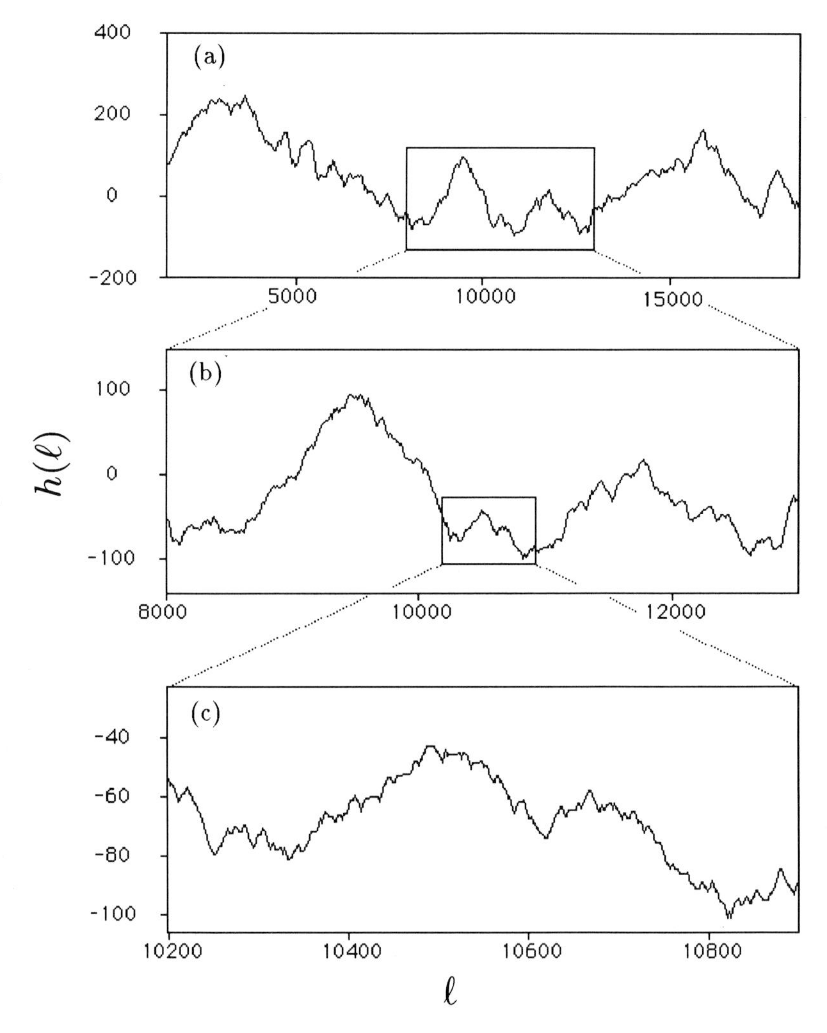

Figure 1.1 Rescaling a self-affine function, in this case the 'DNA walk' introduced in §1.3.2 (cf. Fig. 1.13). Only if the two unequal magnification factors, M_ℓ and M_h, by which the ℓ and h directions are re-scaled, are selected correctly will the enlarged portion have the same statistical properties as the original. (After [416].)

to classify and eventually understand some properties of a crystalline system. Similarly, the symmetry obeyed by self-affine objects will enable us to classify and perhaps better understand some properties of rough interfaces in nature.

We are interested not only in the *morphology* of various 'pre-formed' interfaces, but also in the dynamics of how the morphology develops in time. Some surfaces are formed as a result of a deposition process. Others shrink due to erosion or etching. Some interfaces propagate through inhomogeneous media. An interesting set of questions concerns the *formation, growth, and dynamics* of such interfaces.

In this first chapter, we offer the interested reader an *apéritif* – by exposing a variety of the themes of this book without the baggage imposed by requiring equations or discussing experimental details. The reader whose appetite is already 'up' for the main course is invited to proceed directly to Chapter 2.

1.1 Interface motion in disordered media

1.1.1 Fluid flow in porous media

A familiar scene: at breakfast, your spilled coffee suddenly conquers a segment of the tablecloth. Probably this is not the right moment to contemplate the microscopic forces that balance just when the coffee stops spreading. Nor is it the appropriate moment to start thinking about the shape of the patch, or the roughness of its surface. So wait until the stress subsides, and then examine the large brown patch.

By clipping a paper towel on a stand and immersing one end into a fluid, we can repeat the coffee accident on a laboratory benchtop. Paper is an inhomogeneous material, a prototype of the porous inhomogeneous rock that holds oil! One difference between fluid flow in the paper towel and in oil-bearing rock arises from the length scales at which these phenomena take place. This difference is an advantage: we can use a 20 cm paper towel system (Fig. 1.2) to help develop our understanding of the 20 km oilfield problem. For example, we can characterize the wet–dry interface using scaling laws, whose form is predicted by simple models that capture the essential mechanisms contributing to the morphology. This 'benchtop exercise' is an example of some of the current experiments being carried out on idealized systems which are yielding new insights into practical interface problems.

1.1.2 Propagation of flame fronts

Take a sheet of paper and ignite it at one end. Try to keep it horizontal, so that the entire paper does not take flame at once – if possible, use paper that burns without flame. After burning a part of it, inspect the interface between the burned and unburned parts (see Fig. 1.3). Is it rough, similar to the edge of the coffee droplet on the tablecloth? Is this similarity a coincidence, or there is something in common between these processes?

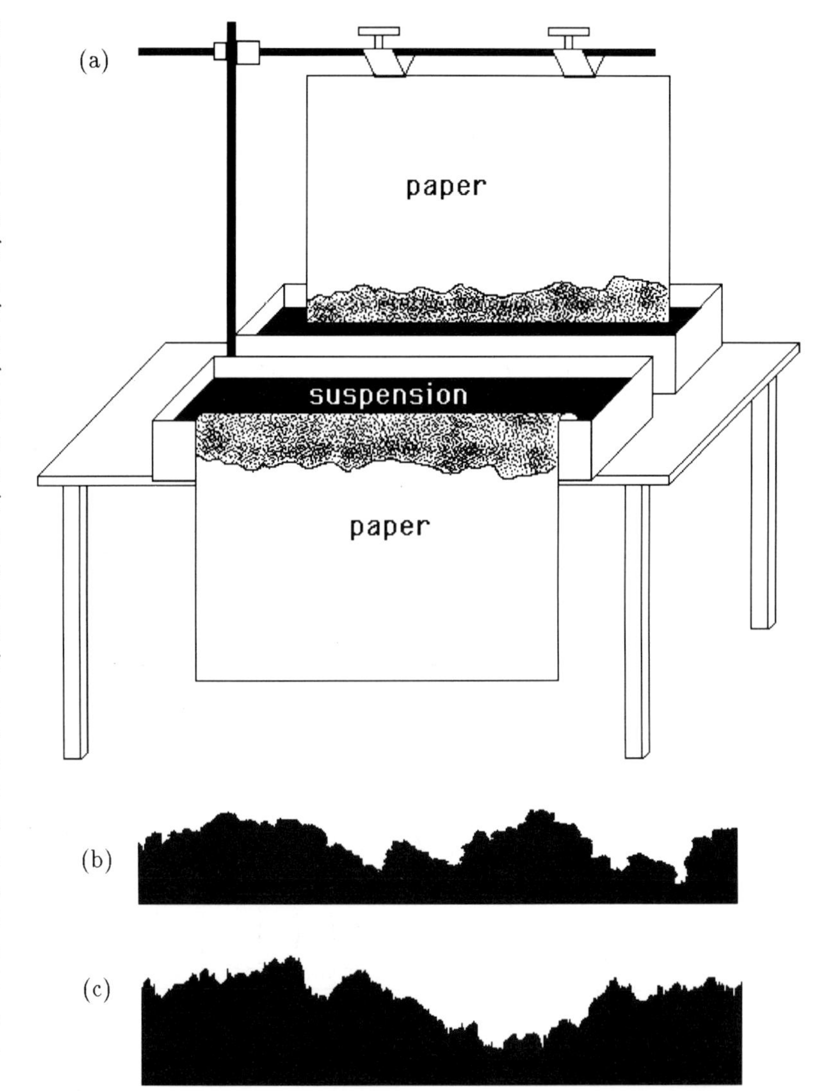

Figure 1.2 (a) Schematic illustration of an experimental setup probing interface motion in random media. Parameters such as type of paper, temperature, humidity, direction of growth and concentration of coffee can be varied systematically. These changes affect the area, the speed of wetting, and the global width of the rough surface, but they do not affect the scaling properties of the surface. (b) Digitized *experimental* interface; the horizontal size of the paper was 20 cm. (c) Typical result of a discrete *model* mimicking interface motion in disordered media (see §10.2). (After [31]).

1.1.3 Flux lines in a superconductor

Suppose we place a superconductor in an external magnetic field. The field penetrates the material by generating flux lines or vortices, each carrying an elementary flux (see Fig. 1.4). If there are no impurities in the superconductor and the temperature is low, the flux lines form an array of straight lines. If there are impurities in the superconductor, the flux lines stretch to get close to the impurity sites. These individual flux lines are rough, resembling the surface of the coffee drop, or the fire front. However, there is one important difference between a flux line and a fire or fluid interface: a firefront is a topologically one-dimensional object moving in a two-dimensional plane, while a

Figure 1.3 An 8 cm segment of paper in which fire propagates from below. (After [504]).

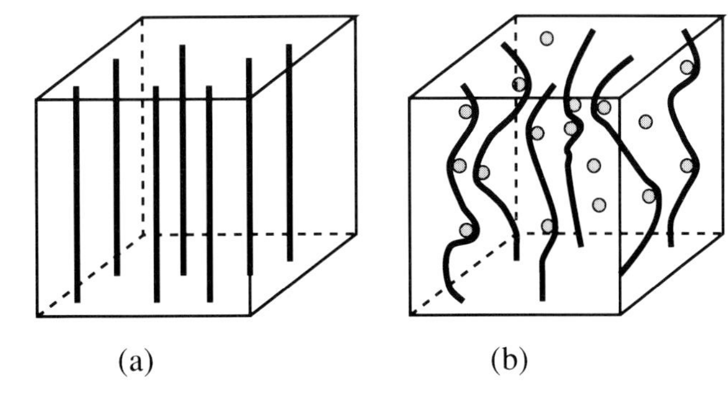

Figure 1.4 Vortex lines in a type II superconductor. (a) In a perfectly clean superconductor at low temperatures the flux lines form an ordered lattice. (b) When the temperature is increased, thermal fluctuations destroy the ordered lattice and the flux lines become wavy. If there are impurities (shaded circles) in the superconductor, the flux lines are pinned in random positions of the impurities.

flux line is a topologically one-dimensional object fluctuating in a *three-dimensional* space.

One purpose of this book is to discuss apparently different phenomena using a common framework. It is coming to be appreciated that all the systems discussed in this section display essentially the same physics: there is an elastic interface ('elastic' in the sense that it does not break, but tries to remain smooth) which propagates in a disordered material. The randomness of the substrate acts to pin the interface, thereby making the interface rough. In the fluid case, local inhomogeneities may block the fluid flow. Some parts of the paper do not burn as rapidly as others, so the flame is halted. Impurities attract the flux line and pin it down in random positions. We shall see that these systems are described by the same laws, and that they can be studied using a similar set of numerical and analytical methods.

1.2 Deposition processes

Winter. Look out of your car window – snow is falling. The larger snowflakes slide down the window slowly, and form the aggregate shown in Fig. 1.5(a). Notice that at length scales comparable to the size of the snowflake the aggregate is very rough. Why? Snowflakes are deposited randomly. Once they arrive at the aggregate, they stick. The *randomness* in the deposition process apparently leads to a *rough* surface. A similar deposition process is illustrated in Fig. 1.5(b). The resulting bulk is nearly homogeneous, but the surface is quite rough. These are but two of the many examples for which random deposition of some material occurs, and we witness the dynamic growth of a rough surface.

1.2.1 Atom deposition

A deposition process of greater technological importance than snowfall takes place during the growth of thin films by molecular beam epitaxy (MBE), a technology used to manufacture computer chips and other semiconductor devices, indispensable in today's technological world. The most common element used in computer chips is silicon. An example of a very clean Si surface is shown in Fig. 1.6, obtained using a scanning tunneling microscope (STM).

Now imagine that you start depositing new atoms on this Si surface. In contrast to the snowflakes that stick on the first contact point, the Si atom does not stick, but diffuses. When a Si atom reaches the edge of a step, it forms covalent bonds with neighboring atoms, and sticks with a high probability. Such bonds may be broken again, but with a low probability. If the incoming flux is large, there is a large number

(a)

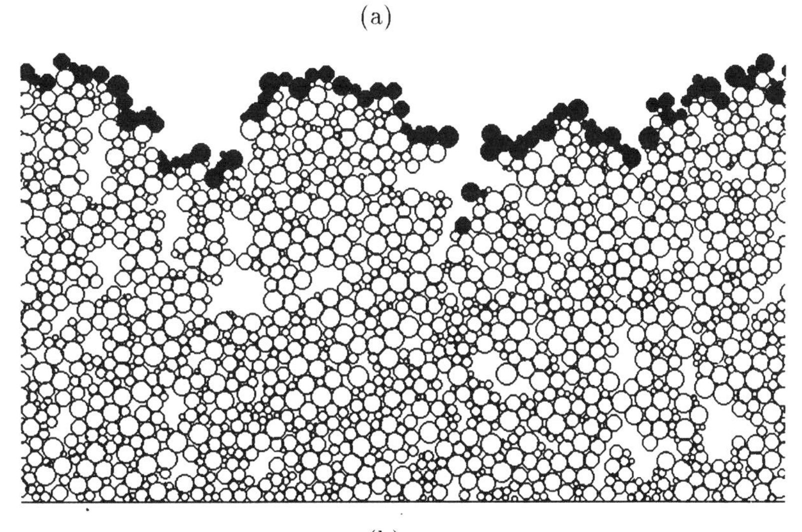

(b)

Figure 1.5 (a) Snow particles falling on a slanted glass window (After [270]); (b) The interface generated by a simple deposition model, in which spherical particles with uniformly distributed random diameters arrive on the surface and roll until they make contact with at least two other discs. (After [89]).

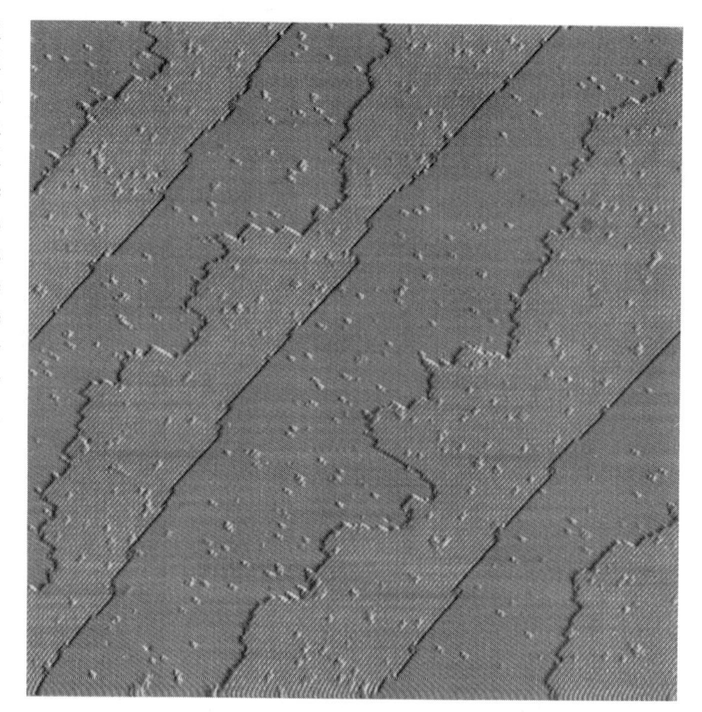

Figure 1.6 A Si surface, as examined by STM. One can distinguish both individual atoms and vacancies. The rugged lines correspond to steps on the surface, where the height of the interface increases by one atom. (Courtesy of M. Lagally).

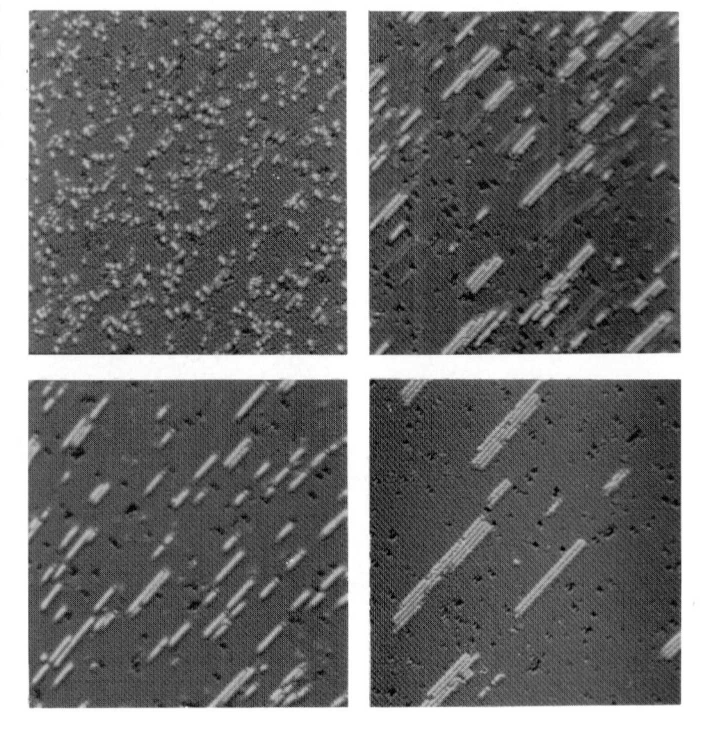

Figure 1.7 Formation of islands by atoms deposition on a Si surface. (Courtesy of M. Lagally).

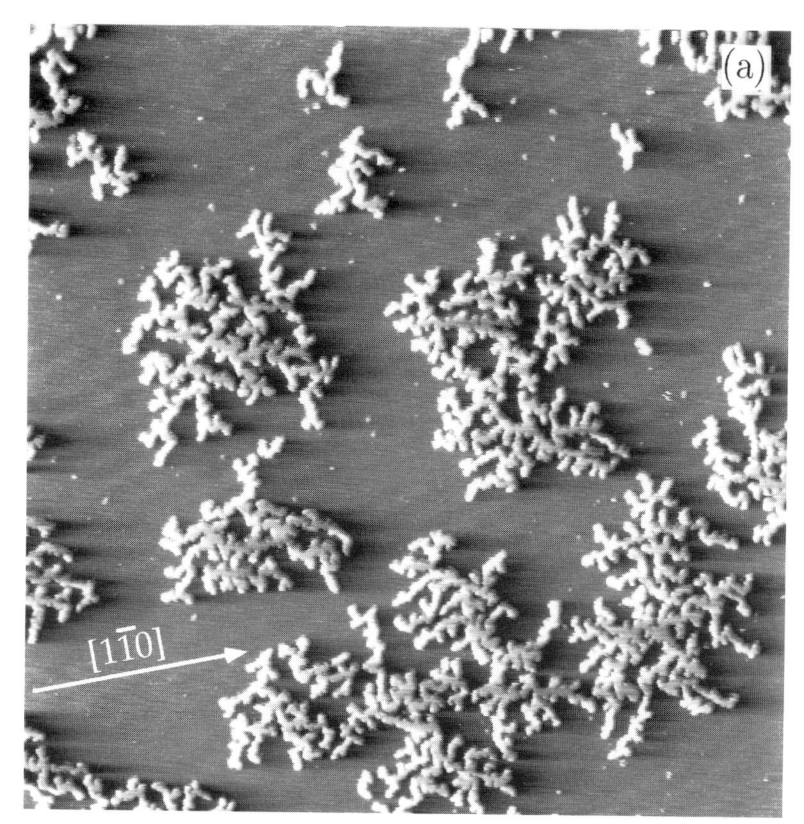

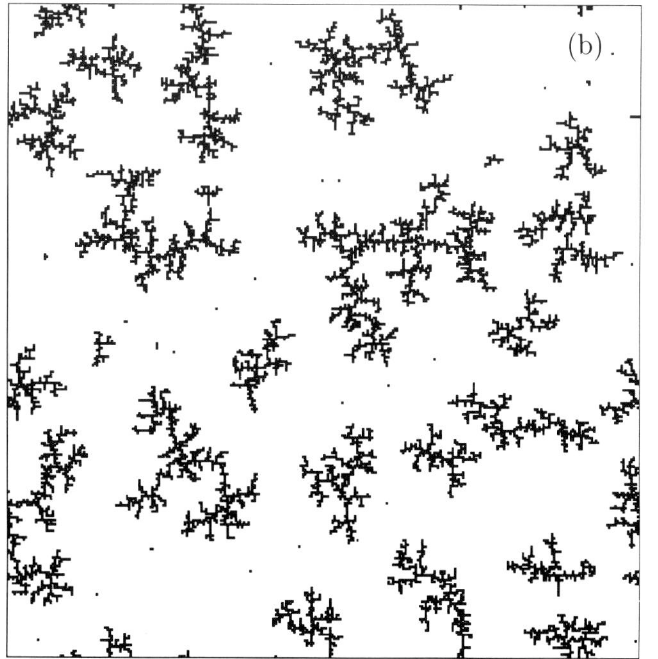

Figure 1.8 (a) Branched island morphologies obtained by Ag atom deposition on Pt at 110 K. (After [57]). (b) Island formation in a model incorporating the three basic processes taking place during MBE: deposition, diffusion, and aggregation. The deposited atoms aggregate due to diffusion, generating branched fractal islands, similar to those observed in (a). (After [197]).

of wandering atoms on the surface, which meet and 'glue' together, forming islands (see Fig. 1.7).

The islands do not always have a regular shape. As Fig. 1.8 illustrates, Ag deposition on Pt surfaces leads to branched structures, which are known to be self-similar, or fractal. What is the difference between the Si and the Pt that accounts for the difference in island structure? Is it related to the difference in the material properties, or only to the different temperatures and deposition rates?

1.2.2 Roughening in MBE

If we continue to deposit atoms, smaller islands will form on the top of the larger islands. The interface eventually becomes quite rough (see Fig. 1.9). Our discussion so far suggests that we may be happy whenever we see a rough surface. However, engineers usually wish to make a *smooth* film, since rough surfaces have poor contact properties and cannot be used in most applications. In order to avoid roughness, one first must understand the basic mechanism leading to roughness, and the processes affecting the morphology in general. This understanding then may be explored to grow films in regimes where roughening is reduced or absent.

Another common method used in film deposition experiments is sputtering. During *sputter erosion* the material is bombarded with an ion beam that hits the surface and kicks out atoms. This process is used to clean a surface, by eroding a few layers – or, by guiding

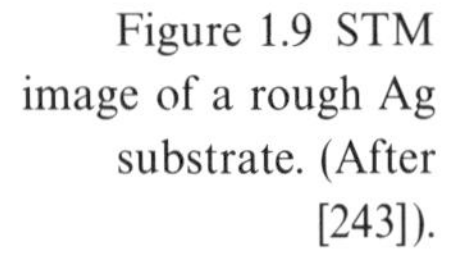

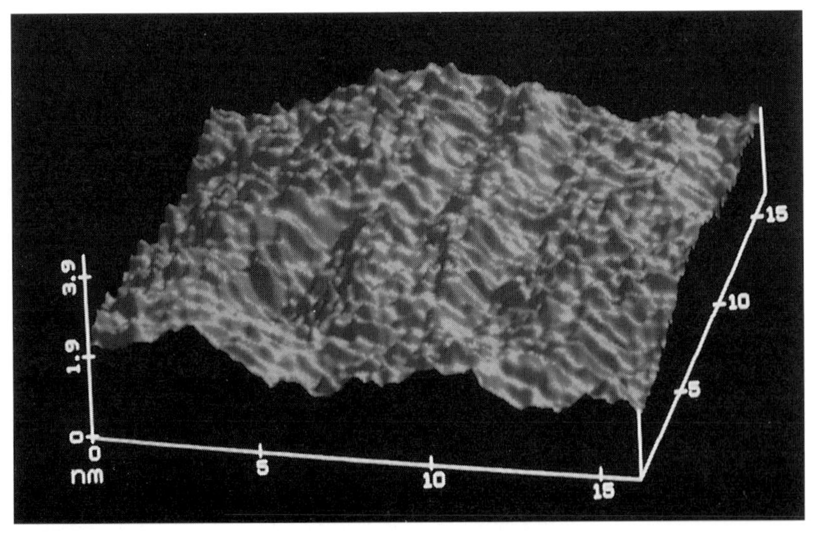

Figure 1.9 STM image of a rough Ag substrate. (After [243]).

the eroded atoms towards a sample, to grow another surface by a process called *sputter deposition*. Films grown by sputter deposition sometimes develop interesting 'cauliflower' structures, as is shown in Fig. 1.10. Why do these manmade films resemble so well the natural cauliflower?

The actual surface morphology in sputter erosion depends on the experimental conditions; some experiments lead to rough interfaces, others to periodic ripple structures (Fig. 1.11). What is the mechanism of the roughening process? Is erosion simply the inverse of deposition, or does it involve processes not present during deposition? How do we explain the formation of the ripple structures on the surface?

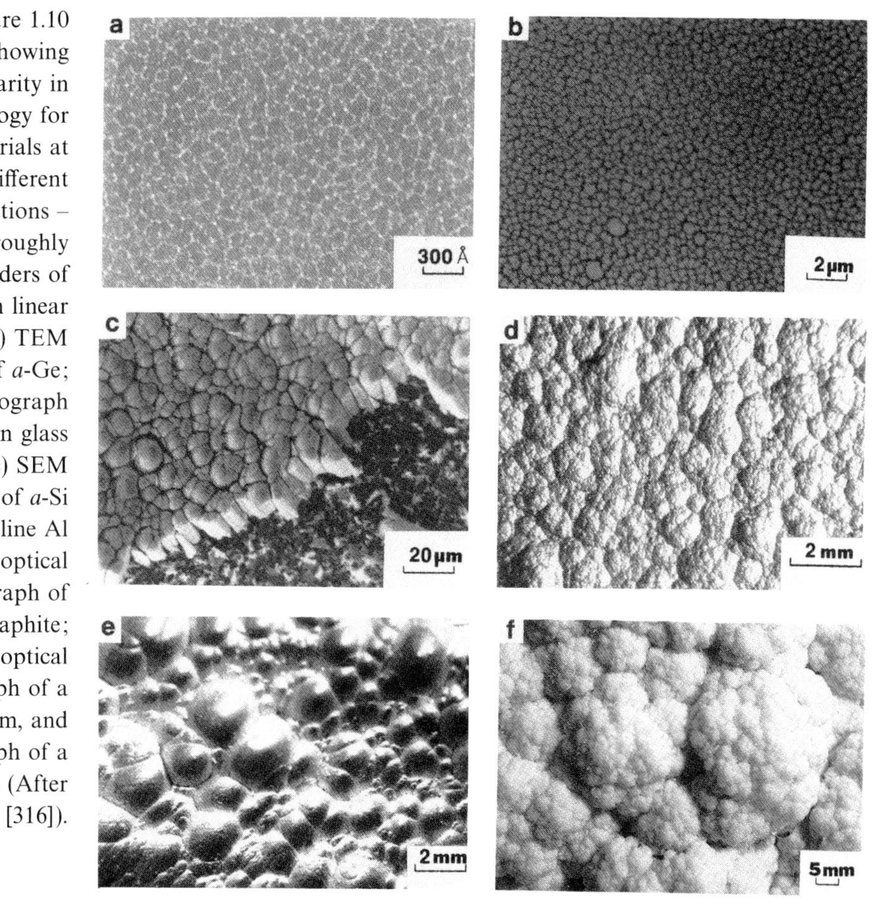

Figure 1.10 Micrographs showing the similarity in morphology for various materials at different magnifications – these span roughly six orders of magnitude in linear dimension. (a) TEM micrograph of *a*-Ge; (b) SEM micrograph of *a*-Si on glass substrate; (c) SEM micrograph of *a*-Si on polycrystalline Al substrate; (d) optical micrograph of pyrolytic graphite; (e) optical micrograph of a thick metal film, and (f) photograph of a cauliflower. (After [316]).

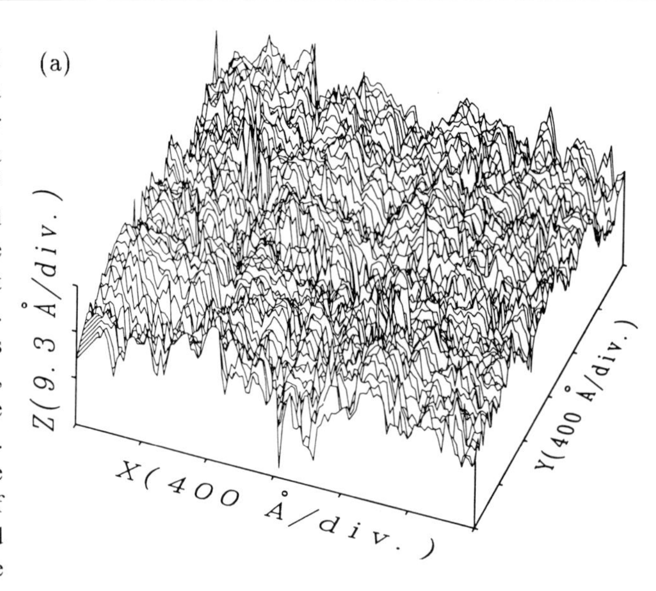

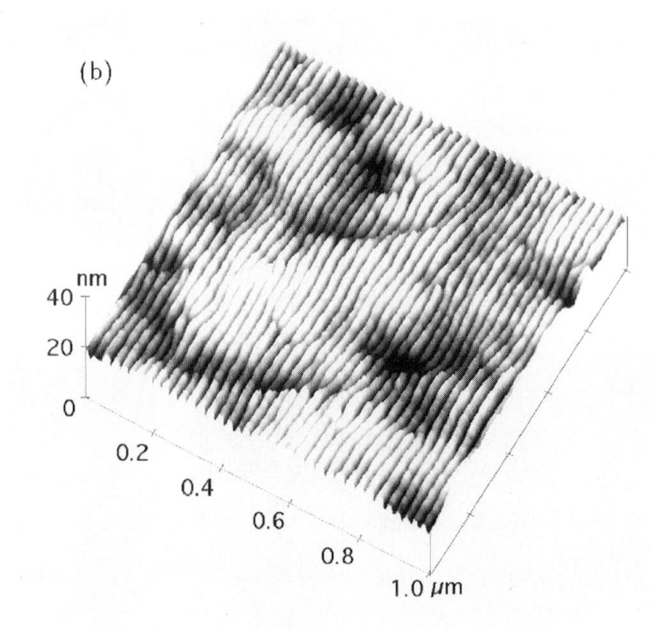

Figure 1.11 (a) STM topograph of a graphite surface after bombarding with a flux $F = 6.9 \times 10^{13}$ ions.cm^{-2}sec^{-1} and an ion fluence $Q = 10^{16}$ ions/cm^2 at room temperature. The sample size is 2400 Å ×2400 Å, and the vertical size 18.6 Å. (After [114]). (b) Atomic force microscope image of a Xe-bombarded SiO_2 film. Note the periodic ripple structure. (After [301]).

1.3 Biological systems

1.3.1 Bacterial growth

The previous examples were selected from the field of materials science. But there are also interesting interfaces in biology. Let us consider a much-studied problem, the growth of bacterial colonies. In a typical experimental setup, agar is prepared in a petri dish. In the middle of the agar a bacterium is inoculated, whereupon it multiplies. At microscopic length scales, the bacteria exhibit a random motion. Looking from a distance, however, a range of interesting morphologies can be observed (Fig. 1.12). The actual morphology depends on the nutrient concentration and on other experimentally-controllable parameters. Some colonies have a compact shape, with a rough surface, similar to the morphology we confronted with spilt coffee. Others are branched, reminiscent of the islands observed in atomic deposition. Are there some general principles common to bacterial growth, island formation, and fluid flow?

1.3.2 DNA

Many systems develop well-defined interfaces. However, there are processes that do not have a surface at all, but with a natural choice of variables can be mapped to a mathematical function or 'landscape' whose roughness is amenable to analysis using the same methods used for real interfaces.

An example that at first sight has nothing to do with interfaces is one-dimensional Brownian motion. Consider a drunk in an elevator of a skyscraper. Let us imagine that the elevator has only two buttons, up and down: the up button takes the elevator one level up, the down button one level down. The drunk punches the buttons randomly. If we plot the position of the elevator as a function of time, we obtain a jagged landscape, called the trail of the random walk, which may be described by many of the same methods used to quantify interface morphology.

Genomic DNA sequences code for protein structure. The human genome contains information for approximately 100 000 different proteins, which define all inheritable features of an individual – it is likely the most sophisticated information database, created entirely through the dynamic process of evolution.

The building blocks for coding this information, called *base pairs*, form two classes, purines and pyrimidines. In order to study the correlations of a DNA sequence, one can introduce a graphical rep-

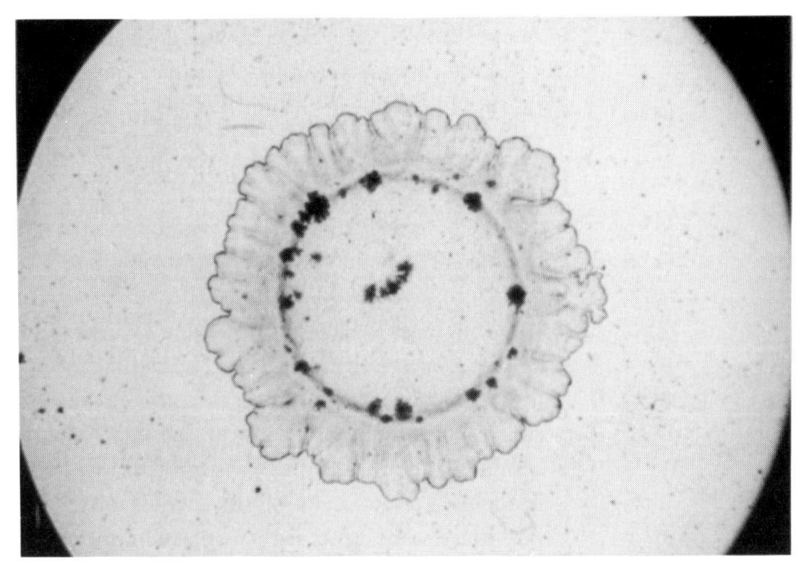

(a)

(b)

Figure 1.12 Examples of bacteria colonies. (a) A colony with roughly compact shape. (After [39].) (b) A colony with a branched morphology, resembling the DLA growth model described in Chapter 19. (After [296]).

resentation of the base pair sequence. This 'DNA walk' allows one to visualize directly the fluctuations of the purine-pyrimidine content in DNA sequences: an 'uphill drift' in a region of the landscape corresponds to high local concentration of pyrimidines, while a 'downhill drift' corresponds to high local concentration of purines.

Figure 1.13 shows a typical example of a gene that contains a significant fraction of base pairs that do *not* code for amino acids. The advantage of the DNA walk representation is that it can be quantitatively studied using the methods developed for interfaces. The methods reveal the surprising fact that such noncoding DNA,

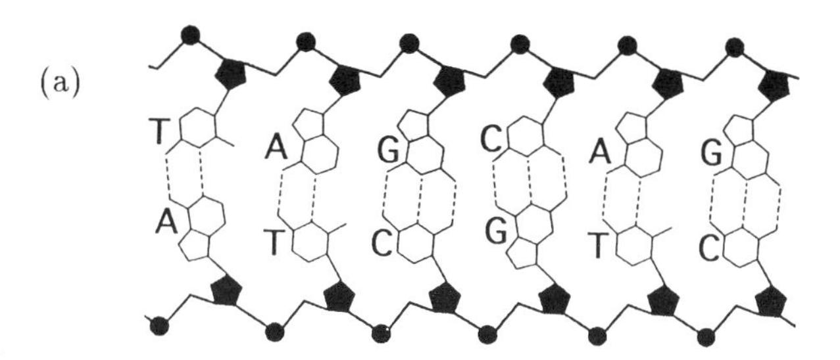

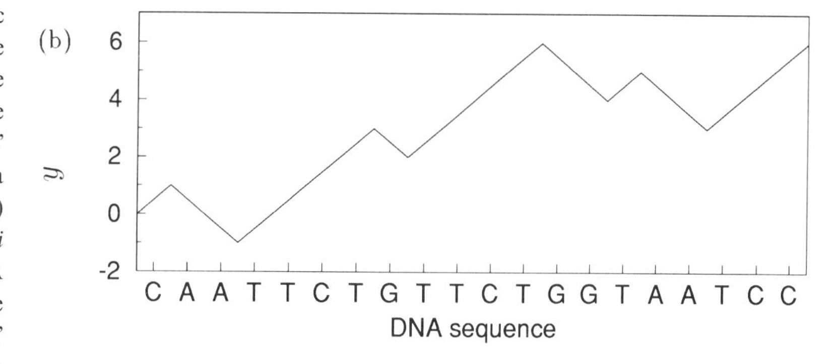

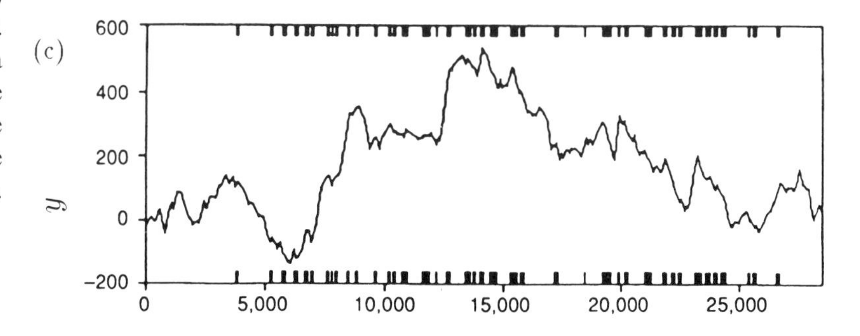

Figure 1.13 (a) The base pairing of two DNA strands to form a double helix. So far as the information content is concerned, a DNA sequence can be represented as a symbolic sequence of four letters: A, C, G and T. (b) Schematic illustration of the definition of the 'DNA walk'. The walker steps 'up' $[u(i) = +1]$ if a pyrimidine (C or T) occurs at position i along the DNA chain, while the walker steps 'down' $[u(i) = -1]$ if a purine (A or G) occurs at position i. (c) DNA walk for a DNA sequence comprising more than 25 000 base pairs. (After [60]).

previously believed to have at most correlations of very short range, in fact displays long-range correlations [60]. The implications of this result for the possible language characterizing the noncoding DNA is a topic under current investigation [290].

1.4 Methods of analysis

Each field has its own methods for treating a given problem. The development of new approaches, with more descriptive and predicting power, is one of the major goals of science. For the field of disorderly surface growth there are a number of standard tools that must be mastered. We briefly discuss four important methods that are developed and used throughout the book.

1.4.1 Scaling concepts

One of the modern concepts used to study various roughening processes is *scaling*. Scaling has a surprising power of prediction, simple manipulations allowing us to connect apparently independent quantities and exponents.

We shall see that many measurable quantities obey simple *scaling relations*. For example, for a large number of systems we shall find that the interface width, $w(t)$, increases as a power of time, $w(t) \sim t^{\beta}$. The width eventually saturates at a value that increases as a power law of the system size, $w(L) \sim L^{\alpha}$.

Studying such scaling relations will allow us to define *universality* classes. The universality class concept is a product of modern statistical mechanics, and codifies the fact that there are but a few essential factors that determine the exponents characterizing the scaling behavior. Thus different systems which at first sight may appear to have no connection between them, behave in a remarkably similar fashion.

The values of the exponents α and β are independent of many 'details' of the system. For example, α and β do not depend on whether we immerse the paper in ink or coffee, or if we use a paper towel or a tablecloth. In fact, we shall see that the scaling exponents obtained for the fluid flow problem coincide with the scaling exponents obtained for the burning front, despite the rather different mechanisms leading to the actual interface.

1.4.2 Experiments

In many branches of physics, experiments are the driving force that leads to new problems for theoretical investigation. In surface roughening the reverse sometimes occurs. For example, the roughening process was an unwanted experimental artifact that one attempted to minimize. It was only recently that a number of experimentalists became interested in this 'ugly-duckling,' and now they are taking the lead – showing that real life is much richer in mechanisms leading to roughening than theoretically anticipated.

We will discuss two different classes of experiments. First, there are experiments – such as paper wetting or burning – that study interface motion in a disordered medium, where impurities play a major role in shaping the morphology of the interface. Second, there are experiments relating to atomic deposition – such as MBE – for which the mechanisms leading to roughening turn out to be rather different.

1.4.3 Discrete models

Theoretical modeling provided a substantial fraction of the driving force behind early investigations of the interface morphology. Many models can afford to be elementary, because the phenomena being modeled are themselves rather simple. In Fig. 1.5, e.g., we show interfaces developing as a result of a simple deposition process. Similarly, 'island formation' in MBE can be modeled successfully using elementary computer algorithms. Figure 1.8 shows the result of such a computer simulation reproducing the morphologies observed experimentally in the Pt system. Moreover, simulation studies represent an essential link between theory and experiments, and can allow us to separate the essential ingredients determining the morphology from the unnecessary details.

1.4.4 Continuum equations

Lately a very successful tool for understanding the behavior of the various growth processes has been introduced, namely stochastic differential equations. Such equations typically describe the interface at large length scales, which means that we neglect the short length scale 'details' and focus only on the asymptotic coarse-grained properties.

One of our initial tasks on investigating a new interface problem is to attempt to derive a continuum growth equation. If we already have a discrete model for the system, then we would like to be able to derive the corresponding continuum equation directly from this model.

There is an alternative method for deriving the growth equation: exploiting symmetry principles. Every growth process obeys some simple symmetries. For example, the laws of physics are independent of where we define the zero height of the interface, so the continuum growth equation must be independent as well. In most cases the scaling properties of the system and the continuum equations are unambiguously determined by the symmetries of the system.

Once we obtain a growth equation we must find its predictions, determining, e.g., the scaling exponents and scaling functions. If the growth process is described by a sufficiently simple equation, we can solve it exactly. However, for many equations no exact solutions exist, so we must apply various approximations to uncover the scaling behavior. One particularly powerful tool, called the *renormalization group method*, can be successfully used in many cases.

1.5 Discussion

There are some aspects of growth processes that make them both interesting and difficult. One is the effect of randomness. Growth phenomena involve noise or randomness, which plays an essential role in shaping the final morphology of the interface. The origin of randomness may depend on the process being studied. In fluid flow problems, the origin of randomness is the disordered nature of the medium through which the interface advances. In superconductors the pinning forces of the quenched disorder, together with the thermal fluctuations, determine the actual dynamics of the flux lines. In deposition processes, there is the non-uniform nature of the incoming flux: atoms reach the surface at random positions, with random time intervals between them. There is also the random nature of adatom diffusion on the surface since atoms usually follow Brownian trajectories while looking for the edge of an island or step on which to stick.

Thus one must deal systematically with the different sources of randomness. Throughout this book, we pay special attention to the effect of various forms of noise, distinguishing thermal noise from quenched noise, conserved from non-conserved, correlated from non-correlated, Gaussian from power-law distributed noise. Due to the central role that randomness and noise play in roughening phenomena, all these distinctions will lead to new universality classes and scaling exponents.

Suggested further reading:

[1, 19, 64, 65, 125, 224, 420, 421, 422, 456]

2 Scaling concepts

The formation of interfaces and surfaces is influenced by a large number of factors, and it is almost impossible to distinguish all of them. Nevertheless, a scientist always hopes that there is a small number of basic 'laws' determining the morphology and the dynamics of growth. The action of these basic laws can be described in microscopic detail through discrete growth models – models that mimic the essential physics but bypass some of the less essential details.

To this end, we introduce a simple model, *ballistic deposition* (BD), which generates a nonequilibrium interface that exemplifies many of the essential properties of a growth process. We shall use the BD model to introduce scaling concepts, a central theme of this book.

2.1 Ballistic deposition

Ballistic deposition was introduced as a model of colloidal aggregates, and early studies concentrated on the properties of the porous aggregate produced by the model [434, 466, 467]. The nontrivial surface properties became a subject of scientific inquiry after the introduction of vapor deposition techniques [20, 124, 306].

It is simpler to define and study the BD model on a lattice, as in Fig. 2.1, but off-lattice versions have been investigated as well. A particle is released from a randomly chosen position above the surface, located at a distance larger than the maximum height of the interface. The particle follows a straight vertical trajectory until it reaches the surface, whereupon it sticks. In the simplest version of the model, particles are deposited onto a surface oriented perpendicular to the particle trajectories.

The deposited particles form a cluster or an aggregate with a very particular geometry. Figure 2.2 shows the structure produced by the model after the deposition of 35 000 particles.† The unit time is chosen to correspond to the deposition of L particles on the interface, where L is the system size – the number of columns.

2.2 Roughening

We define the 'surface' to be the set of particles in the aggregate that are highest in each column. Figure 2.2 highlights the position of the interface at successive time intervals, so that the growth process can be followed qualitatively. To describe the growth quantitatively, we introduce two functions.

- (i) The *mean height* of the surface, $\bar{h}$, is defined by

$$\bar{h}(t) \equiv \frac{1}{L}\sum_{i=1}^{L} h(i,t), \tag{2.1}$$

where $h(i,t)$ is the height of column i at time t. If the deposition rate (number of particles arriving on a site) is constant, the mean height increases linearly with time,

$$\bar{h}(t) \sim t. \tag{2.2}$$

† The BD model for a two-dimensional square lattice can be readily simulated on a personal computer. At time t, the height of the interface of site i is $h(i,t)$. At $t=0$, the interface is flat, so $h(i,t)=0$ for $i=1,...,L$. At any moment t we choose randomly a site i on the lattice and increase $h(i,t)$ to $h(i,t+1)=\max[h(i-1,t), h(i,t)+1, h(i+1,t)]$.

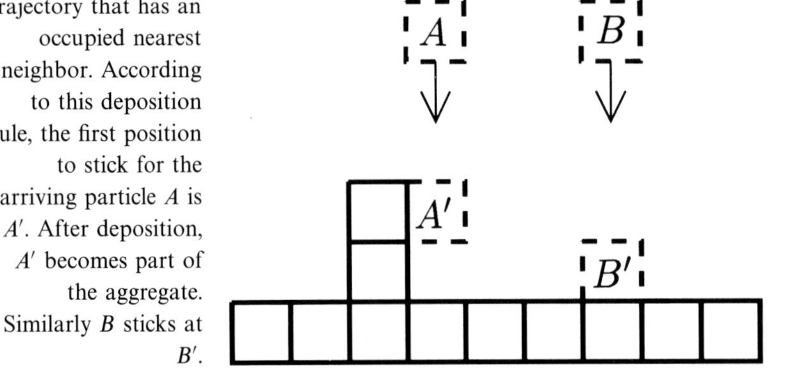

Figure 2.1 The BD model with the nearest-neighbor (NN) sticking rule, illustrating two sticking possibilities for the newly-deposited particles. Shown is the case $L=9$, where L is the number of columns. We choose a random position above the surface, and allow a particle to fall vertically toward it. The particle sticks to the first site along its trajectory that has an occupied nearest neighbor. According to this deposition rule, the first position to stick for the arriving particle A is A'. After deposition, A' becomes part of the aggregate. Similarly B sticks at B'.

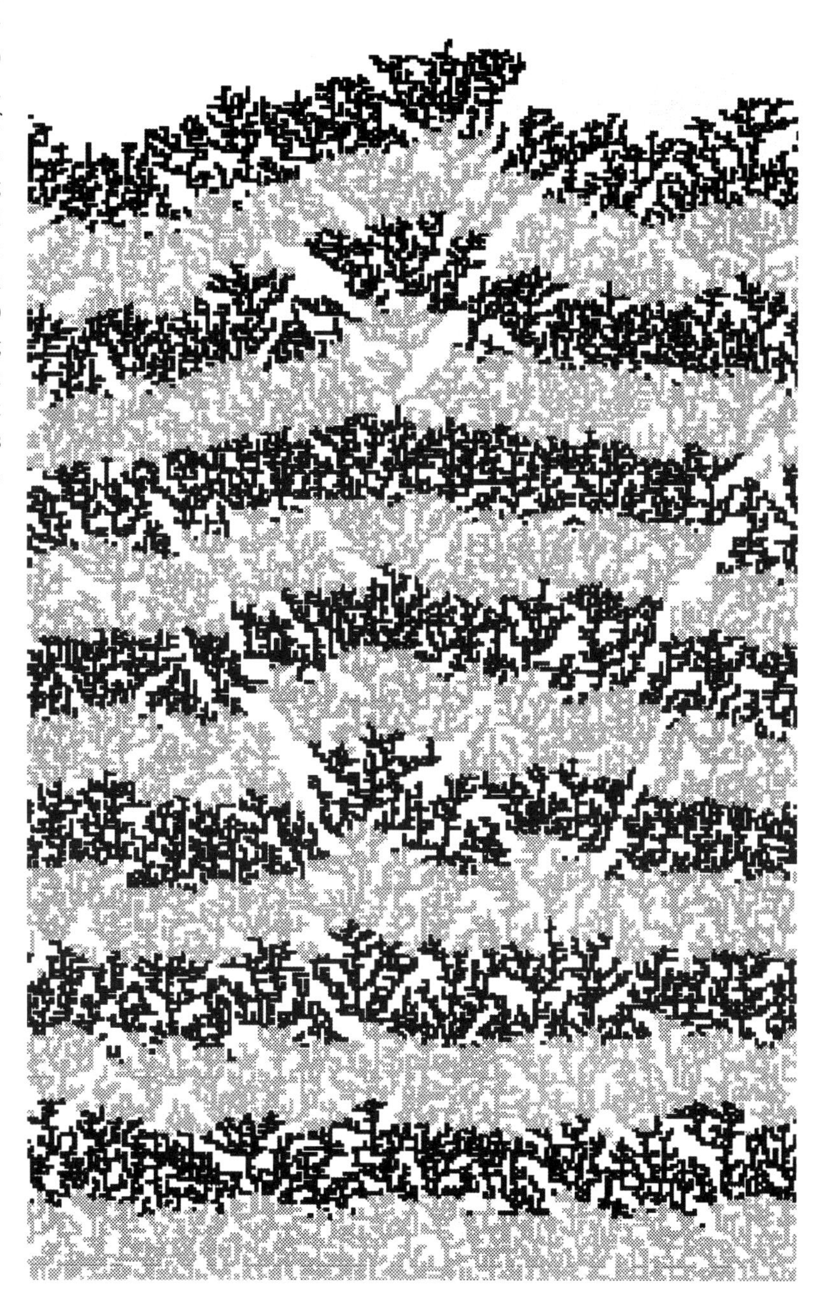

Figure 2.2 A BD cluster obtained by depositing 35 000 particles on a substrate of horizontal size $L = 200$. The shading reflects the arrival time of the particles: after the deposition of each set of 2500 particles, the shading changes. The eye can readily recognize that the roughness increases in time.

- (ii) The *interface width*, which characterizes the *roughness* of the interface, is defined by the rms fluctuation in the height,

$$w(L,t) \equiv \sqrt{\frac{1}{L}\sum_{i=1}^{L}[h(i,t)-\bar{h}(t)]^2}. \qquad \rightarrow (2.3)$$

To monitor the roughening process quantitatively, we measure the width of the interface as a function of time. By definition, the growth starts from a horizontal line; the interface at time zero is simply a straight line, with zero width. As deposition occurs, the interface gradually roughens.

A typical plot of the time evolution of the surface width has two regions separated by a 'crossover' time $t_\times$ (see Fig. 2.3)

(i) Initially, the width increases as a power of time,

$$w(L,t) \sim t^{\beta} \qquad [t \ll t_\times]. \qquad \rightarrow (2.4)$$

The exponent β, which we call the *growth exponent*, characterizes the time-dependent dynamics of the roughening process.†

(ii) The power-law increase in width does not continue indefinitely, but is followed by a saturation regime (the horizontal region of Fig.

† Technically, there occurs an initial transient for $t < 1$ (less than one layer) for which simple Poisson growth dominates, and $\beta = 1/2$.

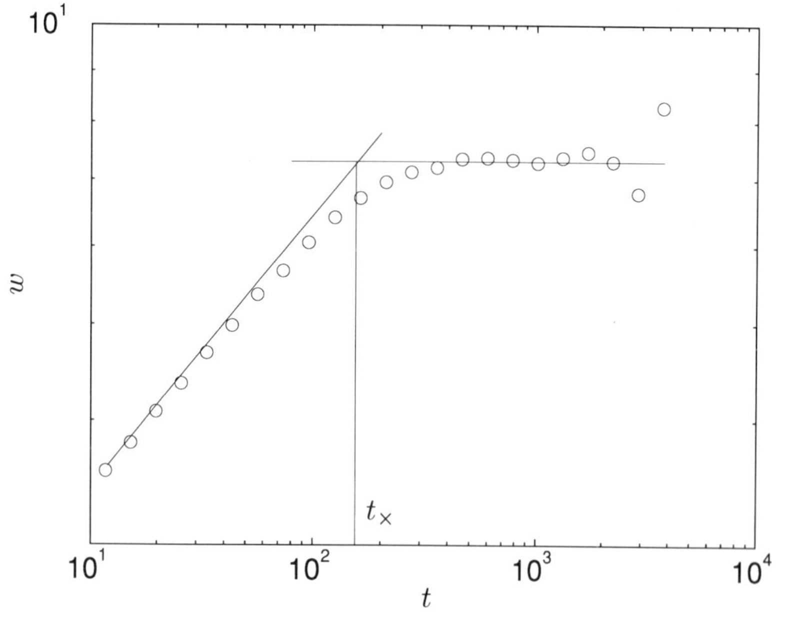

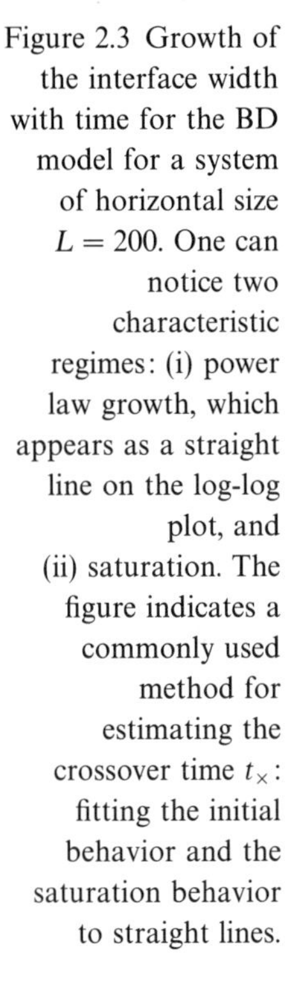
Figure 2.3 Growth of the interface width with time for the BD model for a system of horizontal size $L = 200$. One can notice two characteristic regimes: (i) power law growth, which appears as a straight line on the log-log plot, and (ii) saturation. The figure indicates a commonly used method for estimating the crossover time $t_\times$: fitting the initial behavior and the saturation behavior to straight lines.

2.3) during which the width reaches a *saturation value*, w_{sat}. In Fig. 2.4, four different curves correspond to the time evolution of the width obtained by simulating systems with four different system sizes L. As L increases, the saturation width, w_{sat}, increases as well, and the dependence also follows a power law,

$$w_{\text{sat}}(L) \sim L^{\alpha} \qquad [t \gg t_{\times}]. \qquad \rightarrow (2.5)$$

The exponent α, called the *roughness exponent*, is a second critical exponent that characterizes the roughness of the saturated interface.

(iii) The crossover time $t_{\times}$ (sometimes called *saturation* time) at which the interface crosses over from the behavior of (2.4) to that of (2.5) depends on the system size,

$$t_{\times} \sim L^{z}, \qquad (2.6)$$

where z is called the *dynamic exponent*. The construction illustrated in Fig. 2.3 is a simple way to estimate $t_{\times}$.

2.3 Dynamic scaling

The scaling exponents α, β, and z are not independent, and there exists a simple way to 'collapse' the data of Fig. 2.4 onto a single curve [124].

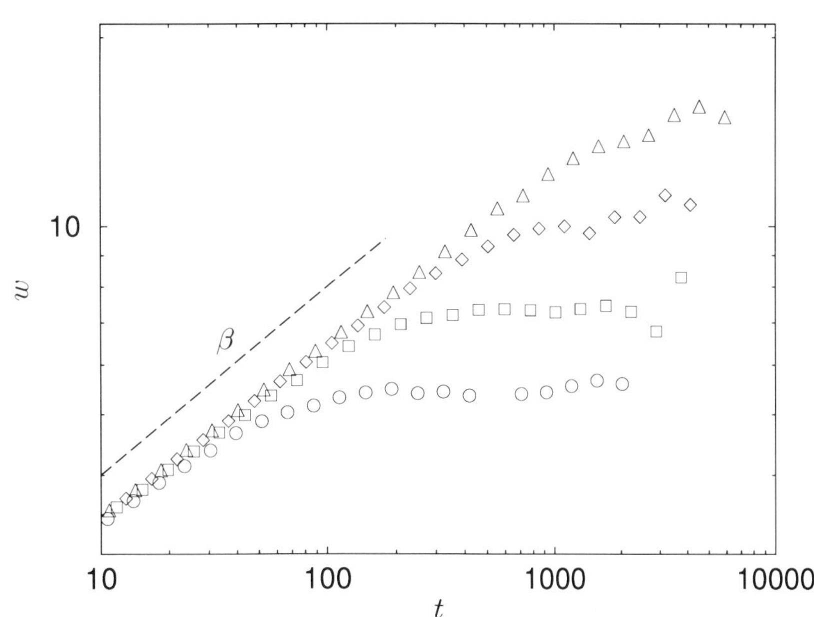

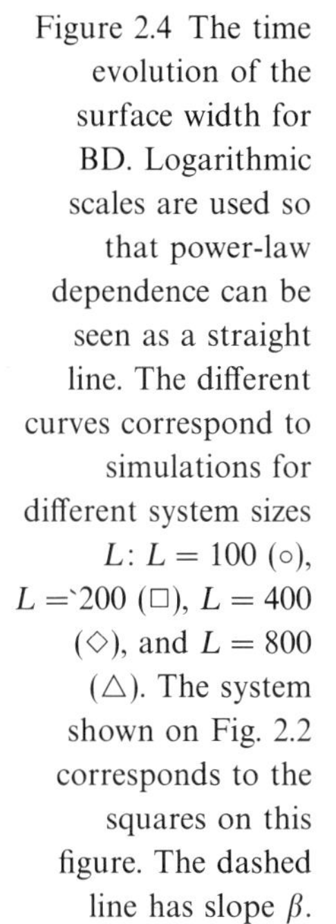

Figure 2.4 The time evolution of the surface width for BD. Logarithmic scales are used so that power-law dependence can be seen as a straight line. The different curves correspond to simulations for different system sizes L: $L = 100$ (○), $L = 200$ (□), $L = 400$ (◇), and $L = 800$ (△). The system shown on Fig. 2.2 corresponds to the squares on this figure. The dashed line has slope β.

(A) Plotting $w(L,t)/w_{sat}(L)$ as a function of time will result in curves that saturate at the same value, independent of the system size L (see Fig. 2.5).

(B) Plotting the width as a function of $t/t_\times$ will cause the curves to saturate at the same characteristic time (see Fig. 2.5).

These two observations suggest that $w(L,t)/w_{sat}(L)$ is a function of $t/t_\times$ only, i.e.,

$$\frac{w(L,t)}{w_{sat}(L)} \sim f\left(\frac{t}{t_\times}\right), \tag{2.7}$$

where $f(u)$ is called a *scaling function.* Replacing w_{sat} and $t_\times$ in (2.7) with their scaling forms (2.5) and (2.6), we obtain the *Family–Vicsek scaling relation* [124]

$$w(L,t) \sim L^\alpha f\left(\frac{t}{L^z}\right). \qquad \rightarrow (2.8)$$

The general form of the scaling function $f(u)$ can be read off from Fig. 2.5. There are two different scaling regimes depending on its argument $u \equiv t/t_\times$.

(a) For small u, the scaling function increases as a power law. Since during the rescaling shown in Fig. 2.5 we did not *rotate* the curves, but only shifted them, it can be seen that in this regime we have

$$f(u) \sim u^\beta \qquad [u \ll 1]. \tag{2.9}$$

(b) As $t \to \infty$, the width saturates. Saturation is reached for $t \gg t_\times$, i.e. the argument of the scaling function $u \gg 1$. In this limit we have

$$f(u) = \text{const} \qquad [u \gg 1]. \tag{2.10}$$

The validity of the scaling relation (2.8) can be tested numerically by replotting the data of Fig. 2.4 in a fashion suggested by (2.8) and

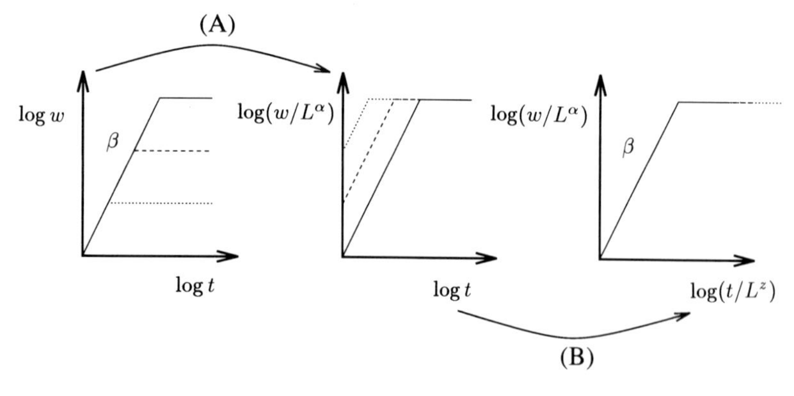

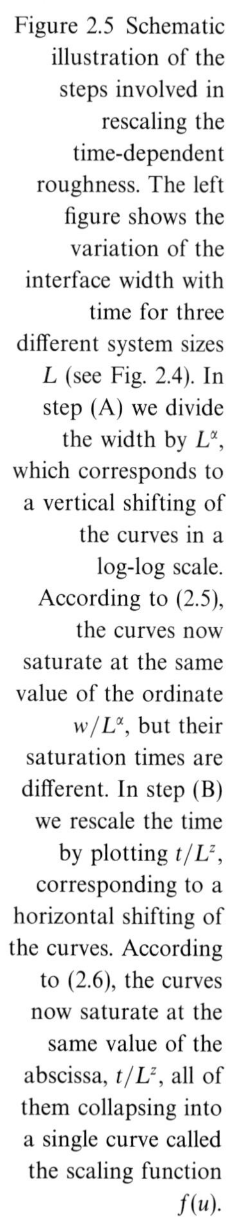
Figure 2.5 Schematic illustration of the steps involved in rescaling the time-dependent roughness. The left figure shows the variation of the interface width with time for three different system sizes L (see Fig. 2.4). In step (A) we divide the width by L^α, which corresponds to a vertical shifting of the curves in a log-log scale. According to (2.5), the curves now saturate at the same value of the ordinate w/L^α, but their saturation times are different. In step (B) we rescale the time by plotting t/L^z, corresponding to a horizontal shifting of the curves. According to (2.6), the curves now saturate at the same value of the abscissa, t/L^z, all of them collapsing into a single curve called the scaling function $f(u)$.

Fig. 2.5. Plotting on the horizontal axis t/L^z, and on the vertical $w(L,t)/L^\alpha$, the different curves of Fig. 2.4 should collapse onto one curve that exhibits the properties (2.9) and (2.10) of the scaling function $f(u)$. The rescaled curves are shown in Fig. 2.6, and the 'data collapse' found indeed supports the scaling hypothesis (2.8).

We next show that the exponents α, β and z are not independent. In Fig. 2.3 if we approach the crossover point $(t_\times, w(t_\times))$ from the left, we find, according to (2.4), that $w(t_\times) \sim t_\times^\beta$. However, approaching the same point from the right, we have, from (2.5), $w(t_\times) \sim L^\alpha$. From these two relations follows $t_\times^\beta \sim L^\alpha$ which, according to (2.6), implies that

$$z = \frac{\alpha}{\beta}. \qquad \rightarrow (2.11)$$

Equation (2.11), a scaling law linking the three exponents, is valid for any growth process that obeys the scaling relation (2.8).

The results thus far are summarized in Table 2.1.

2.4 Correlations

Why does the interface width saturate for the BD model? This question may appear elementary, but it has no simple answer. The

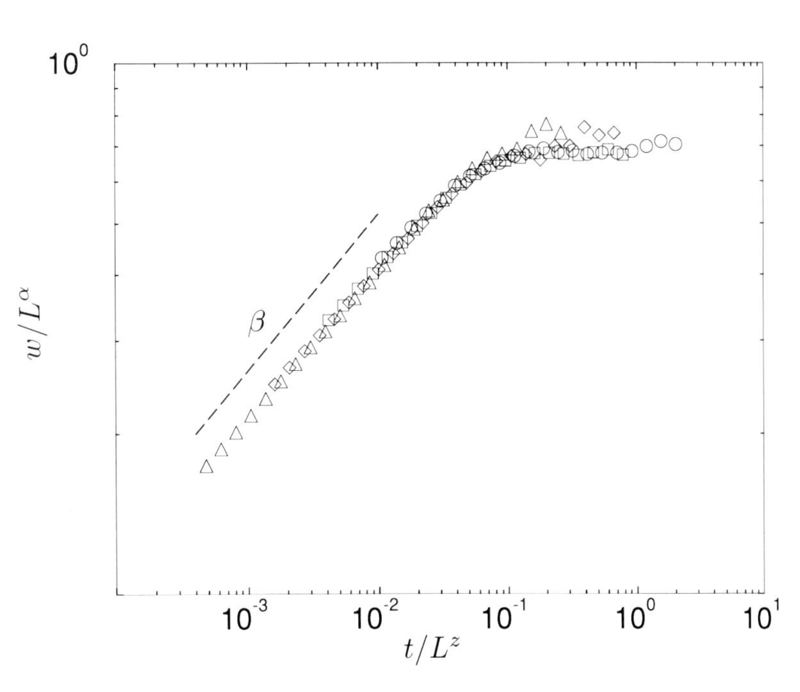

Figure 2.6 The BD simulations of Fig. 2.4 rescaled according to (2.8). The obtained curve is in fact the scaling function $f(u)$, with the properties (2.9) and (2.10). The different symbols correspond to runs with different system sizes $L = 100$ (○), $L = 200$ (□), $L = 400$ (◇), and $L = 800$ (△). The slope of the dashed line is β.

Table 2.1 *Summary of the key definitions.*

Mean height	$\bar{h}(t) \equiv \frac{1}{L}\sum_{i=1}^{L} h(i,t)$		(2.1)
Interface width	$w(L,t) \equiv \sqrt{\frac{1}{L}\sum_{i=1}^{L}[h(i,t)-\bar{h}(t)]^2}$		(2.3)
Growth exponent	$w(L,t) \sim t^{\beta}$	$[t \ll t_\times]$	(2.4)
Roughness exponent	$w_{\rm sat}(L) \sim L^{\alpha}$	$[t \gg t_\times]$	(2.5)
Dynamic exponent	$t_\times \sim L^z$		(2.6)
Scaling relation	$w(L,t) \sim L^{\alpha} f\left(t/L^z\right)$		(2.8)
Scaling law	$z = \alpha/\beta$		(2.11)

fact that the saturation time, $t_\times$, and the saturation width, w_{sat}, increase with the system size suggests that saturation phenomena constitute a *finite size effect*.† But what is the mechanism that leads to saturation? How does the system 'know' *when* to saturate? To answer these questions, we must discuss a very important property of most surfaces, the existence of *correlations* in the system.

One remarkable property of the BD growth process is that *correlations* develop along the surface, which imply that the different sites of the surface are not completely independent, but depend upon the heights of neighboring sites. What is the microscopic origin of these correlations? The answer is not known, but we can gather some clues about it through a close examination of the growth process. The newly-arriving particle sticks to the first nearest-neighbor site it encounters. Thus the height of the new particle will be equal to or larger than that of its neighbors. The height fluctuation will spread *laterally*, because the next particle deposited near it must have a height equal or larger as well (see Fig. 2.7). Although the growth process is *local*, through this lateral growth the 'information' about the height of each of the neighbors spreads globally. The typical distance over which the heights 'know about' each other – the characteristic distance over which they are correlated – is called the *correlation length*, and is denoted by $\xi_\parallel$.

† Note that according to (2.5) an infinite system ($L = \infty$) does not saturate.

At the beginning of the growth, the sites are uncorrelated. During deposition $\xi_\parallel$ grows with time. For a finite system $\xi_\parallel$ cannot grow indefinitely, because it is limited by the size of the system, L. When $\xi_\parallel$ reaches the size of the system, the entire interface becomes correlated, resulting in the saturation of the interface width. Thus at saturation

$$\xi_\parallel \sim L \qquad [t \gg t_\times]. \tag{2.12}$$

According to (2.8), saturation occurs at a time $t_\times$ given by (2.6). Replacing L with $\xi_\parallel$, we obtain $\xi_\parallel \sim t_\times^{1/z}$, which in fact holds for $t < t_\times$ as well,

$$\xi_\parallel \sim t^{1/z} \qquad [t \ll t_\times]. \tag{2.13}$$

A *perpendicular* correlation length, $\xi_\perp$, characterizes the fluctuations in the growth direction, and displays the same scaling behavior as the surface width, $w(L,t)$.

We made the plausible assumption above that when $\xi_\parallel$ becomes equal to the system size, the interface width saturates. Continuum equations, to be discussed in the following chapters, will confirm the existence of such saturation phenomena.

2.5 Discussion

The surface generated by the ballistic deposition model is called a 'nonequilibrium interface.' The word nonequilibrium is used to stress

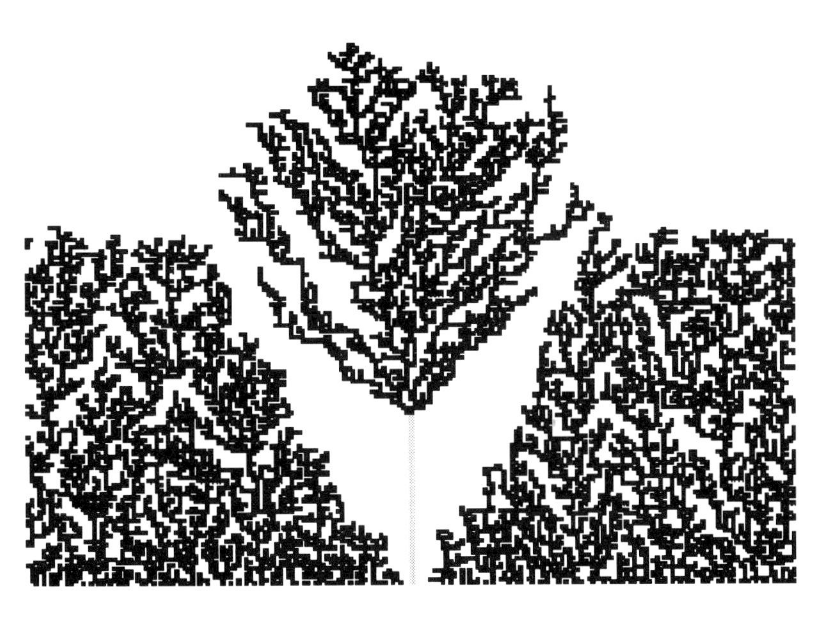

Figure 2.7 Illustration of lateral growth in BD. Growth started from a flat interface, with one column (gray) much higher than its neighborhood. This large height fluctuation will capture the arriving atoms, building up a tree which grows both laterally and vertically. This example illustrates the lateral spread of the 'information' during growth: the excess height spreads along the surface, influencing the height of neighboring columns. Under normal growth conditions, it is unlikely that such a high column would develop, but the mechanism of sending information laterally is still valid.

the far-from-equilibrium nature of the growth process. We neglect thereby the possibility that deposited particles 'restructure' themselves to reach the lowest energy state.

Ballistic deposition is but one of several models under current investigation with the goal of understanding the generic properties of different growth processes. However, the observed roughening process is more general than suggested by this particular model and occurs for a large number of seemingly unrelated growth phenomena.

It may seem unusual that we have defined the exponents α, β and z, but have postponed providing numerical values. The reason is simple: to calculate α, β and z analytically, we must first develop a number of essential tools.

Suggested further reading:

[126, 287, 414, 456]

Exercises:

2.1 Write a computer program to simulate the ballistic deposition model. Calculate the interface width and average height for various system sizes. What is the largest system size in which you can reach saturation in a reasonable computer time? Estimate the time that would be necessary to reach saturation in a system ten times larger.

2.2 An alternative way of collapsing (2.4) and (2.5) into a single scaling relation is to use

$$w(L,t) \sim t^{\beta} g\left(\frac{L}{t^{\varphi}}\right). \tag{2.14}$$

Calculate the exponent φ and derive the scaling properties of the scaling function $g(u)$.

3 Fractal concepts

In the previous chapter, we introduced a simple growth model that exhibits generic scaling behavior. In particular, the interface width w increases as a power of time [Eq. (2.4)], and the saturated roughness displays a power law dependence on the system size [Eq. (2.5)]. There exists a natural language for describing and interpreting such scaling behavior, and this is the language of fractals. In this chapter, we introduce the concepts of fractal geometry, which provide a language in terms of which to better understand the meaning of power laws. Isotropic fractals are *self-similar*: they are invariant under isotropic scale transformations. In contrast, surfaces are generally invariant under anisotropic transformations, and belong to the broader class of *self-affine* fractals. We will therefore also discuss the basic properties of self-affine fractals, as well as numerical methods for calculating the critical exponents α, β, and z.

3.1 Self-similarity

An object can be self-similar if it is formed by parts that are 'similar' to the whole. One of the simplest self-similar objects is the Cantor set, whose iterative construction at successive 'generations' is shown in Fig. 3.1. If we enlarge the box of generation 3 by a factor of three, we obtain a set of intervals that is identical to the generation 2 object. In general, at generation k we can enlarge part of the object by a factor of three and obtain the object of generation $(k-1)$.

For the Cantor set, the enlarged part overlaps the original object exactly; we call such an exactly self-similar object a *deterministic fractal*. However, many objects existing in nature are random. Despite this randomness, such natural objects may be self-similar in a *statistical* sense. One classic example is the coastline of a continent. If we study two maps with different magnifications representing a typical coastline, they look similar. There is no way to distinguish between them, if we are not already familiar with the particular coastline. In fact, we cannot even determine which map has the higher magnification. Unlike the case of deterministic fractals, the two maps at two different magnifications do *not* overlap, but nonetheless, their statistical properties are the same. Objects with inherent randomness that are self-similar only in a statistical sense are called *random fractals* or *statistical fractals*.†

3.2 Fractal dimension

To characterize quantitatively a self-similar system, we must introduce some definitions. By *embedding dimension*, d_E, we understand the smallest Euclidean dimension of the space in which a given object can be embedded.

To decide on the fractality of an object, we must measure its Hausdorff dimension. The volume $V(\ell)$ of an arbitrary object can be measured by covering it with balls of linear size ℓ, and volume ℓ^{d_E}. We need $N(\ell)$ balls to cover it, so

$$V(\ell) = N(\ell)\ell^{d_E}. \tag{3.1}$$

One might at first expect that for any object, $N(\ell) \sim \ell^{-d_E}$, since the volume of an object does not change if we change the unit of

† Self-similarity is a symmetry property of the system. By self-similarity we mean invariance under an isotropic transformation, namely a simple dilation. If we consider an object S formed by a set of points $R = (x_1, x_2, x_3, \ldots)$, a dilation, or similarity transformation with a scaling factor b, changes the coordinates according to $bR = (bx_1, bx_2, bx_3, \ldots)$. The set S formed by the particles of coordinates R is self-similar if it is invariant under this transformation. For a deterministic fractal, scale invariance means that the rescaled system bS is identical with a part of the original system S. For a random fractal we have only statistical identity, which means that every *statistical* quantity should be the same for the rescaled and the original system. [287, 468]

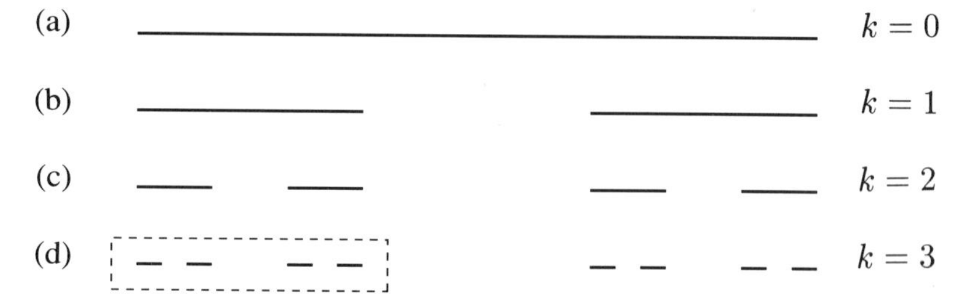

Figure 3.1 Construction of the Cantor set. (a) Cut the unit interval [0,1] into three equal segments, and remove the middle one. (b) What remains is the generator of the fractal to be constructed. (c) Divide the remaining two parts in three again, and remove the middle segments. (d) If we continue this process indefinitely, we obtain the Cantor set, or Cantor dust. To each iteration we may associate a generation number, denoted by k.

measurement ℓ. For fractals we have in general

$$N(\ell) \sim \ell^{-d_f}. \qquad \rightarrow (3.2)$$

Objects with $d_f < d_E$ are called *fractals* [287], where d_f is the *fractal dimension*.† From (3.2) we have

$$d_f = \lim_{\ell \to 0} \frac{\ln N(\ell)}{\ln(1/\ell)}. \qquad (3.3)$$

For the Cantor set, the natural unit to measure the length of the set at iteration k is the length of the smallest interval $\ell_k = (1/3)^k$. The number of intervals of length ℓ_k at level k is $N(\ell_k) = 2^k$. Thus from (3.3) we have for the fractal dimension $d_f = \ln(2)/\ln(3) = 0.639....$ Since $d_f < d_E = 1$, the Cantor set is a fractal.

The Sierpinski gasket, the construction of which is shown in Fig. 3.2, is a generalization to two dimensions of the Cantor set. To observe its *self-similarity*, we increase the size of the dotted box of Fig. 3.2(d) by a factor of two. The new object is identical to the triangle in Fig. 3.2(c). We can cover the Sierpinski gasket at level k with $N(\ell) = 3^k$ triangles of linear size $\ell = (1/2)^k$. The fractal dimension is $d_f = \ln(3)/\ln(2) = 1.585$, smaller than the embedding dimension $d_E = 2$.

We can use (3.3) to measure the fractal dimension of a random fractal. If, e.g., we measure the total length of a coastline, $\ell N(\ell) =$

† Strictly speaking, a fractal is defined by $d_f > d_T$, where d_T is the *topological dimension* of the studied object. $d_T = 0$ for a set of disconnected points, $d_T = 1$ for a curve, $d_T = 2$ for a surface and $d_T = 3$ for a solid.

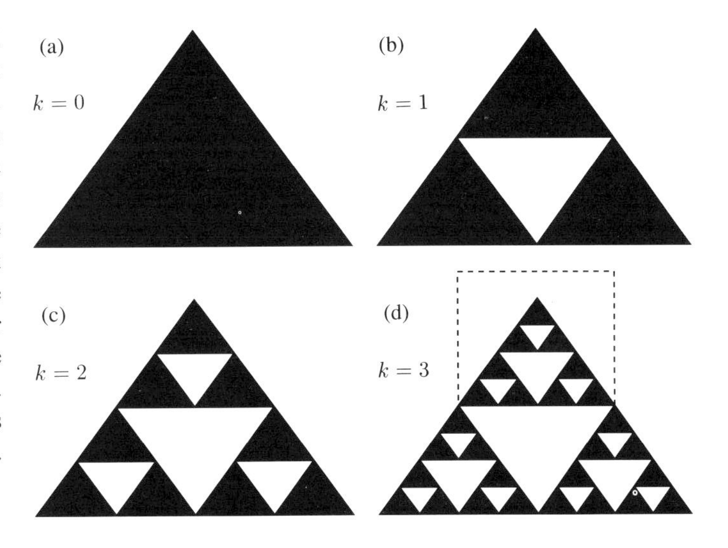

Figure 3.2 Sierpinski gasket. (a) We start from the filled triangle, (b) remove from the middle a triangle whose area is one fourth of the total. (c) In the next step, we repeat the same procedure for the remaining three filled triangles. (d) This process is iterated indefinitely.

ℓ^{1-d_f}, we would find that it *increases* according to (3.2) as the length ℓ of the measuring sticks *decreases*. Plotting $N(\ell)$ as a function of $1/\ell$ on log-log paper, we obtain a straight line whose slope is the fractal dimension. There are many methods for measuring the fractal dimension of an arbitrary random fractal. Best known are the box counting method and the measurement of the correlation function [456]. The choice of the best method is often determined by the particular experimental or numerical situation, or by the nature of the available data, and the careful researcher should use more than one method to confirm consistency.

3.3 Self-affinity

The scale transformation we described for self-similar fractals is isotropic, which means that dilation increases the size of the system *uniformly* in every spatial direction. Fractal objects that must be rescaled using an *anisotropic* transformation are called *self-affine* fractals. The concept of an anisotropic rescaling† is shown in Fig. 3.3. The construction of a simple deterministic self-affine fractal is presented in Fig. 3.4. If we enlarge isotropically the part shown in the box of Fig. 3.4(c), it does not overlap with Fig. 3.4(b). However, it will overlap if we rescale 'anisotropically.'

Scaling – For quantifying disorderly surfaces, we are interested in a special subclass of anisotropic fractals, described by single-valued

† An *anisotropic* rescaling has different scaling factors in different spatial directions, so $b \cdot R = (b_1 x_1, b_2 x_2, b_3 x_3, \ldots)$. For a deterministic self-affine fractal, invariance means that the rescaled system $b \cdot R$ is identical with a part of the original system R. Again, for a random self-affine fractal we have only statistical identity [288, 289].

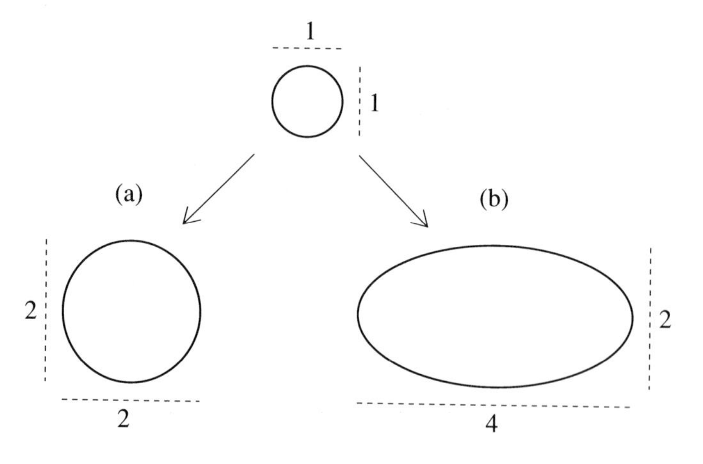

Figure 3.3 The effect of isotropic and anisotropic rescalings on a simple object, the circle. (a) The diameter is enlarged by a factor of two in the case of *isotropic* change of scale. (b) In case of the *anisotropic* rescaling, the East–West diameter is enlarged by a factor of four, while the North–South diameter is enlarged by a factor of two, resulting in an ellipse. While self-similar objects are invariant under an isotropic transformation, for self-affine objects we must perform an anisotropic transformation.

functions called *self-affine functions*. The model of Fig. 3.4 introduces such a single-valued function, $h(x)$, where x is defined on the interval [0,1]. The analog of the scaling relation (3.2) for a self-affine function can be formulated as

$$h(x) \sim b^{-\alpha} h(bx), \tag{3.4}$$

where α (in some texts denoted by H or χ) is called the Hölder or self-affine exponent and gives a quantitative measurement of the 'roughness' of the function $h(x)$. Relation (3.4) formulates in general terms the fact that a self-affine function must be rescaled in a different way horizontally and vertically: if we 'blow up' the function with a factor b horizontally ($x \to bx$), it must be 'blown up' with a factor b^{α} vertically [$h \to b^{\alpha} h$] in order that the resulting object overlaps the object obtained in the previous generation. For the special case $\alpha = 1$, the transformation is isotropic and the system is self-similar.

We may use (3.4) to determine the roughness exponent α for the deterministic model. In order to overlap the box of Fig. 3.4c with Fig. 3.4b, we must enlarge the box *horizontally* by a factor $b_1 = 4$, and

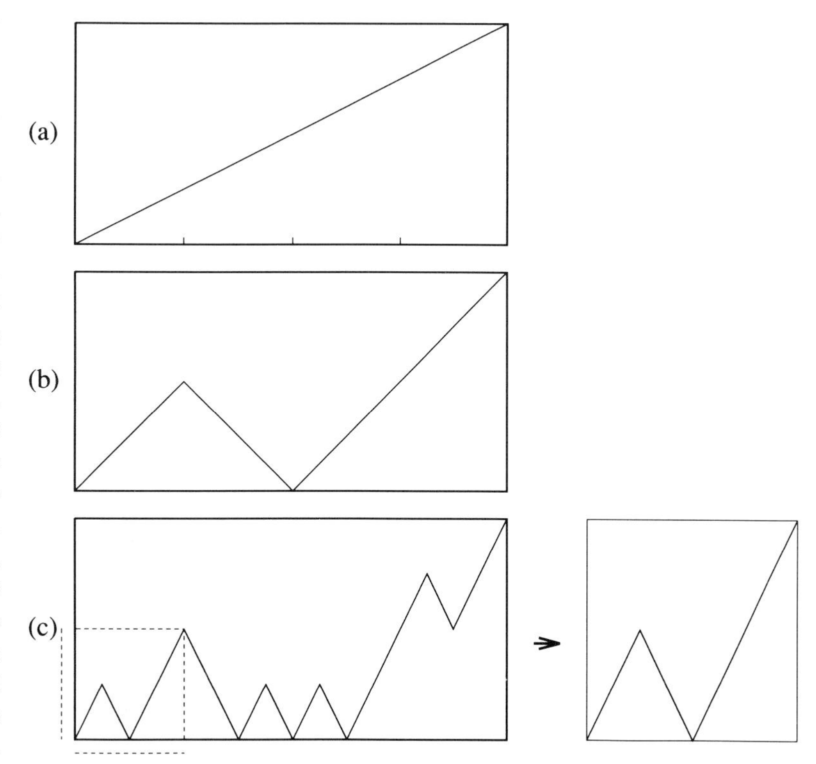

Figure 3.4 Construction of a deterministic self-affine object [288]. The diagonal in (a) is divided in four equal parts horizontally, and replaced by the structure shown in (b). In the next iteration we repeat the same procedure on all four segments, leading to the structure (c). If we were to rescale the object in the dotted box isotropically (I), we would obtain the object shown in the right, different from the object obtained in the previous iteration.

vertically by a factor $b_2 = 2$. From (3.4) follows

$$\alpha = \frac{\log b_2}{\log b_1} = \frac{1}{2}. \tag{3.5}$$

An important consequence of (3.4) concerns the scaling of the height difference $\Delta(\ell) \equiv |h(x_1) - h(x_2)|$ between two points separated by a distance $\ell \equiv |x_1 - x_2|$. For self-affine systems $\Delta(\ell)$ obeys (3.4). The solution of the 'functional equation' (3.4) is a power law [414],

$$\Delta \sim \ell^{\alpha}. \tag{3.6}$$

Fractal Dimension – In addition to the roughness exponent α, it is possible to associate a fractal dimension d_f with a self-affine function. Consider a self-affine function defined on the interval [0,1] which we cover with boxes of size ℓ. We first divide the horizontal domain of the function into N_s segments, so the width of each segment is $\ell = 1/N_s$. In a horizontal interval of size ℓ, the height changes according to (3.6), so we require $\Delta/\ell \sim \ell^{\alpha-1}$ boxes to cover the function. Since we need $\ell^{\alpha-1}$ boxes to cover the variation in one segment, for N_s segments the total number of boxes required is

$$N(\ell) \sim N_s \times \ell^{\alpha-1} \sim \ell^{\alpha-2}. \tag{3.7}$$

Thus from (3.2),

$$d_f = 2 - \alpha. \tag{3.8}$$

The astute reader will perceive a paradox at this point: Eq. (3.8) suggests that the interface is *self-similar*, not *self-affine*. The resolution of the paradox arises from the fact that the argument leading to Eq. (3.7) holds only if $\Delta \ll \ell$. For large enough ℓ, $\Delta > \ell$ and the number of boxes required to cover the function scales as $N(\ell) \sim 1/\ell$, not as $\ell^{\alpha-2}$. The same argument then leads to $d_f = 1$. Thus on very short length scales $d_f = 2 - \alpha$, but on long length scales $d_f = 1$.

Random Walk – The simplest random self-affine fractal is generated by a one-dimensional random walk on a lattice. Consider a particle at $x = 0$. At every moment, the particle moves one site randomly up or down ($|x(t+1) - x(t)| = 1$) with equal probability $p = 1/2$. Record the coordinate $x(t)$ as a function of time for a particular realization of the random walk (see Fig. 3.5). Since all steps are independent, we can calculate the probability $P(x,t)$ that the particle can be found at a given distance x from the origin at time t. In the continuum approximation, we obtain a simple Gaussian distribution

$$P(x,t) = \frac{1}{\sqrt{2\pi t}} \exp\left(-\frac{x^2}{2t}\right). \tag{3.9}$$

A distribution function is determined by its mean value and its standard deviation. The mean is zero, since there is an equal probability to find the particle in the (x) or $(-x)$ position. For the standard deviation, we obtain from (3.9)

$$\langle [x(t_2) - x(t_1)]^2 \rangle^{1/2} \sim |t_2 - t_1|^{1/2}. \tag{3.10}$$

The standard deviation scales with time exactly as does the height difference function, defined in (3.6), giving $\alpha = 1/2$ for the random walk. According to (3.8), the 'fractal dimension' of the random walk

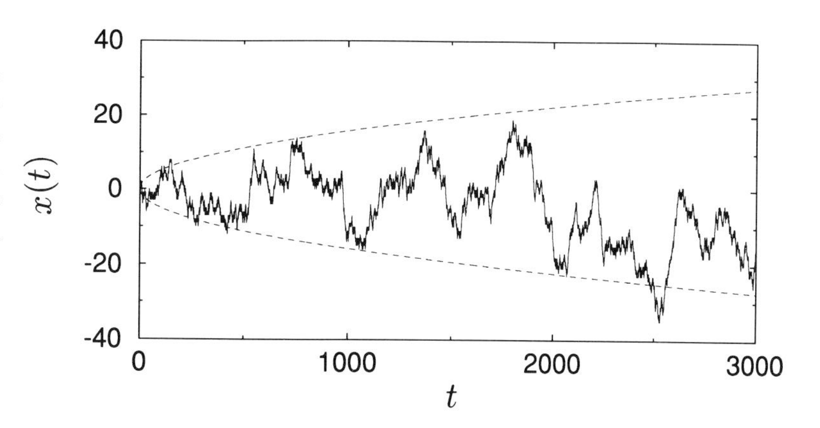

Figure 3.5 A typical trajectory of a random walker. The dashed lines are proportional to $\pm t^{1/2}$, where $t^{1/2}$ is the standard deviation (3.10) of the walker.

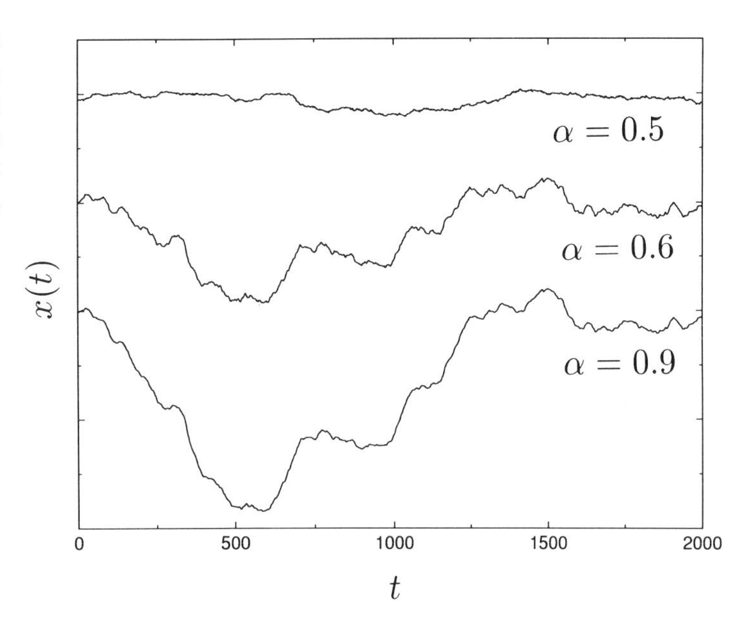

Figure 3.6 The eye can distinguish landscapes with different values of the roughness exponent α. (After [281]).

is

$$d_f = 2 - \alpha = 1.5. \tag{3.11}$$

In general it is possible to generate fractional Brownian motion with a variance

$$\langle [x(t_2) - x(t_1)]^2 \rangle^{1/2} \sim |t_2 - t_1|^\alpha, \tag{3.12}$$

resulting in a landscape with *tunable* roughness exponent α. Interfaces with different roughness exponents α appear different to even the untrained eye, as illustrated in Fig. 3.6.

So far we have concentrated on self-affine functions of a one-dimensional variable x, corresponding to a one-dimensional substrate. In general the substrate may be d-dimensional, and can be described by a function $h(\mathbf{x})$, where $\mathbf{x} \equiv (x_1, x_2, \ldots, x_d)$. The generalization to any dimension is straightforward.

3.4 Discussion

In this chapter, we have discussed the geometric basis of various scaling laws. In particular, we argued that interfaces produced by growth processes are self-affine, which means that rescaling a part of the interface anisotropically we obtain an interface that is *statistically* indistinguishable from the whole. The rescaling factors are b and b^α. These results imply that in order to characterize the morphology of a rough interface, it is sufficient to give its roughness exponent α. However, there are some interfaces for which this characterization is too restrictive, and we require an infinite number of exponents for a complete description. The properties of such 'multi-affine' interfaces are discussed in Chapter 24.

Before commencing our systematic investigation of growth processes, we must resolve a potential ambiguity in notation. The dimension of an interface will be denoted by d. Thus $d = 1$ denotes a one-dimensional interface embedded in a two-dimensional plane (e.g., the wet front during paper wetting experiments), $d = 2$ denotes a two-dimensional interface embedded in a three-dimensional space (e.g., the surface of a table or a mountain). It is sometimes useful to contemplate – or even simulate – higher dimensional 'interfaces,' with $d = 3, 4, \ldots$.

Suggested further reading:

[3, 49, 126, 156, 287, 298, 323, 361, 425, 436, 456]

Exercises:

3.1 Calculate the fractal dimension for the Vicsek fractal, the construction of which is shown in Fig. 3.7.

3.2 Generalize Eq. (3.8) to higher-dimensional interfaces.

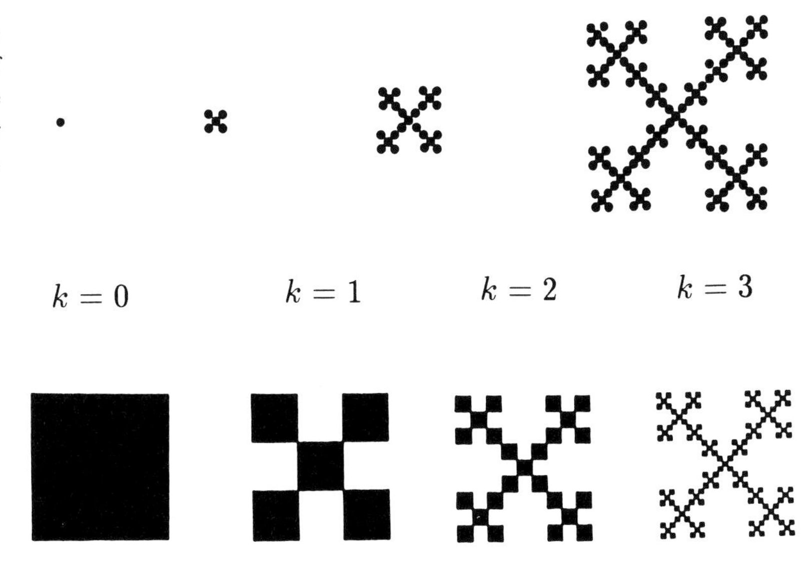

Figure 3.7 Two different methods of constructing the Vicsek fractal. (After [456]).

4 Random deposition

In this chapter, we begin the systematic investigation of different growth processes and the corresponding universality classes. To this end, we introduce the simplest growth model, called random deposition (RD). The simplicity of this model allows us to determine the scaling exponents exactly, and to construct a continuum growth equation that leads to the same scaling exponents. The idea of associating a continuum equation with a discrete growth model will be useful in the following chapters, when we must deal with more complicated processes, such as ballistic deposition (BD).

4.1 Definition

Random deposition is the simplest of the growth models that we shall discuss in this book. It is defined in Fig. 4.1: From a randomly chosen site over the surface, a particle falls vertically until it reaches the top of the column under it, whereupon it is deposited. Thus the simulation

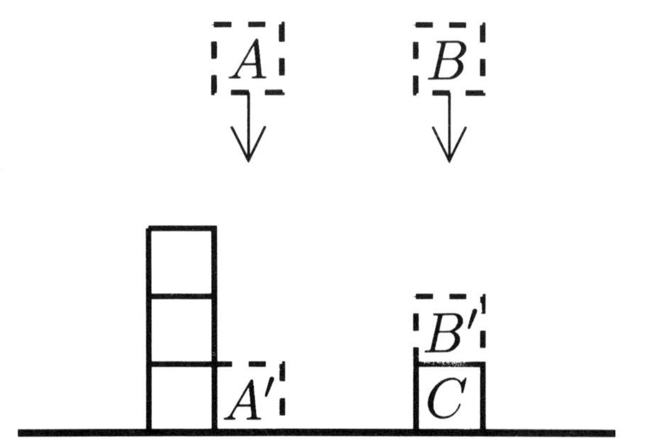

Figure 4.1 The RD model. Particles A and B are dropped from random positions above the surface and are deposited on the top of the column under them. In contrast to BD, in RD the height of the interface at a given point *does not depend* on the height of the neighboring columns.

algorithm could hardly be simpler: we choose a column i randomly, and increase its height $h(i,t)$ by one. Figure 4.2 shows the dynamics of a typical interface generated by this model.

The most important difference between RD and BD is that the RD interface is *uncorrelated*. The columns grow independently, as there is

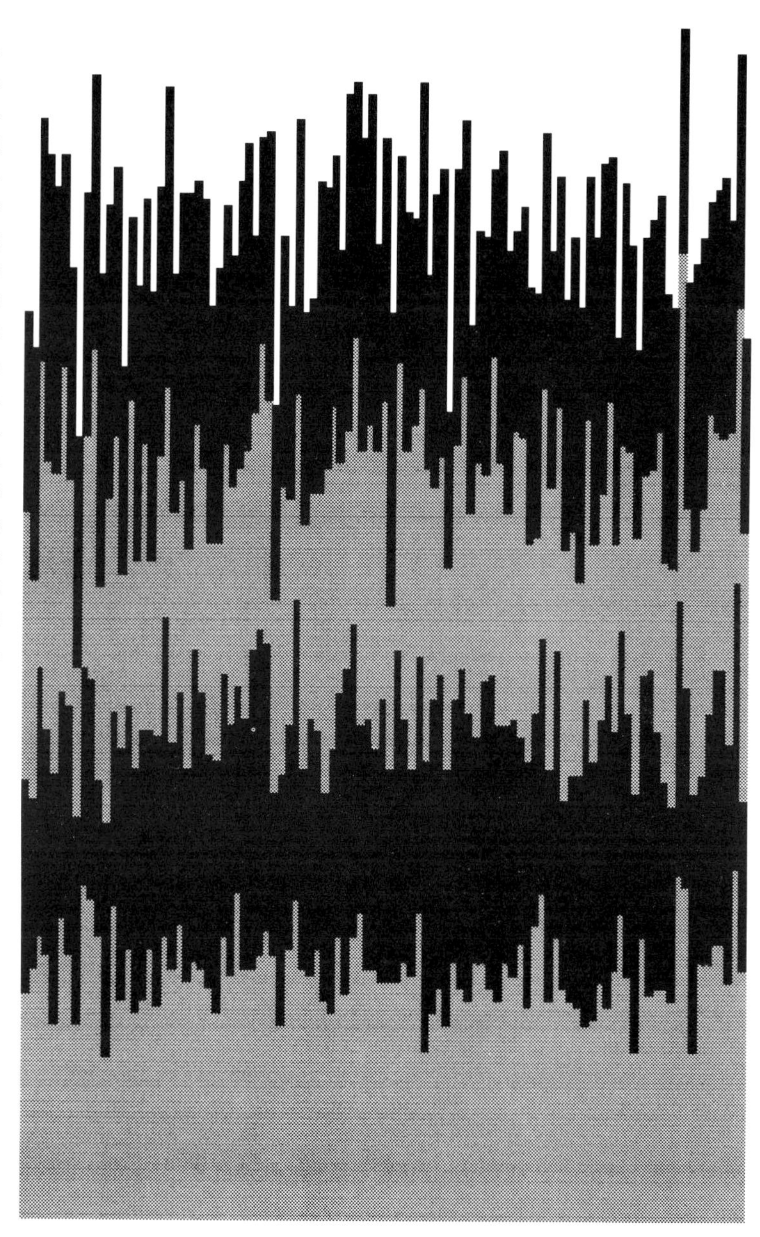

Figure 4.2 A typical interface generated by the RD model after depositing 16 000 particles on a substrate of size $L = 100$. The shading code reflects the arrival time of the particles: after the deposition of each set of 4000 particles, the shading changes. Note the rapid increase in the roughness and the uncorrelated nature of the surface, in contrast to the case of BD (Fig. 2.2).

no mechanism that can generate correlations along the interface. In BD, the fact that particles are capable of sticking to the edge of the neighboring columns leads to lateral growth, allowing the spread of correlations along the surface.

4.2 Exact solution

Since there are no correlations between the columns, every column grows independently with a probability $p = 1/L$, where L is the system size. The probability that a column has height h after the deposition of N particles is

$$P(h, N) = \binom{N}{h} p^h (1-p)^{N-h}. \tag{4.1}$$

We define the time to be the mean number of deposited layers, $t = N/L$. Then the average height grows linearly in time,

$$\langle h \rangle \equiv \sum_{h=1}^{N} h P(h, N) = Np = \frac{N}{L} = t. \tag{4.2}$$

The second moment is also straightforward to calculate,

$$\langle h^2 \rangle \equiv \sum_{h=1}^{N} h^2 P(h, N) = Np(1-p) + N^2 p^2. \tag{4.3}$$

The width of the interface is given in terms of the first and second moments by

$$w^2(t) \equiv \langle (h - \langle h \rangle)^2 \rangle = \langle h^2 \rangle - \langle h \rangle^2 = Np(1-p) = \frac{N}{L}\left(1 - \frac{1}{L}\right). \tag{4.4}$$

Since $t = N/L$, from (4.4) we have $w(t) \sim t^{1/2}$, so

$$\beta = \frac{1}{2}. \qquad \text{[RD]} \qquad \rightarrow (4.5)$$

The RD model allows the interface width to grow indefinitely with time, i.e., without saturation. This brings us back to the question addressed in §2.4, where we argued that the correlated nature of the interface is the origin of the saturation phenomena in the BD model. Since there are no correlations in the RD model, the correlation length $\xi_\parallel$ is always zero, the interface does not saturate, and the roughness exponent α is not defined.† Moreover, since the columns are not correlated, the interface is not self-affine.

4.3 Stochastic growth equations

Because the RD model is simple, we can calculate exactly all the relevant quantities using the microscopic growth rules [476]. This is not the case for other growth models, such as BD. In order to study analytically such growth models, alternative methods have been introduced. One particularly successful approach is to associate a stochastic growth equation with the given growth process.

To illustrate this approach, we now introduce a differential equation to describe RD. The goal is to find the variation with time of the interface height $h(\mathbf{x}, t)$ at any position $\mathbf{x}$, where $\mathbf{x}$ belongs to a d-dimensional substrate. In general, the growth can be described by the continuum equation‡

$$\frac{\partial h(\mathbf{x}, t)}{\partial t} = \Phi(\mathbf{x}, t), \tag{4.6}$$

where $\Phi(\mathbf{x}, t)$ is the number of particles per unit time arriving on the surface at position $\mathbf{x}$ and time t.

The particle flux is not uniform, since the particles are deposited on random positions. The randomness can be incorporated into the theory by decomposing Φ into two terms, so that (4.6) becomes

$$\frac{\partial h(\mathbf{x}, t)}{\partial t} = F + \eta(\mathbf{x}, t). \qquad \rightarrow (4.7)$$

The first, F, is the average number of particles arriving at site $\mathbf{x}$. The second, $\eta(\mathbf{x}, t)$, reflects the random fluctuations in the deposition process and is an uncorrelated random number that has zero configurational average

$$\langle \eta(\mathbf{x}, t) \rangle = 0. \qquad \rightarrow (4.8)$$

The second moment of the noise is given by

$$\langle \eta(\mathbf{x}, t)\eta(\mathbf{x}', t') \rangle = 2D\delta^d(\mathbf{x} - \mathbf{x}')\delta(t - t'). \qquad \rightarrow (4.9)$$

Relation (4.9) implies that the noise has no correlations in space and

† It is common to say that $\alpha = \infty$ for the RD model. Indeed, we can obtain this result if we define the roughness exponent by (2.5), since the saturation width is $w_{sat}(L) = \infty$ for any system size. However, if one recalls the interpretation of α in terms of the local width $w_L(\ell)$, the roughness exponent is not defined.

‡ Strictly speaking, the interface generated by the discrete models is not analytic; it consists of discrete jumps and is not differentiable. To obtain a continuum description, we first must coarse-grain the interface. Thus $h(\mathbf{x}, t)$ correctly describes an interface when the length scale is larger than the lattice spacing.

time, since averaging over the product $\langle \eta(\mathbf{x}, t)\eta(\mathbf{x}', t')\rangle$ produces zero, except for the special case in which $t = t'$ and $\mathbf{x} = \mathbf{x}'$. Conditions (4.8) and (4.9) are automatically satisfied if the noise variable η is chosen from a Gaussian distribution. Another type of noise that also satisfies both (4.8) and (4.9) and is often used in numerical simulations is 'bounded noise,' in which $\eta = +1$ and $\eta = -1$ are chosen with equal probability.

The statistical properties of an interface described by (4.7) are the same as those of an interface in RD. Integrating (4.7) over time we have

$$h(\mathbf{x}, t) = Ft + \int_0^t dt' \; \eta(\mathbf{x}, t'). \tag{4.10}$$

For the time dependence of the average height, we find

$$\langle h(\mathbf{x}, t)\rangle = Ft, \tag{4.11}$$

reproducing the result of (4.2). Squaring (4.10), we obtain $\langle h^2(\mathbf{x}, t)\rangle = F^2t^2 + 2Dt$, from which we find

$$w^2(t) - \langle h^2\rangle - \langle h\rangle^2 - 2Dt. \tag{4.12}$$

Thus we obtain the same exponent $\beta = 1/2$ as in (4.4).

4.4 Discussion

Random deposition, described either by the discrete model of §4.3 or by the stochastic growth equation (4.7), defines the simplest growth process and forms a distinct universality class in growth phenomena. Its most important property lies in the *uncorrelated* nature of the interface. The resulting interface is *not* self-affine, so the scaling picture introduced in Chapter 2 does not apply. Despite its simplicity, it is very instructive as a first step toward the more realistic correlated growth processes.

At this point it is not clear whether we gain anything by introducing the continuum growth equation (4.7), since it only reproduces the results of the exact calculation performed directly on the RD model. In fact, by introducing the continuum equation, we lose the microscopic details of the model. But this apparent disadvantage can be an advantage in some situations. We will see in the next chapter that a small modification of the deposition rule makes the discrete model intractable analytically. However, we will be able to derive a generalized form of (4.7), which in turn can be solved exactly, and provides the correct scaling exponents for the modified discrete model.

Later in the book we discuss models that are described by (4.7), but the microscopic rules are too complicated to prove this equivalence by simply comparing with the discrete model.

Exercises:

4.1 Write a program that simulates random deposition. Calculate the time-dependence of the interface width. Compare the results obtained with those of ballistic deposition.

4.2 Complete the derivation of relation (4.12).

5 Linear theory

Random deposition, discussed in the previous chapter, represents the simplest growth model. Due to the uncorrelated nature of the growth process, we were able to solve the model exactly, both in its discrete form and its continuum equation counterpart. In this chapter, we study another growth model that leads to a correlated surface: *random deposition with surface relaxation* [119]. There is no exact solution to the discrete version of the model. We present a general method to construct the growth equation from symmetry principles that will allow us to associate a stochastic equation with the model. The equation is linear, and can be solved exactly for the values of the scaling exponents.

5.1 Random deposition with surface relaxation

In the RD model, each particle falls along a single column toward the surface until it reaches the top of the interface, whereupon it sticks irreversibly. To include surface relaxation, we allow the deposited

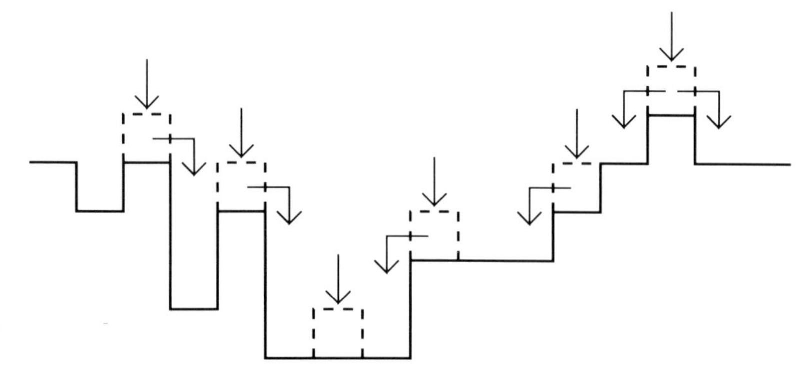

Figure 5.1 Random deposition with surface relaxation. The freshly-deposited atoms do not stick irreversibly to the site they fall on, but rather they can 'relax' to a nearest neighbor if it has a lower height.

particle to diffuse along the surface up to a finite distance, stopping when it finds the position with the *lowest* height (Fig. 5.1) [119]. As a result of the relaxation process, the final interface will be smooth, compared to the model without relaxation. Figure 5.2 shows the time evolution of the interface generated by the model.

Thus the newly-arriving particle compares the heights of nearby columns before 'deciding' where to stick. This process generates *correlations* among the neighboring heights, which lead to the entire interface being correlated. These correlations eventually lead to the saturation of the interface. Simulations in one dimension result in the

Figure 5.2 The interface at 10 successive times for a simulation of random deposition with surface relaxation, obtained by depositing 35 000 particles on a substrate of size $L = 100$. The shading reflects the arrival time of the particles: after the deposition of each set of 3500 particles, the shading changes. Note that the interface is smoother than that shown in Fig. 4.2 for the RD model.

scaling exponents [119]

$$\beta = 0.24 \pm 0.01, \qquad \alpha = 0.48 \pm 0.02. \tag{5.1}$$

These values are quite different from the values $\beta = 1/2$ and '$\alpha = \infty$' that we found for the RD model, presumably reflecting the facts that, unlike in RD, the interface is correlated and the width saturates. Can we calculate analytically these exponents? What are the exponents for higher dimensions? These are among the questions we shall address in this chapter, but first we discuss the continuum equation that describes the model.

5.2 Symmetry principles

In the previous chapter we used simple physical arguments to construct the stochastic growth equation describing the RD model. Those arguments worked well, the continuum equation having reproduced faithfully the results of the exact solution of the discrete model. In order to treat more complicated models, such as random deposition with surface relaxation, we introduce a more systematic method to derive the continuum equation describing a given discrete model. Specifically, we shall see that the key ingredient is judicious consideration of symmetry principles [189]. Our guiding principle is that the equation of motion should be the *simplest possible equation compatible with the symmetries of the problem.*

In this chapter we illustrate the application of symmetry principles by deriving the equation describing the *equilibrium* interface, where by equilibrium we mean that the interface is not driven by an external field. Thus an equilibrium interface separates two domains that are in 'equilibrium' in the sense that one domain is not growing at the expense of the other. Such interfaces are observed in magnetic systems (the boundary between two domains with different magnetization) or in immiscible fluids. The growth equation we derive will describe the random deposition model with surface relaxation as well.

Consider an interface characterized by its height $h(\mathbf{x}, t)$, and assume that $h(\mathbf{x}, t)$ is single valued – i.e., there are no 'overhangs'. Our goal is to generalize (4.7), and derive a growth equation for correlated interfaces. We expect the growth equation to have the form†

$$\frac{\partial h(\mathbf{x}, t)}{\partial t} = G(h, \mathbf{x}, t) + \eta(\mathbf{x}, t). \tag{5.2}$$

† In (5.2) we neglect on the left hand side the 'inertial term' $\partial^2 h/\partial t^2$ and higher order time derivatives, since we are interested in the long-time behavior of the system. Such terms are irrelevant in the asymptotic limit, as we show in this section.

Here $G(h, \mathbf{x}, t)$ is a general function that depends on the interface height, position and time and $\eta(\mathbf{x}, t)$ is the noise term. We are interested in the behavior of the interface shown schematically in Fig. 5.3. As a first step in obtaining the growth equation, we list the basic symmetries of the problem:

(i) *Invariance under translation in time.* The growth equation should not depend on where we define the origin of time, so the equation must be invariant under the transformation $t \to t + \delta_t$. This symmetry rules out an explicit time dependence of G.

Example: Suppose we have the term t^2 in the growth equation. Under translation in time, this term becomes $(t + \delta_t)^2$. Not being invariant, t^2 is excluded from the growth equation. Note that $\partial h/\partial t$ is invariant under translation in time, since $\partial h/\partial(t + \delta_t) = \partial h/\partial t$.

(ii) *Translation invariance along the growth direction.* The growth rule should be independent of where we define $h = 0$, so the growth equation should be invariant under the translation $h \to h + \delta_h$. This symmetry rules out the explicit h dependence of G, so that the equation must be constructed from combinations of $\nabla h, \nabla^2 h, \dots, \nabla^n h$.

Example: Assume that we have the term h^2 in the equation. Under a translation along h it becomes $(h + \delta_h)^2$, so h^2 should be excluded from the equation. Note that $\nabla(h + \delta_h) = \nabla h$, so the ∇h term survives this transformation.

(iii) *Translation invariance in the direction perpendicular to the growth direction.* The equation should not depend on the actual value

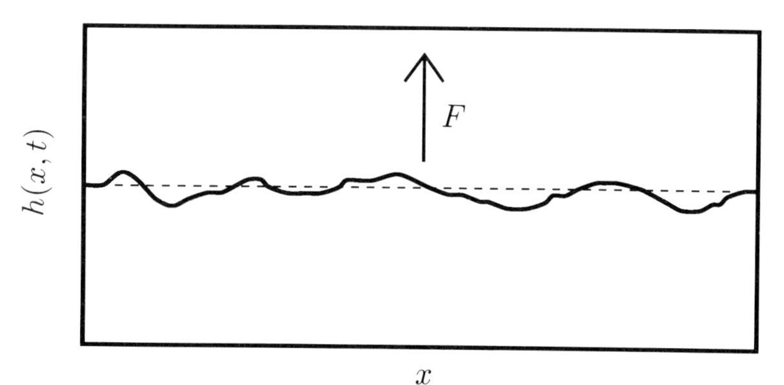

Figure 5.3 Schematic illustration of a typical one-dimensional interface $h(x, t)$ driven by an external force F, where $F = 0$ corresponds to the case of an equilibrium interface studied in this chapter. The interface is single-valued (no overhangs allowed) and has a well-defined orientation (dashed line).

of $\mathbf{x}$, having the symmetry $\mathbf{x} \to \mathbf{x} + \delta_x$. This excludes explicit $\mathbf{x}$ dependence of G.

Example: Assume that we have the term x^2 in the growth equation. Under a translation along x it becomes $(x + \delta_x)^2$, changed under this transformation, thus it should be excluded from the growth equation. Note that $\partial h(x + \delta_x)/\partial x = \partial h/\partial x$, so ∇h survives this transformation. Similarly, the functional form $\partial^n/\partial \mathbf{x}^n$ with $n > 0$ obeys this symmetry.

(iv) *Rotation and inversion symmetry about the growth direction* $\mathbf{n}$. This rules out odd order derivatives in the coordinates, excluding vectors such as ∇h, $\nabla(\nabla^2 h)$, etc.

Example: We saw that ∇h survives both (ii) and (iii). However, it does not survive this last requirement. Under $\mathbf{x} \to -\mathbf{x}$ it becomes $\partial h/\partial(-x) = -\partial h/\partial x$, i.e., $\partial h/\partial x$ changes sign under this transformation, so it should be excluded from the growth equation. Note that $(\nabla h)^2$ and $\nabla^2 h$ both survive this transformation, since they have an even number of derivatives in x.

(v) *Up/down symmetry for h.* The interface fluctuations are similar with respect to the mean interface height. This rules out even powers of h, terms such as $(\nabla h)^2$, $(\nabla h)^4$, etc. This symmetry is intimately connected to the equilibrium nature of the interface, and for nonequilibrium problems this symmetry may be broken (see Chapter 6). (Quick check: look for invariance under $h \to -h$).

Example: The nonlinear term $(\nabla h)^2$ survives all requirements imposed in the previous points. However, it does not survive this last requirement. To see this, we must look at the entire growth equation $\partial_t h = (\nabla h)^2$. Under $h \to -h$ the equation becomes $\partial_t(-h) = (\nabla h)^2$, equivalently $\partial_t h = -(\nabla h)^2$. Since $(\nabla h)^2$ changes sign under this transformation, it should be excluded from the growth equation. Note that $\nabla^2 h$ survives this transformation, since both sides of the growth equation have an odd number of factors of h.

5.3 The Edwards–Wilkinson equation

To find the final form of the growth equation, we consider all terms that can be formed from combinations of powers of $\nabla^n h$. One by one, we eliminate all those that violate at least one of the symmetries listed above. Thus we find

$$\frac{\partial h(\mathbf{x},t)}{\partial t} = (\nabla^2 h) + (\nabla^4 h) + ... + (\nabla^{2n} h) + (\nabla^2 h)(\nabla h)^2 + ... + (\nabla^{2k} h)(\nabla h)^{2j} + \eta(\mathbf{x},t), \tag{5.3}$$

where n, k, j can take any positive integer value. For simplicity of notation we do not explicitly indicate the coefficients in front of the terms.

Since we are interested in the scaling properties, we focus on the long-time ($t \to \infty$), long-distance ($x \to \infty$) behavior of functions that characterize the surface. In this *hydrodynamic limit*, higher order derivates should be less important compared to the lowest order derivatives, as can be confirmed using scaling arguments. A more sophisticated proof is available from the renormalization group treatment (see Chapter 7 and Appendix B).

Example: We show that the $\nabla^4 h$ term is 'irrelevant' in the hydrodynamic limit, compared to the $\nabla^2 h$ term. By 'irrelevant' we mean that this term does not affect the scaling behavior of the growth equation; its inclusion would have no effect on the scaling exponents. We rescale the interface in the $\mathbf{x}$ direction by a factor b, performing the scale transformation $\mathbf{x} \to \mathbf{x}' \equiv b\mathbf{x}$. Since the interface is assumed to be self-affine with a roughness exponent α, the height must be rescaled as well, according to (3.4), as $h \to h' \equiv b^\alpha h$ in order to obtain an interface with similar geometrical properties. As a result of this transformation, the $\nabla^2 h$ and $\nabla^4 h$ terms will be rescaled as

$$\nabla^2 h \to \nabla'^2 h' \equiv b^{\alpha-2} \nabla^2 h \tag{5.4}$$

and

$$\nabla^4 h \to \nabla'^4 h' \equiv b^{\alpha-4} \nabla^4 h \tag{5.5}$$

In the hydrodynamic limit ($b \to \infty$), the term $\nabla^4 h \to 0$ faster than $\nabla^2 h$. Hence $\nabla^4 h$ is irrelevant compared to $\nabla^2 h$ and we neglect the $\nabla^4 h$ term compared to $\nabla^2 h$.

A similar argument can be used to show that $(\nabla^2 h)(\nabla h)^2$ is the most relevant of the possible $(\nabla^{2k} h)(\nabla h)^{2j}$ terms of (5.3), but is irrelevant compared to the $\nabla^2 h$ term.

The noise term, $\eta(\mathbf{x}, t)$ in (5.3) incorporates the stochastic character of the fluctuation process. In the simplest case, it is assumed to be uncorrelated, with the properties (4.8) and (4.9). Thus, the simplest equation describing the fluctuations of an equilibrium interface, called the Edwards–Wilkinson (EW) equation [77, 112], has the form

$$\frac{\partial h(\mathbf{x},t)}{\partial t} = \nu \nabla^2 h + \eta(\mathbf{x},t). \qquad \text{[EW]} \qquad \rightarrow (5.6)$$

Here ν is sometimes called a 'surface tension', for the $\nu\nabla^2 h$ term tends to smooth the interface. Equation (5.6) is valid in the small gradient approximation, i.e., in the limit $(\nabla h) \ll 1$. This approximation is consistent with the requirement $\alpha < 1$, since $\delta h \sim (\delta \mathbf{x})^\alpha$ means that local slopes $\delta h/\delta \mathbf{x} \sim (\delta \mathbf{x})^{\alpha-1}$ decrease as we increase the size $(\delta \mathbf{x})$ of the chosen domain – slopes decrease as the system size increases.

A simple but intuitive geometrical interpretation, shown in Fig. 5.4, illustrates the *smoothing effect* of the Laplacian term $\nu\nabla^2 h$. Its most important property is that it smooths by 'redistributing' the irregularities on the interface, while *maintaining the average height unchanged*. Thus the surface tension acts as a *conservative* relaxation mechanism.

The average velocity of the interface is zero, which can be seen by using (5.6),

$$v \equiv \int_0^L d^d\mathbf{x} \left\langle \frac{\partial h}{\partial t} \right\rangle = 0. \qquad (5.7)$$

The contribution from the Laplacian term is zero, due to the periodic

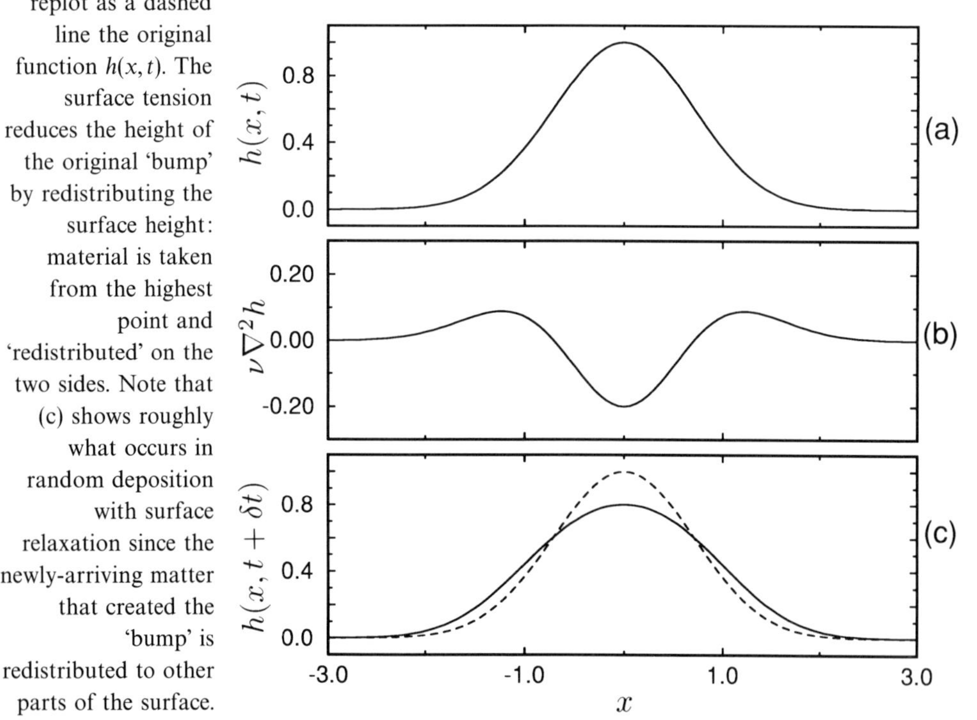

Figure 5.4 The effect of the surface tension ν in (5.6) on interface morphology. (a) Suppose that at time t the interface has a height fluctuation ('bump'). In (b) we show the $\nu\nabla^2 h$ term, which is *negative* at the maximum of $h(x,t)$. (c) At time $t+\delta t$, the height becomes $h(x,t+\delta t) \approx h(x,t)+\delta t \times \nu\, \nabla^2 h(x,t)$ [continuous line], where $\delta t \to 0$ and we neglect the effect of the noise $\eta(x,t)$. For comparison, we replot as a dashed line the original function $h(x,t)$. The surface tension reduces the height of the original 'bump' by redistributing the surface height: material is taken from the highest point and 'redistributed' on the two sides. Note that (c) shows roughly what occurs in random deposition with surface relaxation since the newly-arriving matter that created the 'bump' is redistributed to other parts of the surface.

boundary conditions, and since the noise has zero average according to (4.8).

If the interface is moving with a nonzero velocity, a velocity term can be added to (5.6),

$$\frac{\partial h(\mathbf{x}, t)}{\partial t} = v + \nu \nabla^2 h + \eta. \tag{5.8}$$

The uniform motion generated by the velocity term does not affect the scaling properties of the interface. It can be transformed away by viewing the process from a system of coordinates moving with velocity v by performing the change of variables $h \to h + vt$.

The growth of an interface governed by (5.8) displays the same general features as the BD model. We start from flat initial conditions, $h(\mathbf{x}, 0) = 0$, and the interface advances with a constant velocity v. Its roughness increases in time as $w \sim t^\beta$, and saturates at a size-dependent width $w_{\mathrm{sat}} \sim L^\alpha$. This behavior can be incorporated into the scaling law (2.8), but the growth exponents are different from those of BD. Due to the linear character of the growth equation (5.8), we can obtain the growth exponents exactly – as we shall see in the next section.

5.4 Solving the EW equation

There are two ways to obtain the scaling exponents of the EW equation: we can use scaling arguments, or we can simply solve the equation. We present both methods, since scaling arguments are useful in situations when exact solutions are not available.

5.4.1 Scaling approach

If the interface $h(\mathbf{x}, t)$ is self-affine, then on rescaling it both horizontally

$$\mathbf{x} \to \mathbf{x}' \equiv b\mathbf{x} \tag{5.9}$$

and vertically

$$h \to h' \equiv b^\alpha h, \tag{5.10}$$

we should obtain an interface that is statistically indistinguishable from the original one. The growth equation (5.6) must be invariant under this transformation. Since the interface roughness depends on time t as well, to compare two interfaces obtained at different moments we must also rescale the time,

$$t \to t' \equiv b^z t. \tag{5.11}$$

Substituting (5.9)–(5.11) into (5.6) we find

$$b^{\alpha-z}\frac{\partial h}{\partial t} = \nu b^{\alpha-2}\nabla^2 h + b^{-d/2-z/2}\eta, \tag{5.12}$$

where we used the general property of the delta functions,

$$\delta^d(a\mathbf{x}) = \frac{1}{a^d}\delta^d(\mathbf{x}). \tag{5.13}$$

Since $\eta(\mathbf{x},t) \to b^{-(d+z)/2}\eta(\mathbf{x},t)$, on rescaling the product $\langle\eta(\mathbf{x},t)\eta(\mathbf{x}',t')\rangle$, we find, using (4.9),

$$\langle\eta(b\mathbf{x}, b^z t)\eta(b\mathbf{x}', b^z t')\rangle = 2D\delta^d(b\mathbf{x} - b\mathbf{x}')\delta(b^z t - b^z t')$$

$$= 2Db^{-(d+z)}\delta^d(\mathbf{x} - \mathbf{x}')\delta(t - t'). \tag{5.14}$$

Multiplying both sides of (5.12) with $b^{z-\alpha}$, we obtain

$$\frac{\partial h}{\partial t} = \nu b^{z-2}\nabla^2 h + b^{-d/2+z/2-\alpha}\eta. \tag{5.15}$$

To find the correct exponents α and z, we require that the EW equation must be invariant under the transformation (5.12). Thus to ensure scale invariance, each term on the rhs of (5.15) must be independent of b, which implies

$$\alpha = \frac{2-d}{2}, \qquad \beta = \frac{2-d}{4}, \qquad z = 2. \qquad \text{[EW]} \qquad \to (5.16)$$

5.4.2 Exact solution

Because of the linear character of (5.6), we can obtain both the scaling form and the exponents by solving the growth equation exactly. Fourier transforming (5.6) in space and time, we find

$$h(\mathbf{k},\omega) = \frac{\eta(\mathbf{k},\omega)}{\nu k^2 - i\omega} \tag{5.17}$$

where $\eta(\mathbf{k},\omega)$ is the Fourier transform of $\eta(\mathbf{x},t)$. From (4.8) and (4.9),

$$\langle\eta(\mathbf{k},\omega)\rangle = 0 \tag{5.18}$$

and

$$\langle\eta(\mathbf{k},\omega)\eta(\mathbf{k}',\omega')\rangle = 2D\delta^d(\mathbf{k}+\mathbf{k}')\delta(\omega+\omega'). \tag{5.19}$$

From (5.17), we obtain the correlation function

$$\langle h(\mathbf{k},\omega)h(\mathbf{k}',\omega')\rangle = \frac{\langle\eta(\mathbf{k},\omega)\eta(\mathbf{k}',\omega')\rangle}{(\nu k^2 - i\omega)(\nu k'^2 - i\omega')}. \tag{5.20}$$

With some algebra [340] we find, after the inverse transform back to real space, the result

$$\langle h(\mathbf{x},t)h(\mathbf{x}',t')\rangle = \frac{D}{2\nu}|\mathbf{x}-\mathbf{x}'|^{2-d} \; f\left(\frac{\nu|t-t'|}{|\mathbf{x}-\mathbf{x}'|^2}\right). \qquad (5.21)$$

Here $f(u)$ is a scaling function with the properties $f(u) \to$ const as $u \to \infty$ and $f(u) \to u^{(2-d)/2}$ as $u \to 0$. Comparing (5.21) with (2.8), we obtain the same scaling exponents (5.16).

For $d = 2$, we find $\alpha = \beta = 0$, confirming the scaling result (5.16). In this case the correlations decay logarithmically, i.e., the width scales logarithmically with time at early times, and the saturation width depends on the logarithm of the system size. We find $t_\times \sim L^2$, so $z = 2$, confirming (5.16).

For the EW equation, we find that the scaling exponents depend on

Figure 5.5 *Off-lattice* random deposition with surface relaxation. (After [307]).

the dimension of the interface. For $d > 2$, the roughness exponent α of (5.16) becomes negative, which means that the interface is flat. Every noise-induced irregularity which generates nonzero width is suppressed by the surface tension.

We note that the scaling exponents (5.16) for $d = 1$ give $\beta = 1/4$ and $\alpha = 1/2$, very close to the results (5.1) that have been found to describe the random deposition model with surface relaxation. Indeed, as illustrated in Fig. 5.4, the way the surface tension reorganizes the height of the surface is similar to the local relaxation process used in the model. The similarity of the relaxation mechanism and the coincidence of the exponents in one dimension lead us to conclude that the model and the EW equation belong to the *same universality class*. Further support for this conclusion is given by numerical results in higher dimensions: $2 + 1$ dimensional simulations reveal a very weak divergence of the width, consistent with the logarithmic scaling ($\alpha = 0$) predicted by (5.16) [307].

The RD model we described in §4.1 is a lattice model, which means that particles can have only discrete positions fixed by the lattice. Similar exponent values are found for off-lattice restructuring models (in which discs are falling on the surface and relax until they reach the lowest position in the neighborhood – see Fig. 5.5) [307].

5.5 Discussion

The discrete growth model, RD with surface relaxation, and the continuous Edwards–Wilkinson equation (5.6) define a new universality class, different from RD. The main difference between the two universality classes can be reduced to the question of correlations between neighboring sites, which are absent in RD, and present in the EW model. The construction of the EW equation provides a general procedure that will be useful in the following chapters, where more complicated growth models and processes will be discussed. The main property of the EW theory that renders it linear is the existence of the up-down symmetry in h. This symmetry excludes nonlinear terms such as $(\nabla h)^2$, which, if incorporated, would change the scaling properties.

Suggested further reading:

[112, 120]

Exercises:

5.1 Write a program that simulates random deposition with surface relaxation. Calculate the time-dependence of the average height and of the interface width. Compare the results obtained with those of ballistic deposition and random deposition.

5.2 Show that for the Edwards–Wilkinson growth equation the inertial term, $\partial^2 h/\partial t^2$, is irrelevant in the asymptotic limit in the sense that it does not affect the scaling properties.

5.3 Construct the simplest growth equation that has the symmetry (ii) of §5.2 broken, but obeys symmetries (i), (iii), (iv) and (v).

5.4 Perform the calculations leading to (5.21), and calculate the scaling function $f(u)$ for the Edwards–Wilkinson equation.

6 Kardar–Parisi–Zhang equation

The EW equation, discussed in the previous chapter, was the first continuum equation used to study the growth of interfaces by particle deposition. The predictions of this linear theory change, however, when nonlinear terms are added to the growth equation. The first extension of the EW equation to include nonlinear terms was proposed by Kardar, Parisi and Zhang (KPZ) [217]. The KPZ equation, as it has come to be called, is capable of explaining not only the origin of the scaling form (2.8), but also the values of the exponents obtained for the BD model.

Although the KPZ equation cannot be solved in closed form due to its nonlinear character, a number of exact results have been obtained. Moreover, powerful approximation methods, such as dynamic renormalization group, can be used to obtain further insight into the scaling properties and exponents. In this chapter, we introduce the KPZ equation and present some of its key properties. The discussion will lead us to the exact values of the scaling exponents for one-dimensional interfaces. The renormalization group approach to the KPZ equation is then treated in the following chapter.

6.1 Construction of the KPZ equation

Although one cannot formally 'derive' the KPZ equation, one can develop a set of plausibility arguments using both (i) *physical* principles, which motivate the addition of nonlinear terms to the linear theory (5.6), and (ii) *symmetry* principles, as we did in the case of the EW equation.

First we return to the BD model of Chapter 2, which was used to introduce scaling concepts. What are the values of the scaling exponents characterizing this model? For $d = 1$, detailed numerical simulations [20, 312] result in values that differ from the predictions of the EW and RD models,

$$\alpha = 0.47 \pm 0.02, \qquad \beta = 0.33 \pm 0.006. \qquad \text{[BD]} \tag{6.1}$$

Our goal in this chapter is to calculate α and β analytically for $d = 1$.

The first step is to construct a continuum equation describing the model. To this end, we will attempt to generalize the EW equation. What is the difference between random deposition with surface relaxation and the BD model? In random deposition with surface relaxation, particles arrive on the surface and then relax, while for BD they stick to the first particle they meet. The BD rule generates lateral growth (Fig. 2.7) – implying that growth occurs in the direction of the *local* normal to the interface.

To include lateral growth into the growth equation, we first add a new particle to the surface. Growth occurs locally normal to the interface, generating an increase δh along the h axis which, by the Pythagorean theorem (Fig. 6.1), is

$$\delta h = [(v\delta t)^2 + (v\delta t \nabla h)^2]^{1/2} = v\delta t[1 + (\nabla h)^2]^{1/2}. \tag{6.2}$$

If $|\nabla h| \ll 1$, we can expand† (6.2),

$$\frac{\partial h(\mathbf{x}, t)}{\partial t} = v + \frac{v}{2}(\nabla h)^2 + \ldots, \tag{6.3}$$

suggesting that a nonlinear term of form $(\nabla h)^2$ must be present in the growth equation to reflect the presence of lateral growth. Adding this

† In general one can derive growth equations without using the small slope approximation. However, in the large scale limit these equations reduce to the KPZ equation [220, 292].

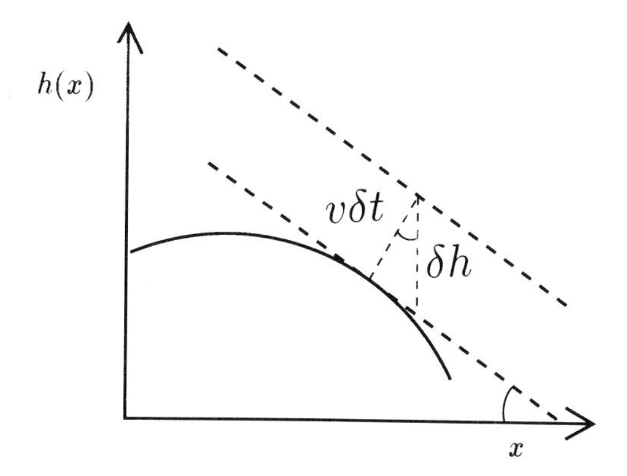

Figure 6.1 The origin of the nonlinear term in the KPZ equation (6.4). Growth occurs along the local normal v.

term to the EW equation, we obtain the KPZ equation [217]

$$\frac{\partial h(\mathbf{x},t)}{\partial t} = \nu \nabla^2 h + \frac{\lambda}{2}(\nabla h)^2 + \eta(\mathbf{x},t). \qquad \text{[KPZ]} \qquad \rightarrow (6.4)$$

The first term on the rhs describes relaxation of the interface caused by a surface tension ν. The noise η satisfies (4.8) and (4.9).

Before discussing the properties of the KPZ equation, let us show how to derive it more systematically, using symmetry principles. The KPZ equation is the simplest growth equation which has the symmetries (i)–(iv) of the linear theory discussed in §5.2, but the up-down symmetry of the interface height $h(\mathbf{x},t)$ is broken. The source of the symmetry breaking is the existence of a driving force, F, perpendicular to the interface, which selects a particular growth direction for the interface (see Fig. 5.3). In BD, this symmetry is broken due to the lateral growth property of the model. Relaxing the up-down symmetry means that the interface is *not* invariant under the transformation

$$h \rightarrow -h. \tag{6.5}$$

In the linear theory this up-down symmetry excludes terms such as $(\nabla h)^{2n}$ from the growth equation. The lowest order term of this sort is $(\nabla h)^2$ which, if added to the EW equation (5.6), results in the KPZ equation.

In general, the existence of a driving force is a *necessary*, but not *sufficient* condition for the broken up-down symmetry in h, and hence for the appearance of the nonlinear term. As an example, we recall the random deposition model with surface relaxation of §5.1, for which particle deposition generates a driving force that makes the interface grow. Deposition apparently breaks the up-down symmetry of the growth, but the model is nevertheless described by the EW equation, so in fact no symmetry breaking occurs. If we transform to a system of coordinates moving together with the interface, the model generates an interface which is invariant under the transformation $h \rightarrow -h$. Lateral growth usually implies the presence of nonlinearity, which is present in the BD model. Indeed, we shall see that the BD scaling properties are described by the KPZ equation (6.4).

6.2 Excess velocity

We noted in the previous chapter that in the absence of a driving force ($F = 0$), the mean velocity of the interface is zero [Eq. (5.7)]. In the *presence* of a nonlinear term, the mean velocity

$$v = \frac{\lambda}{2} \int_0^L d^d\mathbf{x} \, \langle (\nabla h)^2 \rangle \tag{6.6}$$

is nonzero – unless the interface is flat, with $h(x,t) = \text{const}$. Thus an interface governed by the KPZ equation has nonzero velocity even in the absence of an external driving force. This property will prove useful in the following chapters, when we try to obtain information on the existence of the nonlinear terms affecting growth processes.

It is instructive to consider a geometrical interpretation of the nonlinear term. Figure 6.2 shows a bump on the interface at time t. At time $t + \delta t$ the height of the interface is

$$h(x, t + \delta t) = h(x, t) + \frac{\lambda}{2} (\nabla h)^2 \delta t, \tag{6.7}$$

where we neglect the linear term and the noise. Since $(\nabla h)^2$ is positive, it generates an increase in height by *adding* material to the interface (or taking away material, if $\lambda < 0$), as shown in Fig. 6.2. This situation may be contrasted to the effect of the linear term, which *reorganizes* the interface height such that the total mass remains unchanged (see

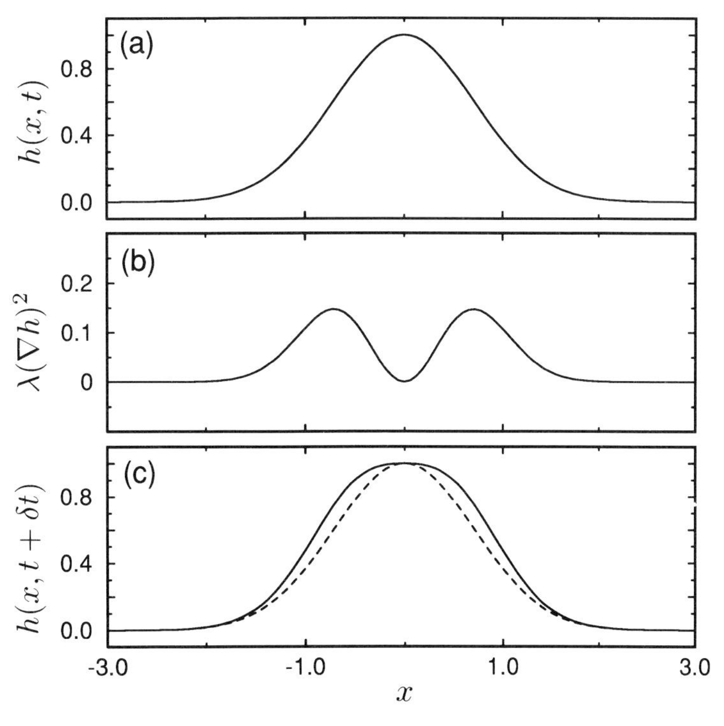

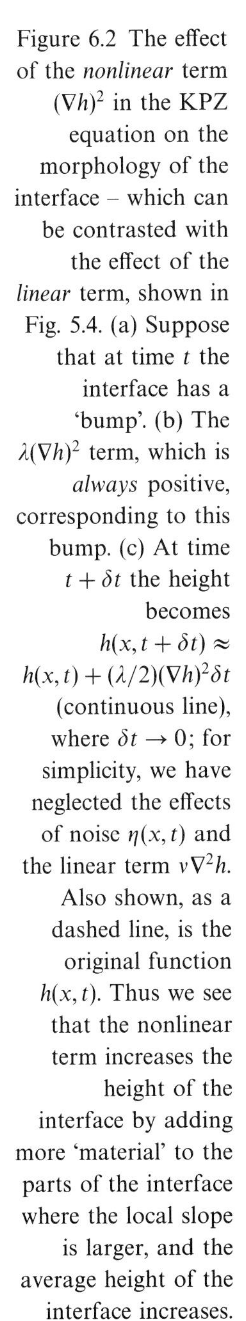
Figure 6.2 The effect of the *nonlinear* term $(\nabla h)^2$ in the KPZ equation on the morphology of the interface – which can be contrasted with the effect of the *linear* term, shown in Fig. 5.4. (a) Suppose that at time t the interface has a 'bump'. (b) The $\lambda(\nabla h)^2$ term, which is *always* positive, corresponding to this bump. (c) At time $t + \delta t$ the height becomes $h(x, t + \delta t) \approx h(x,t) + (\lambda/2)(\nabla h)^2 \delta t$ (continuous line), where $\delta t \to 0$; for simplicity, we have neglected the effects of noise $\eta(x,t)$ and the linear term $\nu \nabla^2 h$. Also shown, as a dashed line, is the original function $h(x,t)$. Thus we see that the nonlinear term increases the height of the interface by adding more 'material' to the parts of the interface where the local slope is larger, and the average height of the interface increases.

Fig. 5.4). The material added by the nonlinear term generates the 'excess velocity' of Eq. (6.6).

6.3 Scaling arguments

We now show that the nonlinear term $(\nabla h)^2$ is in fact relevant, and therefore affects the scaling exponents. The scale transformation $\mathbf{x} \to b\mathbf{x}$, together with the corresponding rescaling in the height $h \to b^\alpha h$ and time $t \to b^z t$, transform (6.4) to

$$b^{\alpha-z}\frac{\partial h}{\partial t} = \nu b^{\alpha-2}\nabla^2 h + \frac{\lambda}{2} b^{2\alpha-2}(\nabla h)^2 + b^{-d/2-z/2}\eta. \tag{6.8}$$

If we compare $b^{\alpha-2}\nabla^2 h$ with $b^{2\alpha-2}(\nabla h)^2$, we see that, in the limit $b \to \infty$, the nonlinear term dominates over the surface tension term, provided $\alpha > 0$. Multiplying both sides of (6.8) with $b^{z-\alpha}$, we obtain

$$\frac{\partial h}{\partial t} = \nu b^{z-2}\nabla^2 h + \frac{\lambda}{2} b^{\alpha+z-2}(\nabla h)^2 + b^{-d/2+z/2-\alpha}\eta. \tag{6.9}$$

If we apply the scaling arguments that lead to the correct exponents for the EW equation, we run into a problem. To insure scaling invariance, naively one would expect that the rhs of (6.9) must be independent of b. However, this procedure provides *three* scaling relations for the *two* exponents, α and z, thereby overdetermining them. We proceed to argue that since the nonlinear term dominates over the surface tension, we can neglect the $\nu\nabla^2 h$ term. We thereby obtain $\alpha = (2-d)/3$ and $\beta = (2-d)/(4+d)$, which cannot be correct since for $d = 1$ they predict $\alpha = 1/3$ and $\beta = 1/5$, quite different from (6.1).

There are two possible resolutions: either (i) the KPZ equation does not describe BD, or (ii) something went wrong with the scaling arguments. Choice (ii) is correct. The reason is that, rescaling our system, the different terms (ν, λ and D) in the growth equation do not renormalize independently – being coupled to each other. Thus, in order to have scale invariance, we cannot simply assume that the exponents of b are zero in (6.9), since ν, D and λ may also change under rescaling. A more precise discussion of this problem will be possible in the next chapter, where we discuss the renormalization group approach. In this chapter we continue by discussing some exact results that lead us to the correct scaling exponents in one dimension.

6.4 Exponents

To obtain the scaling exponents we proceed in two steps. First, we discuss Galilean invariance and mapping to the Burgers equation which, combined with the scaling arguments of §6.3, will lead us to a scaling relation between the two independent exponents. This scaling relation will leave only one independent exponent. Second, we discuss the fluctuation–dissipation theorem, that will allow us to calculate the roughness exponent in one dimension.

6.4.1 Burgers equation and Galilean invariance

Let us consider for the moment the Burgers equation with noise for a vorticity-free velocity field, a paradigm of turbulence studies [66]

$$\frac{\partial \mathbf{v}}{\partial t} + \lambda(\mathbf{v} \cdot \nabla)\mathbf{v} = \nu\nabla^2\mathbf{v} - \nabla\eta(\mathbf{x}, t). \tag{6.10}$$

Here $\mathbf{v}(\mathbf{x}, t)$ is the velocity of the fluid, ν is the viscosity and $\nabla\eta(\mathbf{x}, t)$ is a random force. The Burgers equation describes a randomly stirred vorticity-free ($\nabla \times \mathbf{v} = 0$) fluid. The noisy Burgers equation (6.10) can be mapped into the KPZ equation by using the change of variables

$$\mathbf{v} = -\nabla h. \tag{6.11}$$

Thus the Burgers equation and the KPZ equation should have related scaling exponents.

The left hand side of the Burgers equation originates from the *total* derivative

$$\frac{D\mathbf{v}}{Dt} = \frac{\partial \mathbf{v}}{\partial t} + (\mathbf{v} \cdot \nabla)\mathbf{v}. \tag{6.12}$$

Hence we should have $\lambda = 1$. We include the coefficient λ only for convenience. After any rescaling we expect that the total derivative (6.12) remains unchanged,† so $\lambda = 1$. When we rescale the KPZ equation – arriving at (6.9) – the coefficient of the nonlinear term should therefore remain unchanged under rescaling, leading to the scaling relation

$$\alpha + z = 2. \tag{6.13}$$

Equation (6.13) provides a relation between the two unknown exponents α and $z = \alpha/\beta$ characterizing the growth. The scaling relation (6.13) is valid in any dimension, and is confirmed by the renormalization group calculation (see Chapter 7).

† More precisely, the total derivative $D\mathbf{v}/Dt$ may change, but both terms in the rhs of (6.12) will have the same scaling factor.

This relation is also a consequence of an additional symmetry of the KPZ equation, called Galilean invariance. The Burgers equation is invariant under the Galilean transformation

$$\mathbf{v}(\mathbf{x},t) \to \mathbf{v}_0 + \mathbf{v}'(\mathbf{x} - \mathbf{v}_0 t, t), \tag{6.14}$$

which is an exact symmetry of the microscopic dynamics. Physically (6.14) means that viewing the fluid from a system of coordinates that moves with velocity $\mathbf{v}_0$ does not change the laws of physics, so the Burgers equation should be invariant. This invariance transforms into an invariance under tilting of the interface by an infinitesimal angle ϵ for the KPZ equation. With the transformation

$$h' \equiv h + \epsilon \mathbf{x}, \quad \mathbf{x}' \equiv \mathbf{x} - \lambda \epsilon t, \quad t' = t, \tag{6.15}$$

the tilted KPZ equation satisfies

$$\frac{\partial h'(\mathbf{x}',t')}{\partial t'} = \nu \nabla'^2 h' + \frac{\lambda}{2}(\nabla' h')^2 + \eta(\mathbf{x}' + \lambda \epsilon t', t'). \tag{6.16}$$

Equation (6.16) displays Galilean invariance, since the uncorrelated noise can be shown to be invariant as well [314].

6.4.2 Fokker–Planck equation and fluctuation–dissipation theorem

Let us consider in general the Langevin equation

$$\frac{\partial h}{\partial t} = G(h) + \eta(t), \tag{6.17}$$

where the noise has the correlations $\langle \eta(t)\eta(t')\rangle = 2D\delta(t-t')$. We can associate with (6.17) a Fokker–Planck equation describing the time evolution of the probability $\Pi(h,t)$ of having height h at time t, [212]

$$\frac{\partial \Pi}{\partial t} = -\frac{\partial}{\partial h}\left[G(h)\Pi\right] + D\frac{\partial^2 \Pi}{\partial h^2}. \tag{6.18}$$

The situation with the KPZ equation is slightly more complicated because the height depends on $\mathbf{x}$. Thus we have a set of continuous variables parametrized by the coordinate $\mathbf{x}$, with respect to which we must integrate the Fokker–Planck equation [144, 159]

$$\frac{\partial \Pi}{\partial t} = -\int d^d\,\mathbf{x}\frac{\delta}{\delta h}\left[\left(\nu\nabla^2 h + \frac{\lambda}{2}(\nabla h)^2\right)\Pi\right] + D\int d^d\,\mathbf{x}\frac{\partial^2 \Pi}{\partial h^2}. \tag{6.19}$$

Here we replaced $G(h)$ with its form given by the KPZ equation (6.4).

Since (6.4) cannot be directly obtained from a Hamiltonian (see Appendix C), it is not *a priori* true that (6.19) has a stationary solution. Nevertheless, we can check that for $d = 1$ the probability distribution

$$\Pi = \exp\left(-\int dx \left[\frac{\nu}{2D}(\partial_x h)^2\right]\right) \tag{6.20}$$

is a solution of (6.19). It is instructive to check that for $d > 1$, (6.20) is *not* a solution of the Fokker–Planck equation (6.19) [389]. However, for $\lambda = 0$ it is a solution for *any* dimension.

We can obtain the roughness exponent directly from (6.20), which tells us that the local slopes, ∇h, follow a Gaussian distribution – i.e. they are random and uncorrelated. Summing up local random slopes we should obtain the interface, which is exactly Brownian motion. Thus the roughness exponent of the interface compatible with (6.20) is†

$$\alpha = \frac{1}{2}. \qquad \text{[KPZ]} \qquad \rightarrow (6.21)$$

Using (6.13) and (2.11) we obtain for the other two exponents

$$z = \frac{3}{2}, \qquad \beta = \frac{1}{3}. \qquad \text{[KPZ]} \qquad \rightarrow (6.22)$$

Comparing with the numerically-obtained exponents for the BD model (6.1), we find remarkable agreement, suggesting that indeed the KPZ equation and the BD model belong to the same universality class.

6.5 Discussion

In this chapter we introduced the KPZ equation, and we presented some of its properties. Additional important properties of the KPZ equation are discussed in Chapter 25. We emphasize that the scaling relation (6.13) is exact, and is valid in *any* dimension. However, the fluctuation–dissipation theorem, leading to $\alpha = 1/2$, is valid only in one dimension. Thus we cannot determine the higher-dimensional exponents at this point. In the next chapter we discuss the dynamic renormalization group, and its application to the KPZ equation. We will be able to recover again the one-dimensional scaling exponents, and to obtain important results regarding the higher dimensional behavior.

Suggested further reading:

[159, 217]

† Another way to obtain (6.21) is to observe that (6.20) is a solution of the Fokker–Planck equation associated with both the KPZ and the EW equations. Thus, since the two interfaces have the same height distribution, they should have the same roughness exponent.

Exercises:

6.1 Find a geometrical interpretation for (6.15).

6.2 Show that (6.20) is a solution to the Fokker–Planck equation (6.19) if $d = 1$, but not for $d > 1$.

6.3 Consider a growth process, described by the KPZ equation, in which the $x \to -x$ symmetry is broken and a term $a_1 \nabla h$ thus allowed into the continuum equation. Show that with a suitable coordinate transformation this term can be eliminated from the growth equation, thereby demonstrating that it does not affect the scaling exponents.

7 Renormalization group approach

The exact results presented in the previous chapter allow us to obtain the scaling exponents for $d = 1$, and reduce the number of independent scaling exponents to one. The same results can be obtained using the dynamic renormalization group method, which we now develop and use to study the scaling properties of the KPZ equation. In particular, we analyze the 'flow equations' and extract the exponents describing the KPZ universality class for $d = 1$. We also discuss numerical results leading to the values of the scaling exponents for higher dimensions.

7.1 Basic concepts

So far, we have argued that we can distinguish between various growth models based on the values of the scaling exponents α, β and z. The existence of universal scaling exponents and their calculation for various systems is a central problem of statistical mechanics. A main goal for many years has been to calculate the exponents for the Ising model, a simple spin model that captures the essential features of many magnetic systems. A major breakthrough occurred in 1971, when Wilson introduced the renormalization group (RG) method to permit a systematic calculation of the scaling exponents [479]. Since then the RG has been applied successfully to a large number of interacting systems, by now becoming one of the standard tools of statistical mechanics and condensed matter physics. Depending on the mathematical technique used to obtain the scaling exponents, the RG methods can be partitioned into two main classes: real space and **k**-space (Fourier space) RG. The real space calculations are more intuitive, but **k**-space methods are more controlled mathematically. In this section, we illustrate the basic ideas of the RG on the Ising model using a real space construction [67, 423].

Consider a triangular lattice, and classical spins with values $s_i = \pm 1$ (up or down) on each site i. The spins interact with one another, the interaction being described by the Ising Hamiltonian

$$\mathcal{H}(\{s_i\}) \equiv -J \sum_{\langle i,j \rangle} s_i s_j. \tag{7.1}$$

Here J is a coupling constant characterizing the strength of the interaction between the spins. The first sum extends over nearest-neighbor pairs only, so only nearest-neighbor spins interact.

If $J > 0$, the neighboring spins attempt to align parallel to each other, since that minimizes the total energy of the system. In this case, the model is called the ferromagnetic Ising model. The lowest energy state of the model (the minimum of the Hamiltonian (7.1)) corresponds to the configuration in which *all* spins are parallel. The probability that the system is in an arbitrary state, described by the spin configuration $\{s_i\}$, is

$$P(\{s_i\}) = \frac{1}{Z} \exp\left[-\mathcal{H}(\{s_i\})/k_B T\right]. \tag{7.2}$$

Here k_B is the Boltzmann constant, and Z is the partition function

$$Z \equiv \sum_{\{s_i\}} \exp\left[-\mathcal{H}(\{s_i\})/k_B T\right]. \tag{7.3}$$

The sum is performed over all spin configurations $\{s_i\}$. According to (7.2), at high temperatures ($T \to \infty$) all configurations are equally probable, and the system is completely disordered. However, on lowering the temperature, configurations with lower energies, corresponding to mostly parallel spins, dominate the partition function, having probability much higher than for a random state. Experiments and numerical simulations indicate that there exists a critical temperature, T_c, at which long-range correlations spread across the entire system and suddenly more than half the spins have the same orientation.

Next we outline the basic steps of the RG transformation on the triangular lattice. The RG transformation has two major steps, which we illustrate in Fig. 7.1.

Step 1 – We replace every three spins with a single block spin. If there were N spins s_i in the original system, we now have $N/3$ block spins σ_α. The direction of the block spin σ_α depends on the direction of the spins s_i comprising the block. There are a number of ways to define this dependence. The simplest is to use the majority rule: the direction of the block spin σ_α is decided by the majority of the spins s_i forming the block, so $\sigma_\alpha \equiv \text{sign}(s_i^{(1)} + s_i^{(2)} + s_i^{(3)})$. For example, if we have two down spins and one up, then the block spin will be down.

Step 2 – Since the distance between the block spins is larger than it was for the original spin system, we must rescale the lattice with a factor $b = 1/\sqrt{3}$. After this transformation the new lattice will be statistically indistinguishable from the old.

Steps 1 and 2 together define the *renormalization transformation*, which is analogous to the rescaling we applied to the continuum equations in §6.3. Here we rescale the spins instead of the height (Step 1: $s_i \to \sigma_\alpha$ is the analog of $h \to hb^\alpha$), and we rescale the entire lattice instead of only the horizontal coordinate (Step 2: changing the size of the lattice is the same as $x \to bx$). For the interface problem, we argued that on choosing the right exponents, the renormalized system is indistinguishable from the original one – i.e., we required that the equation be invariant under the rescaling. For the Ising model the Hamiltonian plays the role of the continuum equation, so we study how the Hamiltonian changes under the scale transformation performed in Steps 1 and 2.

Having the transformation rule from the spins s_i to the block spins σ_α, we can calculate the probability of finding a given spin block configuration $\{\sigma_\alpha\}$, provided we know the probability (7.2) of each spin configuration comprising the block spins. Thus we calculate

$$P(\{\sigma_\alpha\}) = \frac{1}{Z} \exp\left[-\frac{\mathscr{H}(\{\sigma_\alpha\})}{k_B T}\right], \tag{7.4}$$

which is used to define the effective Hamiltonian $\mathscr{H}(\{\sigma_\alpha\})$ describing the interaction between the block spins σ_α.

If the system is to be scale invariant, then the effective Hamiltonian for the block spins defined by (7.4) must have the same form as the original – i.e., must have a product between the neighboring blocks, and a sum over all blocks. The parameter J will be different: the starting Hamiltonian may have coupling parameters

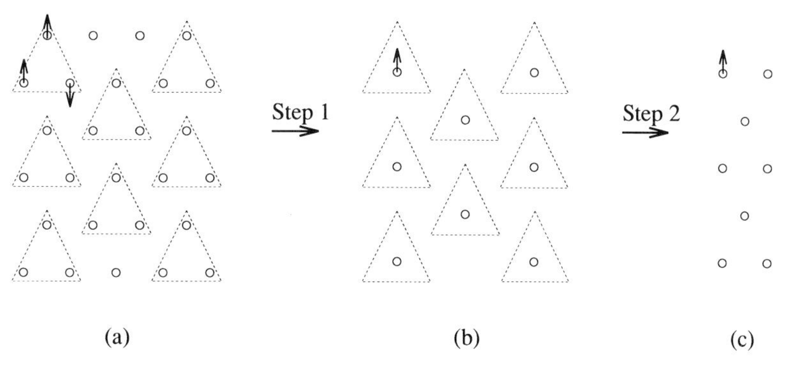

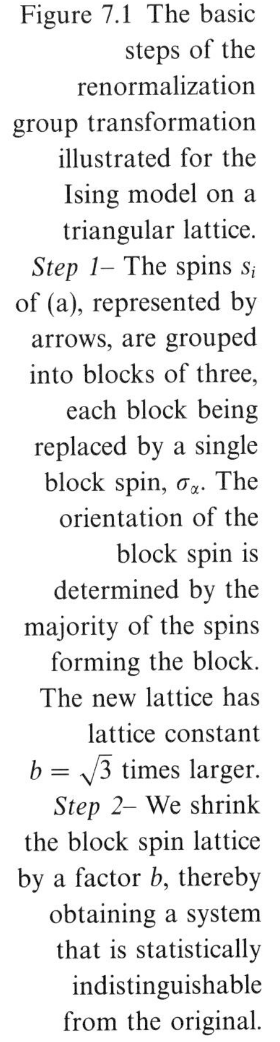
Figure 7.1 The basic steps of the renormalization group transformation illustrated for the Ising model on a triangular lattice. *Step 1*– The spins s_i of (a), represented by arrows, are grouped into blocks of three, each block being replaced by a single block spin, σ_α. The orientation of the block spin is determined by the majority of the spins forming the block. The new lattice has lattice constant $b = \sqrt{3}$ times larger. *Step 2*– We shrink the block spin lattice by a factor b, thereby obtaining a system that is statistically indistinguishable from the original.

$\mathbf{J}^{(0)} \equiv (J_1^{(0)}, J_2^{(0)}, \ldots, J_m^{(0)})$, which after the RG transformation became $\mathbf{J}^{(1)} \equiv (J_1^{(1)}, J_2^{(1)}, \ldots, J_m^{(1)})$.

We can apply the RG transformation again to the block spins, resulting in the set of parameters $\mathbf{J}^{(2)}$. In general the RG transformation R will relate the parameters $\mathbf{J}^{(n+1)}$ to $\mathbf{J}^{(n)}$ as

$$\mathbf{J}^{(n+1)} = \mathsf{R}(\mathbf{J}^{(n)}). \tag{7.5}$$

If we define the configurational space of all possible Hamiltonians (which is an m-dimensional space, since there are m parameters characterizing the effective Hamiltonian), every Hamiltonian will be a point in this space. The RG transformation (7.5) defines a 'flow' in this parameter space, telling us the direction in which the Hamiltonian moves under consecutive application of the RG transformation.

The hope behind the RG method is that under successive applications of the RG transformation we will approach an 'invariant Hamiltonian' that does not change further under the RG transformation. Mathematically this means that the transformation (7.5) has a fixed point defined by the *fixed point equation*

$$\mathbf{J}^* = \mathsf{R}(\mathbf{J}^*). \tag{7.6}$$

In general the fixed point $\mathbf{J}^*$ can be attractive, repulsive or mixed (see Fig. 7.2).

The concept of scale invariance implies that we are interested not in the original Hamiltonian, but in the fixed point Hamiltonian defined by the parameters of (7.6). It is this Hamiltonian that is helpful in determining the scaling exponents, since any other Hamiltonian in the parameter space will flow toward it (or away from it).

Our goal in this section has been to sketch the basic structure of the RG method, using the Ising model in a concrete example. We now return to our interface problem.

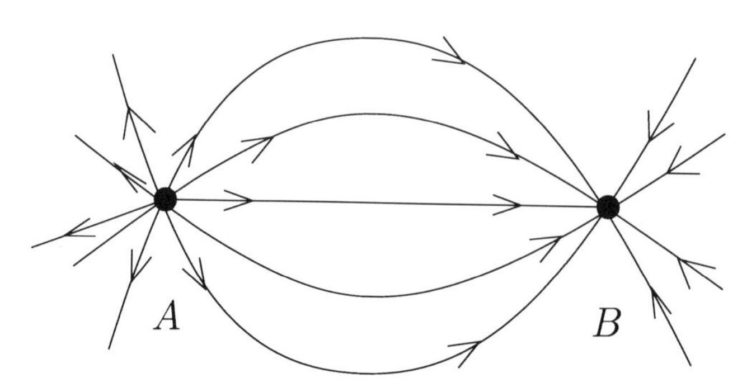

Figure 7.2 Illustration of the flow near (A) a repulsive fixed point and near (B) an attractive fixed point. Every point in this configurational space corresponds to a Hamiltonian, and under the RG transformation the Hamiltonian changes, following one of the trajectories shown as continuous lines. Under successive RG transformations a Hamiltonian in the vicinity of A flows further away from A, while one near B flows toward B.

7.2 Re-scaling in momentum space

The real space RG leads to a fixed point Hamiltonian, and the flow near the fixed point Hamiltonian provides the *equilibrium* properties of the system. However, for growth phenomena we are interested in the *dynamic* properties of the roughening process. Due to the irreversible nature of the growth process, the system has no equilibrium counterpart. Thus in order to apply RG to the KPZ equation, we must generalize the method to handle dynamic properties. The general method suitable for the study of the dynamics of various stochastic systems was introduced soon after Wilson's equilibrium theory, in the context of the dynamic properties of spin systems [179, 277]. The goal was to obtain the scaling exponents for stochastic equations similar to the KPZ equation. In this section, we outline the basic steps of the method directly applied to the KPZ equation.

In order to apply the RG to the KPZ equation, we must put the KPZ equation in a suitable form. The idea is to try to solve perturbatively the KPZ equation, starting from the known solution (5.17) of the linear problem. To this end we Fourier transform (4.19), and write it in the form (see Appendix B)

$$h(\mathbf{k},\omega) = G_0(\mathbf{k},\omega)\eta(\mathbf{k},\omega) + \lambda\mathcal{N}[h(\mathbf{k},\omega)], \tag{7.7}$$

where $G_0(\mathbf{k},\omega) \equiv 1/(\nu k^2 - i\omega)$ and $\mathcal{N}[h(k,\omega)]$ is a *nonlinear* integral functional of the height. The lattice constant a_0 provides a lower cutoff in real space, corresponding to which in momentum space there is an upper cutoff $k < \Lambda$.

If $\lambda = 0$, we have an exact solution to this equation. For $\lambda \neq 0$, we do not have such a solution due to the nonlinear part $\mathcal{N}$. However, we can try to solve the problem *perturbatively*, by assuming that λ is a small parameter. To this end we iterate (7.7), by replacing $h(\mathbf{k},\omega)$ in the rhs of (7.7) with the full form of h given by (7.7), and obtain

$$\tilde{h}(\mathbf{k},\omega) = G_0(\mathbf{k},\omega)\eta(\mathbf{k},\omega) + \lambda\mathcal{N}\left[G_0(\mathbf{k},\omega)\eta(\mathbf{k},\omega) + \lambda\mathcal{N}\right]. \tag{7.8}$$

As discussed in Appendix B, we can continue this procedure of successively iterating Eq. (7.7), and can truncate at the order desired. However the rhs contains integrals over the phase space $(\mathbf{k},\omega)$ that diverge for small $\mathbf{k}$, so the terms of the perturbation expansion are not arbitrarily small. To avoid this problem, the RG procedure effectively 'reorganizes' the perturbation series. This is done in two steps (see Fig. 7.3):

Step 1 – Divide the Brillouin zone $k \in [0,\Lambda]$ into two parts, one with high momenta, $k^> \in [\Lambda/b,\Lambda]$ and one with low momenta, $k^< \in$

$[0, \Lambda/b]$. The divergences are now contained in the low momenta part. Hence, we can integrate out the high momenta components, since the integrals in this region of **k** space are analytic.

Step 2 – The resulting integrals have a cutoff Λ/b, i.e., the system is different from the original one in that it has a different lattice spacing. We can remove this difference by rescaling our system using $k \to kb$. Since the interface is self-affine, in order to obtain a system similar to the original, we must rescale both the height and the time differently, using $h \to b^{\alpha}h$ and $t \to b^{z}t$.

Step 1 integrates out the fast modes from the growth equation, corresponding to the replacement of the spins with the larger units, the block spins. Step 2 is the same in both the KPZ and spin models, and amounts to rescaling the lattice $a_0 \to a_0/b$ to obtain back the original lattice spacing a_0. Thus the basic steps involved in the dynamic RG are fully analogous to those used for the Ising model.

The new equation of motion contains parameters that are functions of the old parameters $\mathbf{J} \equiv (\nu, \lambda, D)$. We obtain the flow equations describing the change in the parameters under the RG transformation

$$\frac{d\nu}{dl} = \nu \mathsf{R}_\nu(\nu, \lambda, D) \tag{7.9}$$

$$\frac{dD}{dl} = D\mathsf{R}_D(\nu, \lambda, D) \tag{7.10}$$

$$\frac{d\lambda}{dl} = \lambda \mathsf{R}_\lambda(\nu, \lambda, D). \tag{7.11}$$

Here the parameter l is related to the rescaling parameter by $e^l \equiv b$, so that $dl = d \log b$. Thus, instead of solving the growth equation perturbatively, we apply the RG procedure that leads us to the flow diagrams describing the variation of the parameters under rescaling.

In principle it is easy to obtain the exponents once we have these flow equations: we must adjust the exponents such that the flow

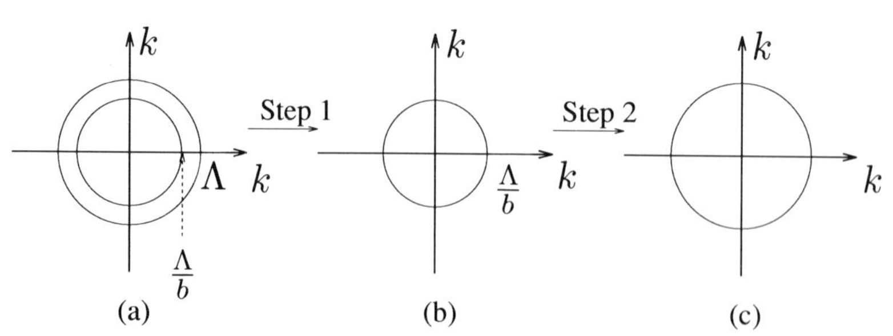

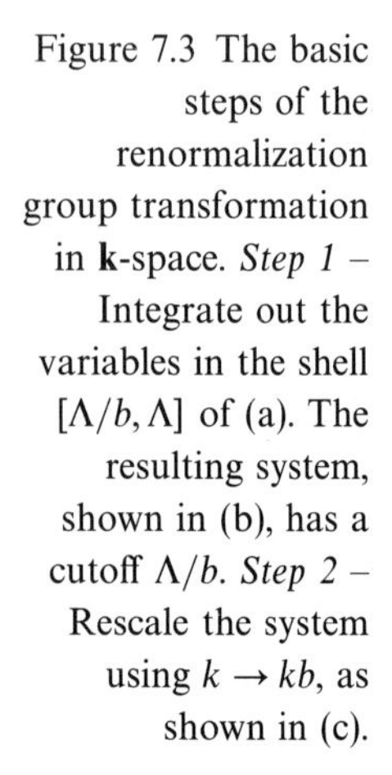
Figure 7.3 The basic steps of the renormalization group transformation in **k**-space. *Step 1* – Integrate out the variables in the shell $[\Lambda/b, \Lambda]$ of (a). The resulting system, shown in (b), has a cutoff Λ/b. *Step 2* – Rescale the system using $k \to kb$, as shown in (c).

equation has a fixed point. The fixed point means that upon further application of the RG transformation, the parameters of the KPZ equation do not change – i.e., the system is scale-invariant. The fixed point can be obtained from $(d/dl)\mathbf{J} = 0$ which implies $\mathsf{R}_\mathbf{J} = 0$. The details of this last step are discussed in the next section, and are directly applied to the KPZ equation.

7.3 Flow equations for the KPZ equation

Calculating the flow equations is a simple task 'in principle', but in fact Step 1 is quite involved. The calculational details are presented in Appendix B. The concrete form of (7.9)–(7.11) for the KPZ equation is given by (B.24), (B.26) and (B.27), which we reproduce here†

$$\frac{d\nu}{dl} = \nu\left[z - 2 + K_d g^2 \frac{2-d}{4d}\right], \tag{7.12}$$

$$\frac{dD}{dl} = D\left[z - d - 2\alpha + K_d \frac{g^2}{4}\right], \tag{7.13}$$

$$\frac{d\lambda}{dl} = \lambda[\alpha + z - 2]. \tag{7.14}$$

Here the coupling constant is defined through

$$g^2 \equiv \frac{\lambda^2 D}{\nu^3}, \qquad \rightarrow (7.15)$$

while $K_d \equiv S_d/(2\pi)^d$ and S_d is the surface area of the d-dimensional unit sphere. The exponents can be obtained by searching for the fixed points of the flow equations, i.e., by considering

$$\frac{d\nu}{dl} = \frac{dD}{dl} = \frac{d\lambda}{dl} = 0. \tag{7.16}$$

Before turning to the discussion of the KPZ problem, we examine the predictions of the flow equations for the simpler linear EW equation. Taking $\lambda = 0$, to satisfy (7.16), we must have $z-2 = 0$ and $z-d-2\alpha = 0$. Thus $z = 2$ and $\alpha = (2-d)/2$, in agreement with (5.16). Thus the parameters of the EW equation rescale *independently* (ν does not depend on D), so the scaling arguments used in §5.4.1 were correct.

Next we calculate the exponents for the KPZ problem. In this case, the parameters are not independent, indicating why the scaling arguments of §6.3 give incorrect exponents. Equation (7.14) leads

† Note that (7.12)–(7.14) were obtained using a one-loop approximation, which means that in the perturbation expansion we neglected terms of order λ^4 or higher. For higher order expansions see [135, 433].

to the scaling relation (6.13). Since there is no correction to λ, this relation is valid in any dimension. Furthermore, as a consequence of Galilean invariance, we expect that higher loop corrections will vanish as well – making (6.13) exact.

Using (7.15) and (7.12)–(7.14), we can calculate the flow of the coupling constant

$$\frac{dg}{dl} = \frac{2-d}{2}g + K_d \frac{2d-3}{4d} g^3. \tag{7.17}$$

The fixed point g^* of this equation is obtained by taking $dg/dl = 0$.

(i) For $d = 1$, there are two fixed points:

$$g_1^* = 0 \qquad g_2^* = \left(\frac{2}{K_d}\right)^{1/2}. \tag{7.18}$$

The g_2^* fixed point is attractive, which means that if we start the flow with a nonzero g different from g_2^*, it flows toward g_2^*, independent of its initial value g. On the other hand, the g_1^* fixed point is repulsive, i.e., if we start with $g \neq g_1^*$, then the system flows away from the fixed point, as illustrated in Fig. 7.4(a). If we insert the nonzero fixed point in (7.12) and (7.13), we find for the

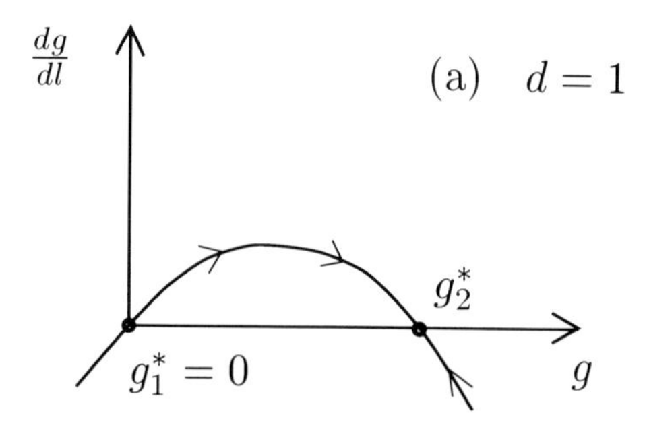

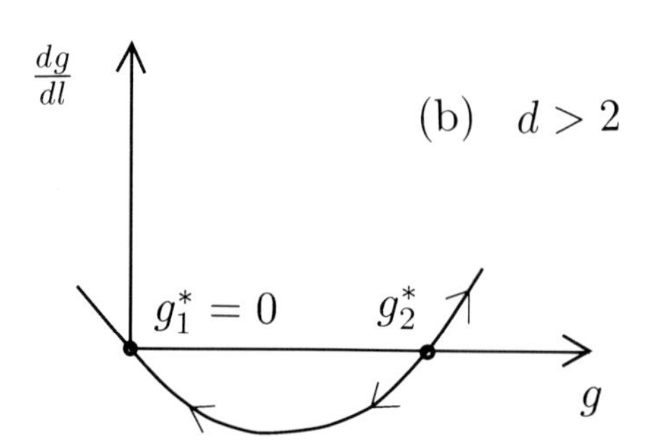

Figure 7.4 (a) The flow of the coupling constant g for $d = 1$. There is an attractive nonzero fixed point g_2^*, which determines the scaling exponents. (b) The flow of g for $d > 2$. The $g_2^*(d)$ fixed point is repulsive, separating two distinct scaling regimes: one is dominated by the linear theory (for which $\lambda = g = 0$), and the other is the strong coupling regime, for which $\lambda \to \infty$ in the one-loop approximation.

scaling exponents

$$z = \frac{3}{2} \qquad \alpha = \frac{1}{2}. \qquad \text{[KPZ (d=1)]} \qquad (7.19)$$

Thus for $d = 1$ we have obtained the *exact* values of the scaling exponents. The reason that the RG provides exact exponents is a consequence of the simultaneous validity of the Galilean invariance (6.15) and the fluctuation-dissipation theorem (6.20).

(ii) $d = 2$ is the critical dimension of the EW equation. The flow equation (7.17) indicates that the coupling constant g is marginally relevant and grows under rescaling. There is no fixed point at a nonzero g. The fixed point is determined by a strong-coupling behavior which is not accessible by perturbation theory.

(iii) For $d > 2$ the situation is different. There is a nonzero fixed point, but it is repulsive, as indicated in Fig. 7.4(b). If $g < g_2^*(d)$, the *weak coupling* regime, the coupling constant flows to zero. With zero coupling constant, the growth is described by the EW equation (5.6), with the exponents of the linear theory (5.16). If $g > g_2^*(d)$, the *strong coupling* regime, the coupling constant flows to infinity – signaling the existence of a new scaling behavior, whose exponents cannot be obtained perturbatively. The fixed point controlling the scaling behavior is not accessible in the one-loop approximation, so the scaling exponents in this regime can be obtained only from numerical simulations (see §7.5 and Chapter 8).

7.4 Phase transition in the KPZ equation

The flow for $d > 2$ is represented schematically in Fig. 7.5. The existence of a nonzero repulsive fixed point signals the existence of a phase transition between the linear problem ($g = 0$) and a strong-coupling regime ($g > g_2^*(d)$).

How can we detect experimentally, or by simulations, such a phase transition?

Suppose e.g. that we have a system described by the three-dimensional KPZ equation, and that we are able to tune some parameters (such as λ, ν, or D) that control the coupling constant g – which is a combination of these according to (7.15). If we change g from small to very large values, and measure the scaling exponents, we should observe a jump in the exponents. For $g < g_2^*$, we should find the exponents of the linear theory – which for $d = 3$ are negative, so we have a flat 'interface.' For $g > g_2^*$, the measurement should give the exponents of the strong-coupling phase.†

The main implication of this phase transition for $d > 2$ is that by the measurement of the scaling exponents alone we cannot decide on the equation describing the growth process. An experimental or numerical result might indicate the absence of roughening, a clear indication of the EW scaling exponents. On the other hand, changing some parameters determining the growth might trigger the roughening process, resulting in a 'morphological' phase transition.

Such a phase transition has been observed in the numerical simulations of a modified BD model and in the numerical integration of the KPZ equation [5, 226, 330, 362, 363, 494].

7.5 Exponents for $d > 1$

Only for $d = 1$ can we obtain the scaling exponents exactly. In higher dimensions, we have only the scaling relation (6.13) relating the two exponents, but their values are not given by the dynamic RG analysis. Since perturbative methods fail in the strong coupling regime, considerable effort has been invested to determine the exponents using numerical methods. Several different models in the KPZ universality class have been studied. The models are discussed in the following chapter, and a summary of results is provided in Tables 8.1 and 8.2. Here we summarize only the results of the numerical simulations.

Based on the simulation results, different conjectures have been made regarding the dimension dependence of the exponents. Studies

† The exponents of the strong coupling phase are positive, as we shall see in the following section.

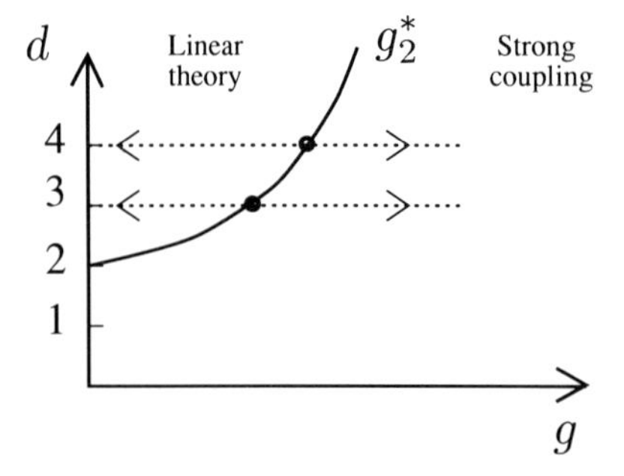

Figure 7.5 Phase diagram for $d > 2$. For $g < g_2^*(d)$ the coupling constant g renormalizes to zero, the scaling behavior being described by the linear theory (5.6). For $g > g_2^*(d)$, g flows to infinity, and the exponents in this regime cannot be obtained analytically.

on the Eden model led Wolf and Kertész (WK) to conjecture [486]

$$\alpha = \frac{1}{d+1} \qquad \beta = \frac{1}{2d+1}, \quad z = \frac{2d+1}{d+1} \qquad \text{[WK]}. \tag{7.20}$$

On the other hand for the restricted solid-on-solid model, Kim and Kosterlitz (KK) conjectured [231]

$$\alpha = \frac{2}{d+3} \qquad \beta = \frac{1}{d+2}, \quad z = 2\,\frac{d+2}{d+3} \qquad \text{[KK]}. \tag{7.21}$$

Both conjectures reproduce the exact results for $d = 1$, and both predict no upper critical dimension.

Recent large scale simulations provide more reliable estimates for the growth exponents, which are different from both conjectures. For example, Forrest and Tang calculated $\beta = \alpha/z = 0.240 \pm 0.001$ for $d = 2$, in between the two predictions [131]. Figure 7.6 compares the conjectures with some of the available numerical results. It is now believed that neither of the two conjectures is exact, but that both provide reasonable approximate values for the exponents.

The standard perturbative methods fail for $d > 2$, but recently new approaches to determine the exponents analytically have been introduced, and provide encouraging results [48, 170, 328, 392, 453]. Schwartz and Edwards developed a perturbative method to solve the Fokker–Planck equation associated with the KPZ equation [392]. They obtained the exact values for the exponents for $d = 1$, and for $d = 2$ they obtained $\alpha = 0.29$. Bouchaud and Cates proposed a self-consistent screening approximation to handle the perturbation

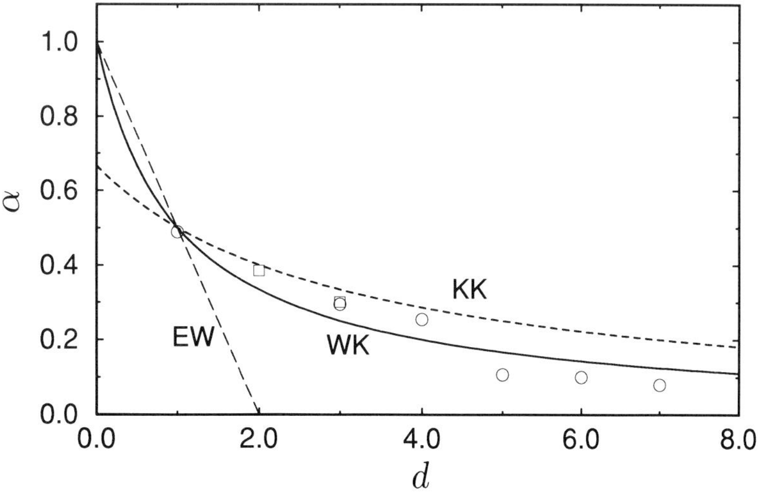

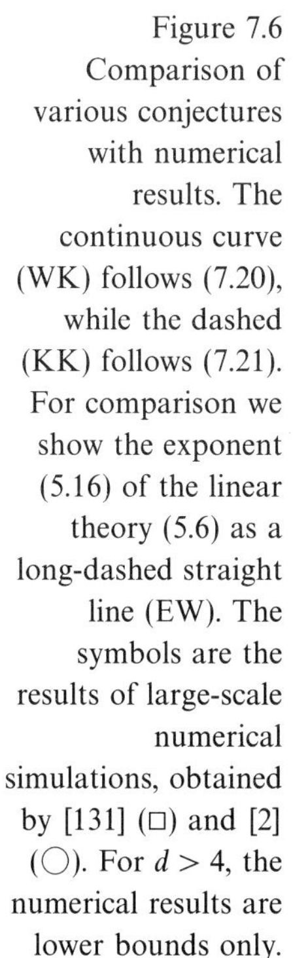
Figure 7.6 Comparison of various conjectures with numerical results. The continuous curve (WK) follows (7.20), while the dashed (KK) follows (7.21). For comparison we show the exponent (5.16) of the linear theory (5.6) as a long-dashed straight line (EW). The symbols are the results of large-scale numerical simulations, obtained by [131] (□) and [2] (○). For $d > 4$, the numerical results are lower bounds only.

Table 7.1 *A summary of the three universality classes discussed so far.*

Name	Equation	Scaling exponents
Random Deposition	$\frac{\partial h}{\partial t} = \eta(\mathbf{x}, t)$ (4.7)	$\beta = 1/2$ α not defined
Edwards–Wilkinson equation (Linear theory)	$\frac{\partial h}{\partial t} = \nu\nabla^2 h + \eta(\mathbf{x}, t)$ (5.6)	$\alpha = \frac{2-d}{2}$ $z = 2$
KPZ equation (Nonlinear theory)	$\frac{\partial h}{\partial t} = \nu\nabla^2 h + \frac{\lambda}{2}(\nabla h)^2 + \eta(\mathbf{x}, t)$ (6.4)	$d = 1:$ $\alpha = 1/2$ $z = 3/2$ $d > 1:$ Not known

expansion [48]. Their method gives $\alpha = 0.36$ for $d = 2$. Both methods predict a finite upper critical dimension between $d = 2$ and $d = 3$, in contrast to the conjectures (7.20) and (7.21).

Other calculations also indicate the existence of an upper critical dimension. The functional renormalization approach of Halpin-Healy predicts $d_c = 4$ [157, 158]. Using a $1/d$ expansion, Cook and Derrida concluded that a finite d_c may exist [82, 83]. For $d > d_c$, the strong coupling fixed point must vanish as the system is dominated by the linear $\lambda = 0$ fixed point. However, there is no evidence for this behavior in numerical simulations. In fact, measurements of Ala-Nissila *et al.* find a positive exponent up to seven dimensions (see Fig. 7.6), suggesting that if a critical dimension exists, it should be above seven [2].

7.6 Discussion

The KPZ equation defines a third universality class that is distinct from the RD and from EW universality classes. BD is one discrete model belonging to the KPZ universality class, but it is not the only model. There are in fact a large number of phenomena and models that are believed to be described in some form by the nonlinear theory. We provide additional examples in the next chapter, where a systematic presentation of the generic models and useful numerical methods is presented.

The three universality classes discussed so far are summarized in Table 7.1. Although we attempted to discuss most of the results on nonlinear theory, important results – such as the effects of different kinds of noise (correlated, power law distributed and quenched) – are either omitted or saved for later chapters. Further properties of the KPZ equation are discussed in Chapter 25.

Suggested further reading:

[132, 277, 314]

Exercises:

7.1 Derive the RG flow equation for the continuum equation (14.2).

7.2 The flow equation of the coupling constant (7.17) allows for the existence of an additional fixed point, $g_3^* = -g_2^*$, not discussed in the text. Discuss the stability of this fixed point, the associated scaling exponents, and its physical relevance to growth processes described by the KPZ equation.

8 Discrete growth models

In this chapter, we describe a few models that have had a key impact on our knowledge about specific aspects of interface roughening. Due to intractable mathematical difficulties, numerical methods are commonly used to determine the scaling exponents for systems with $d > 1$. Most growth models originate from specific physical or biological problems, and only recently have been investigated using the methods described in this book.

8.1 Ballistic deposition

The ballistic deposition model introduced in Chapter 2 is the simplest version – termed the nearest-neighbor (NN) model because falling particles stick to the first nearest neighbor on the aggregate. If we allow particles to stick to a diagonal neighbor as well, we have the next-nearest neighbor (NNN) model (Fig. 8.1). Since the nonlinear term *is* present for both models ($\lambda \neq 0$), the scaling properties for both models are described by the nonlinear theory. These two models therefore belong to the same universality class, since they share the same set of scaling exponents, α, β, and z. Their non-universal parameters, however, are different. For example, for the velocity v_0 (see (A.13)), we find $v_0 = 2.14$, 4.26 for the NN and NNN models, respectively. The coefficient λ of the nonlinear term differs as well, with $\lambda = 1.30$, 1.36, respectively [249].

The origin of the nonlinear term in the model is the lateral sticking rule, leading to the presence of voids. If we now tilt the interface, the resulting aggregate becomes more porous, with an increasing number of voids (Fig. 8.2). The formation of voids increases the growth velocity of the interface, since with the same number of deposited particles the mean height increases faster. Therefore a larger λ term is generated for the tilted case.†

8.2 Eden model

A classic growth model was introduced in 1961 by M. Eden as a model for the formation of cell colonies, such as bacteria or tissue cultures [111]. The model is defined as follows. Consider a lattice and place a seed particle at the origin. A new particle is added on any randomly-chosen perimeter site of the seed, forming thereby a two-site cluster. When iterated, the Eden model generates a cluster with a compact overall shape, but with a rough perimeter. Due to the slight anisotropy of the lattice, the roughly circular cluster is slightly distorted and resembles a diamond for a large enough number of particles [134, 178, 229, 310].

To study the interface properties of the model, it is convenient to start the growth from an entire line of seeds instead of from a single seed (Fig. 8.3). There are three versions of the Eden model, which differ in the microscopic rule of choosing the growing site [207], and result in different crossover behavior. Versions A and B have strong finite-size effects, while better scaling is found for version C. Determination of the scaling exponents can be facilitated by taking into account the 'intrinsic width' according to (A.16) [485].

Moreover, the intrinsic width can be decreased by using *noise reduction* techniques [223, 347, 348, 349, 435, 438], in which a counter is

† Two numerical recipes – measurement of λ and corrections due to an intrinsic width – which are useful for following the details of this chapter, are discussed in Appendix A.

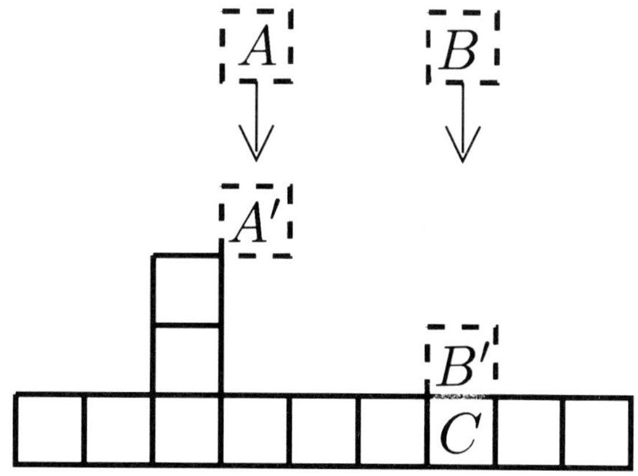

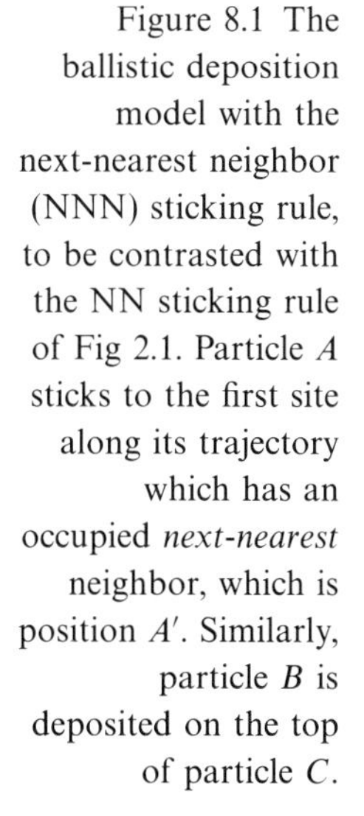

Figure 8.1 The ballistic deposition model with the next-nearest neighbor (NNN) sticking rule, to be contrasted with the NN sticking rule of Fig 2.1. Particle A sticks to the first site along its trajectory which has an occupied *next-nearest* neighbor, which is position A'. Similarly, particle B is deposited on the top of particle C.

used to record how many times a given growth site is chosen randomly from the available ones. The site is occupied only when the counter reaches a prescribed value s, called the noise reduction parameter; $s = 1$ reduces to the original model, while when $s \to \infty$ the growth is eventually deterministic. Noise-reduction algorithms allow a more

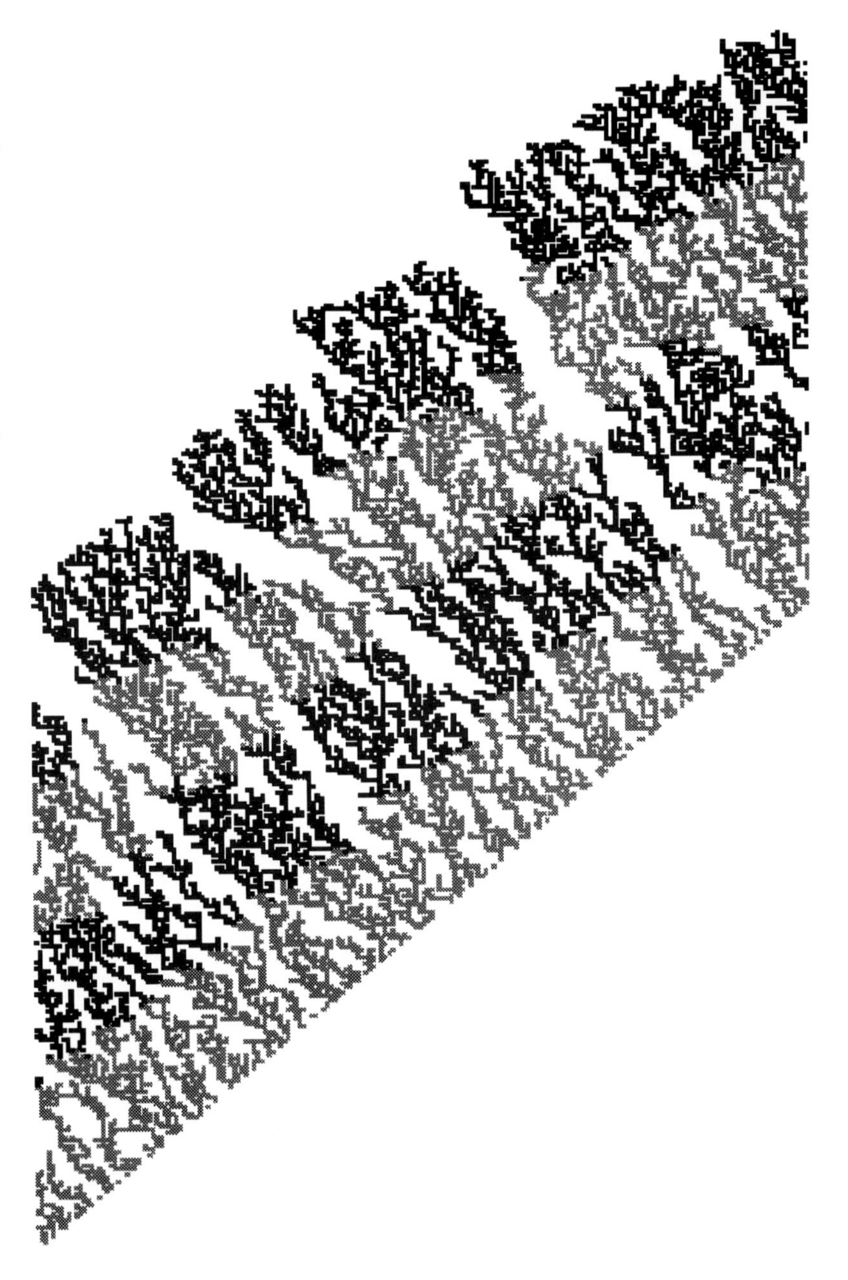

Figure 8.2 A ballistic deposition cluster generated by the NN model using a tilted substrate. A total of 16 000 particles was deposited (slope $m = 1$). Comparing with Fig. 2.2, one can see that the tilted substrate makes the aggregate grow faster, increasing the size and number of voids in the bulk.

reliable estimation of the exponents, and verify the validity of the scaling relation (6.13) [486].

The scaling properties are now believed to be described by the KPZ equation. Early simulations established $\alpha \approx 0.5$ for $d = 1$, in accord with the KPZ predictions. In higher dimensions, however, strong crossover effects give results scattered between 0.2 and 0.4 for $d = 2$ and between 0.08 and 0.33 for $d = 3$. Representative results are summarized in Tables 8.1 and 8.2.

8.3 Solid-on-solid models

Appendix A discusses the fact that many models display pronounced effects of 'corrections-to-scaling'. One class of models, collectively called solid-on-solid (SOS) models, has been introduced in order to minimize these corrections. The SOS models (a) consider a single-valued interface, i.e., do not allow overhangs, and (b) limit the height difference between neighboring sites and thus eliminate large slopes. In this section, we discuss the two SOS-type models that have contributed most to our knowledge by providing extremely accurate values of the scaling exponents in higher dimensions.

8.3.1 Single-step model

The single-step model is useful because one can obtain some of its parameters analytically by, e.g., mapping it to an Ising model [312,

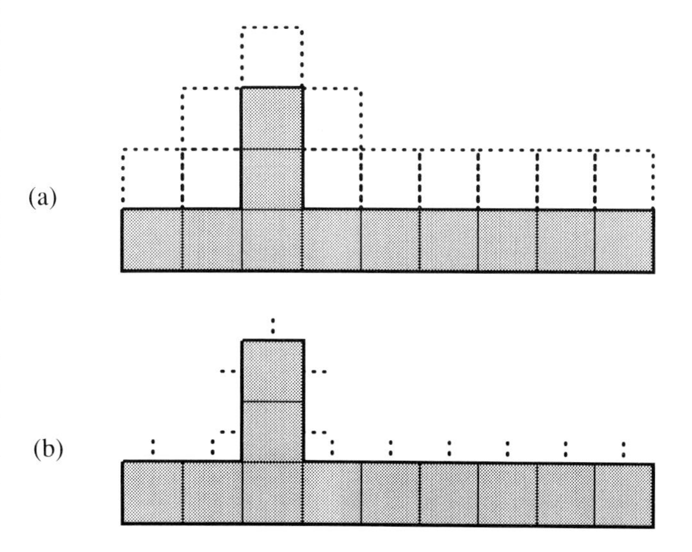

Figure 8.3 The three versions of the Eden model. (a) In version A, a particle is added with equal probability to any unoccupied site (dotted sites) adjacent to the surface (shaded sites). (b) In version B, a bond (dotted segments) linking an occupied site on the cluster with an empty perimeter site is chosen with equal probability. In version C, not shown, an occupied site on the surface is chosen with equal probability, and the new particle is added, with equal probability, to any of the empty sites adjacent to the chosen occupied particle. The three versions produce interfaces that differ microscopically, but belong to the same universality class.

Table 8.1 *Numerical results for exponents in the strong coupling limit, for the case $d = 1$.*

Model	α	β	z	Reference
Ballistic	0.42	0.30		[124]
Ballistic	0.47	0.30		[312]
Ballistic		0.33		[20]
Eden	0.50			[124]
Eden			1.55	[372]
Eden	0.50	0.30		[206]
Eden	0.50	0.30		[310]
Eden	0.51	0.32		[501]
Single Step	0.5	0.33		[312]
Single Step			1.57	[373]
RSOS	0.5	0.33		[231]
RSOS	0.489	0.332		[2]
PNG		0.33		[252]
KPZ	0.5	0.33		[155]
KPZ		0.330		[330]

373]. The model, illustrated in Fig. 8.4, is defined as follows. At time $t = 1$ the interface has a grooved shape, with height $h(2i) = 0$ and $h(2i+1) = 1$, where $i = 0, \ldots, L/2$. Growth occurs with a probability p_+ at each local minimum, resulting in $h_i \to h_i + 2$. The interface can decrease its height ('desorption') at site i, if that site is a local maximum. Desorption occurs at local maxima with probability p_-, resulting in $h_i \to h_i - 2$. The fact that we always choose the local minima (maxima) to grow (desorb) guarantees that at any stage of the growth process the height difference between two neighboring sites must be exactly unity.

There are two methods of updating the model. In *sequential updating* a site on the interface is randomly selected, and growth (or desorption) occurs with probability p_+ (p_-). In *parallel updating*, all eligible growth sites (i.e., all local minima) grow with probability p_+ simultaneously, and all eligible desorption sites desorb with probability p_-. It is possible to separate the lattice into two sublattices (formed by odd and even lattice sites, respectively), and perform the updating on a parallel or vector computer. This makes the simulation of the model very effective. The scaling exponents are independent of the way we update the model, but some parameters may depend on the simulation method.

Table 8.2 *Numerical results for the exponents in the strong coupling regime, for $d > 1$.*

Model	d	α	β	z	Reference
Ballistic	2+1	0.33	0.24		[312]
Ballistic	2+1	0.3	0.22		[20]
Ballistic	2+1	0.35	0.21		[120]
Eden	2+1	0.20			[206]
Eden	2+1	0.33	0.22		[486]
Eden	2+1	0.39	0.22		[106]
Single Step	2+1	0.36		1.58	[312]
Single Step	2+1	0.375		1.63	[276]
Single Step	2+1	0.385	0.240		[131]
RSOS	2+1	0.40	0.25		[231]
KPZ	2+1	0.18	0.10		[71]
KPZ	2+1	0.24	0.13		[155]
KPZ	2+1	0.39	0.25		[4]
KPZ	2+1		0.240		[330]
Eden	3+1	0.08			[206]
Eden	3+1	0.24	0.146		[486]
Eden	3+1	0.22	0.11		[106]
Single Step	3+1	0.30	0.180		[131]
RSOS	3+1		0.20		[231]
RSOS	3+1	0.294	0.180	1.709	[2]
KPZ	3+1		0.17		[330]
RSOS	4+1		0.16		[231]
RSOS	4+1	0.254	0.139		[2]
RSOS	5+1		≥ 0.107		[2]
RSOS	6+1		≥ 0.10		[2]
RSOS	7+1		≥ 0.08		[2]

Kinetic Ising model – To map the single-step model to a kinetic Ising model, we observe that the interface properties can be described in terms of a set of Ising variables $\{s\} = \{s_1 s_2, ..., s_L\}$, where

$$s_i \equiv h(i) - h(i-1) = \pm 1. \tag{8.1}$$

At $t = 0$ we have 'antiferromagnetic' ordering – i.e., up and down spins alternate (Fig. 8.4). Growth can occur if two spins (s_i, s_{i+1}) have a $(-+)$ configuration; growth of the interface at site i exchanges the value of the two spins with a probability $p_+ = 1/2$. The transition $(-+) \to (+-)$ conserves the total 'magnetization' of the system. Similarly, desorption occurs with probability p_-, resulting in the transition $(+-) \to (-+)$.

Lattice gas model – The single-step model can be mapped to a lattice gas if we replace every surface element of slope (-1) with a hard core particle, and every element with slope $(+1)$ with a hole (Fig. 8.4). Thus initially we have $L/2$ particles, placed at every second lattice site. A growth process corresponds to rightward motion of the corresponding particle, and desorption to leftward motion. In the lattice gas model, particles move to the right (left) with a probability p_+ (p_-), with the restriction that only one particle can occupy each site at any given moment. This condition corresponds to the height-difference restriction in the original growth model.

Nonlinear term – If the model is described by the nonlinear theory, an overall tilt should change the velocity of the interface (see Appendix A). The interface velocity is given by the relation

$$v = 2(\Pi_+ p_+ - \Pi_- p_-), \tag{8.2}$$

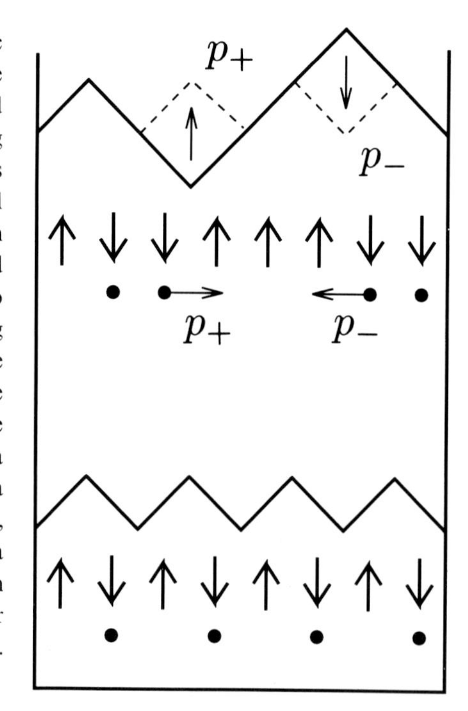

Figure 8.4 Schematic representation of the single-step model and the equivalent Ising and lattice gas models. The initial smooth configuration (bottom) is a grooved surface, equivalent to alternating Ising spins, or to the presence of a particle at every second site (lattice gas). At a later time, we have a rough interface (top), corresponding to a random distribution of up-down spins or diffusing particles.

where Π_+ (Π_-) is the probability of choosing a site eligible for growth (desorption), and the factor 2 arises from the fact that every growth process increases the interface height by 2. For the untilted interface, the number of eligible growth sites equals the number of sites eligible for desorption, so $\Pi_+ = \Pi_-$. In the case $p_+ = p_-$, the interface does not grow. There is an up-down symmetry in h – i.e., desorption and deposition are equally probable. Hence, following the discussion of symmetry principles in § 5.2, the model must be described by the EW equation.

Suppose we elevate the right side of the interface by k lattice spacings, thereby generating an overall tilt of slope $m \equiv k/L$. In the spin representation this will increase (decrease) the initial number of up (down) spins, and

$$N_+ = \frac{L}{2} + \frac{mL}{2}, \quad N_- = \frac{L}{2} - \frac{mL}{2}. \tag{8.3}$$

One essential property of the kinetic Ising model is that in its steady state all configurations $\{s_i\}$ are equivalent. The probability Π_+ of finding a pair $(-+)$ eligible for growth can be calculated as follows. The probability of finding a $(+)$ spin is $N_+/(N_+ + N_-)$, and the probability of finding *next to it* a $(-)$ spin is $N_-/(N_+ + N_-)$. Thus the probability of finding a $(-+)$ combination is

$$\Pi_+ = \frac{N_+ N_-}{(N_+ + N_-)^2}, \tag{8.4}$$

which is the probability of finding a site eligible for growth. The same expression also gives Π_-.

Using (8.2)–(8.4), we thereby obtain for the tilt-dependent velocity

$$v(m) = \frac{(p_+ - p_-)}{2}(1 - m^2) \tag{8.5}$$

which, combined with (A.13), gives the coefficient of the nonlinear term in the KPZ equation

$$\lambda = -(p_+ - p_-). \tag{8.6}$$

This relation expresses quantitatively the fact that for $p_+ = p_-$, the nonlinear term vanishes, and the model is described by the Edwards–Wilkinson equation, (5.6). A second observation is that the growing interface ($p_+ > p_-$) has $\lambda < 0$, so the local velocity decreases with the local slope.

Scaling exponents – A simple argument can provide the roughness exponent as well [312]. In equilibrium, all configurations are equally probable, so neighboring spins are uncorrelated, $\langle s_i s_j \rangle = \delta_{ij}$. The height of the interface at site k is

$$h_k = \sum_{i=1}^{k} s_i + h_1. \tag{8.7}$$

These properties are reminiscent of Brownian motion: the trace of a Brownian particle is given by the sum of uncorrelated random numbers as the particle performs a random walk of k steps, each of length ± 1. Hence the roughness exponent of the interface is $\alpha = 1/2$, which coincides with the predictions of both the EW and KPZ equations.

The previous argument can be used in the equilibrium situation, for which the interface does not grow and $p_+ = p_-$. It is also applicable to the moving interface as well, which is a nonequilibrium problem [312]. The equilibrium state of the kinetic Ising model corresponds to the stationary (saturated) state of the nonequilibrium growth model. While both the KPZ and EW theories predict the same roughness exponent, $\alpha = 1/2$, they predict different dynamic exponents z.

Generalizations of single-step models to $d > 1$ provide the first systematic study of conjectures (7.20) and (7.21) [131, 441]. For $d > 1$, $d+1$-dimensional hypercubes are deposited on (or desorbed from) a d-dimensional interface. The rate of deposition (or desorption) is p_+ (or p_-). If deposition is balanced with desorption, one can expect exponents predicted by the linear theory [see (5.16)].

Such tuning of the deposition and desorption rates allows us to systematically study crossover effects, and thus to locate the scaling region. One source of deviation of the effective roughness exponent from its asymptotic value is a crossover from the scaling behavior of the linear equation to the nonlinear one, if the value of the effective coupling constant $g \sim D\lambda^2/\nu^3$ is small. For parallel updating, the choice $p_+ = 1/2$ and $p_- = 0$ maximizes this quantity, while for sequential updating, $p_+ = 1$ and $p_- = 0$ according to (8.6). Large Monte-Carlo simulations incorporating corrections-to-scaling give exponents different from the predictions of (7.20) and (7.21) (see also Fig. 7.6).

8.3.2 Restricted solid-on-solid model

The restricted SOS model, introduced by Kim and Kosterlitz (KK), permitted the first systematic investigation of growth exponents for higher dimensions [231]. The KK growth algorithm randomly selects a site on a d-dimensional interface, and increases the height of the interface by one, $h_i \to h_i + 1$, provided that at every stage of growth the restricted solid-on-solid condition $|\Delta h| = 0, 1$ is fulfilled between the selected and the neighboring sites.† The model exhibits very good scaling properties, without noticeable intrinsic width or corrections-to-scaling. Tilt-dependent velocity measurements reveal that the coefficient of the nonlinear term is negative: $\lambda = -0.75$ [249].

KK studied their model for $d \leq 4$. They concentrated on the determination of the exponent β, because this exponent describes the early time behavior of the interface width and so does not require the saturation of the system (which for large systems is very time consuming). The results, summarized in Tables 8.1 and 8.2, led KK to conjecture that the dimension dependence of the exponents is given by (7.21). By using a fitting *Ansatz* from the correlation function, Ala-Nissila *et al.* were able to obtain more reliable estimates of the exponents [2] . Their values are systematically larger than those given by (7.21), and are in agreement with the $d > 1$ hypercube-stacking simulations mentioned in the previous section.

8.4 Propagation of interfaces in the Ising model

When the temperature is sufficiently low, a well-defined interface develops between the up and down spins in a simple Ising model during the growth of the stable phase at the expense of the unstable phase. For the Ising model in a field H, Eq. (7.1) is replaced by

$$\mathscr{H} = -J \sum_{\langle ij \rangle} s_i s_j - H \sum_i s_i. \tag{8.8}$$

Here $\langle ij \rangle$ denotes a sum over the nearest-neighbor pairs, J the nearest-neighbor coupling constant, and H the external field. The system follows nonconservative dynamics, such as the one proposed by Metropolis in which spins are flipped with a probability

$$P(s \to s') \propto \exp\left(-\frac{\mathscr{H}(s') - \mathscr{H}(s)}{k_B T}\right). \tag{8.9}$$

We prepare our system such that there is a straight interface separating two domains, one with spins up and the other with spins down. We have one stable domain (with spins oriented parallel to H) and one unstable domain (spins antiparallel to H) on decreasing the temperature to below the critical temperature. The stable phase grows, making the interface advance at the expense of the unstable one. The properties of this evolving interface can be described by the

† In general, one can assume $|\Delta h| = 0, 1, ..., k$ without changing the growth exponents, but most of the simulations are performed for the simplest case $k = 1$.

KPZ equation [105, 254]. An unwanted effect that can disturb the identification of the interface is the nucleation and droplet growth of the stable phase in the bulk of the unstable phase. These 'islands' eventually coalesce with the growing interface, changing its geometrical structure. In simulations, this effect can be eliminated by allowing spins to flip only in the vicinity of the interface, thus neglecting the dynamics of the bulk.

8.5 Numerical integration of the KPZ equation

So far, we have concentrated on the determination of the exponents of the nonlinear theory using discrete models that are believed to belong to the KPZ universality class. Despite the presented analytical and numerical evidence that these models have a nonzero nonlinear term, the reader might reasonably doubt our claim that they can be described using the KPZ equation, or, conversely, that the KPZ equation predicts the same exponents as the models. The situation is clear in one dimension, where exact values of the KPZ exponents can be determined analytically and their values coincide with the numerical results, but is less clear in higher dimensions.

An alternative proof might be possible if we could obtain the scaling exponents directly from the growth equation. For these reasons, many groups have tried to obtain the scaling exponent by integrating numerically the KPZ equation [4, 71, 155, 329, 330]. Early investigations have found very small values of β in two dimensions, compared to the results provided by discrete models. Later works have reported agreement with the lattice models [4, 329, 330]. In addition, a very detailed investigation by Moser *et al.* [329, 330] has produced evidence of a phase transition between the smooth and rough phase in $d = 3$.

To integrate numerically the KPZ equation, one must discretize the continuum equation, (6.4). This is done by using a forward-backward difference method on a cubic grid with a lattice constant Δx, and the Euler algorithm with time increments Δt. Denoting the basis vectors by $\mathbf{e}_1, \mathbf{e}_2, \ldots, \mathbf{e}_d$, and labeling the grid points by $\mathbf{n}$, we find for the discretized growth equation

$$
\begin{aligned}
h(\mathbf{x}, t + \Delta t) = h(\mathbf{x}, t) + \\
\frac{\Delta t}{(\Delta \mathbf{x})^2} \sum_{i=1,d} (\nu\, [h(\mathbf{x} + e_i, t) - 2h(\mathbf{x}, t) + h(\mathbf{x} - e_i, t)] \\
+ (1/8)\lambda [h(\mathbf{x} + e_i, t) - h(\mathbf{x} - e_i, t)]^2) + \sigma (12 \Delta t)^{1/2} \eta(t).
\end{aligned}
\tag{8.10}
$$

Here $\sigma^2 \equiv 2D/(\Delta x)^d$, and the random numbers η are uniformly distributed between $-1/2$ and $1/2$. For practical purposes it is useful to use the dimensionless variables $\tilde{h} \equiv h/h_0$, $\tilde{t} \equiv t/t_0$ and $\tilde{x} \equiv x/x_0$, where the natural units can be obtained from the combination of the coefficients of the growth equation $h_0 = \nu/\lambda$, $t_0 = \nu^2/\sigma^2\lambda^2$, and $x_0 = (\nu^3/\sigma^2\lambda^2)^{1/2}$, while the dimensionless coupling constant is given by $g = (\Delta\tilde{x})^d/2$.

Simulating the KPZ equation in these dimensionless parameters gives $\beta = 0.330 \pm 0.004$, in good agreement with the exact result 1/3 [330]. For two dimensions, the same method gives $\beta = 0.240 \pm 0.005$, coinciding with the value found in the SOS models (Table 8.2).

The results obtained for three dimensions are particularly interesting. As can be seen in Fig. 8.5, at $g^* \approx 32$, there is a qualitative change in the curves. The upward trend of the curves for large coupling constant can be interpreted as a sign of kinetic roughening, g^* representing the transition point between smooth and rough phases. Further studies on the ensemble fluctuations also indicate the presence of a transition at g^*. These results are in good agreement with the predictions of the dynamic RG analysis, that a smooth-to-rough transition is possible in dimensions higher than two (see §7.4).

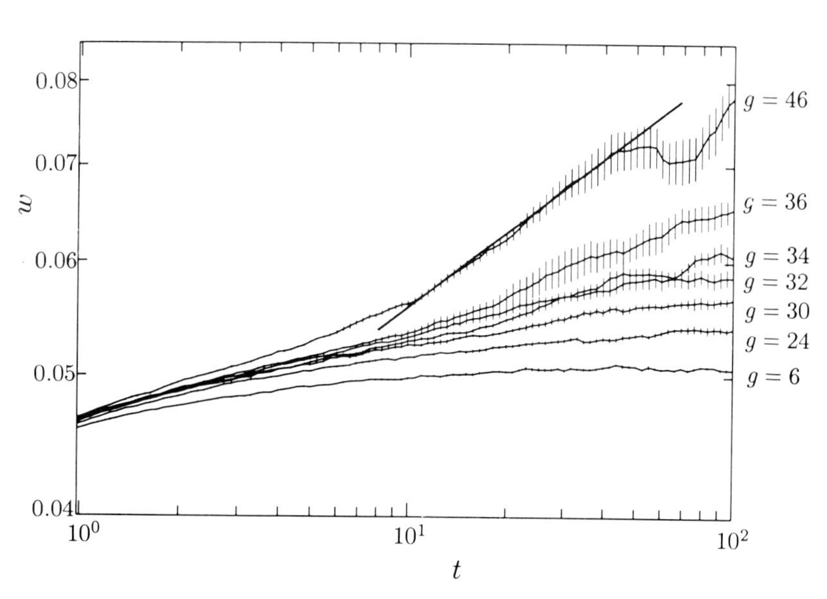

Figure 8.5 Time dependence of the width for the KPZ equation with $d = 3$. The qualitative change in the curves at $g^* = 32$ indicates the possibility of a morphological phase transition from the smooth phase ($g < 32$) to roughening ($g > 32$). (After [330]).

8.6 Discussion

There are various reasons for introducing a discrete growth model. A number of models are designed to faithfully mimic some natural process. In particular, the BD and Eden models were designed to simulate vapor deposition and biological growth, respectively. These models have certain limitations – e.g., they cannot provide accurate scaling exponents. To go beyond these limitations, we need models that will provide answers to specific questions. A good example is the SOS model, which does not exhibit strong corrections-to-scaling and so allows the accurate determination of scaling exponents.

Although we are not able to calculate the KPZ exponents analytically for $d > 1$, the use of discrete models does enable us to obtain good estimates of exponent values. Why is it important for us to have accurate numerical values for the exponents? First, in order to identify the universality class, we need to know the exponents. Second, when we measure an exponent experimentally, we also need to know the value of the exponent in order to determine whether the growth process is indeed described by the KPZ equation.

Suggested further reading:

[254, 304, 330]

Exercises:

8.1 Write a computer program that simulates the NNN ballistic deposition model, and compare the scaling of the interface width with that of the NN model.

8.2 Calculate λ for the NN and NNN ballistic deposition models.

8.3 In the BD model $\lambda > 0$, while in the SOS models $\lambda \leq 0$. In the Eden model, is λ positive or negative? Does the sign vary from version to version of this model?

PART 3 Interfaces in random media

9 Basic phenomena

In previous chapters, we focused on interfaces that grow and roughen due to thermal fluctuations, the origin of the randomness arising from the random nature of the deposition process. For a class of interface phenomena, however, we do not have deposition, but rather we have an interface that moves in a disordered medium. The experiment described in Chapter 1 in which a fluid interface propagates through a paper towel is one example. The velocity of the interface is affected by the inhomogeneities of the medium: the resistance of the medium against the flow is different from point to point; we call this *quenched noise*, because it does not change with time. Fluid pressure and capillary force drive the fluid, and disorder in the medium slows its propagation. If the disorder 'wins' the competition, the interface becomes 'pinned.' Conversely, if the driving forces win, the interface stays 'depinned.' This transition from a pinned to a moving interface – obtained by changing the driving force – is called the depinning transition.

In the following three chapters, we discuss how quenched disorder leads to interface pinning – and depinning. We will show that the same theoretical ideas describe interface motion in a random field Ising model, which is relevant to the problem of domain growth in a disordered magnetic material.

The problem of a moving interface in the presence of quenched noise is a new type of critical phenomena, arising from 'quenched randomness.' The investigation of this problem will lead to the introduction of new critical exponents, which can be studied using numerical simulations and RG calculations. Although many important results have been obtained, the diversity and richness of the underlying phenomena continue to present us with a large number of unanswered questions.

9.1 Depinning transition

The existence of the depinning transition distinguishes interfaces moving in a disordered medium. In order to understand the origin of the depinning transition, we first consider a simple example of depinning [214].

Motion of a particle with friction – In Fig. 9.1(a) we show a particle of mass m on a plane, driven by an external force F. There are two friction forces acting: *static* friction, F_c, and *dynamic* friction $\rho dr/dt$, proportional to the velocity. For $F < F_c$, $v = 0$. For $F > F_c$, the equation of motion is

$$m\frac{d^2r}{dt^2} = F - F_c - \rho\frac{dr}{dt} \qquad [F > F_c]. \tag{9.1}$$

After a transient period, the particle will reach a velocity v and the constant acceleration term can be neglected. There are two regimes as a function of the external force F, as shown in Fig. 9.1(b).

Pinned phase – If $F < F_c$, the driving force cannot overcome the static friction, thus the velocity of the particle is zero.

Moving phase – If $F > F_c$, the velocity is $v = (F - F_c)/\rho$, proportional to the reduced force

$$f \equiv \frac{F - F_c}{F_c}. \tag{9.2}$$

The transition from a pinned to a moving particle is called a *depinning transition*, and takes place at a critical threshold force F_c. In the vicinity of the depinning transition, the average velocity has the form

$$v \sim f^{\theta}, \tag{9.3}$$

where θ is the *velocity exponent*. In the example, $v \propto f$ so $\theta = 1$.

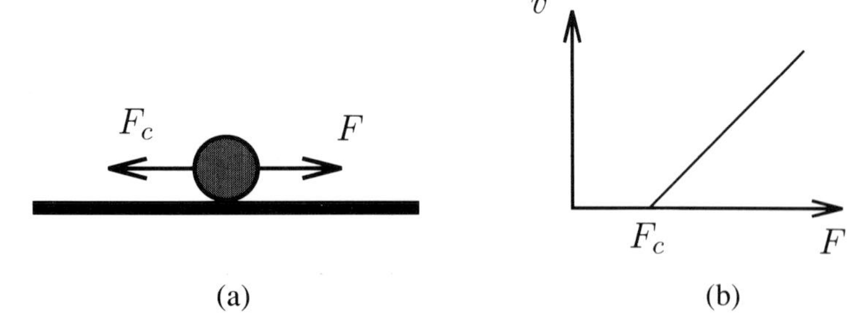

Figure 9.1 Depinning transition of a single particle on a plane. (a) The particle of mass m is driven by the external force, F, to the right, and the friction force, F_c, acts against its motion. (b) The velocity of the particle as a function of the external driving force.

9.2 Interfaces in a disordered medium

Let us consider the general situation of an interface in a porous medium, as shown in Fig. 9.2, with a driving force F acting on the interface in a vertical direction. This is in fact a generalization of the single particle motion shown in Fig. 9.1. Instead of a single particle, we have a spatially extended string that has a surface tension and so tries to remain straight. The disorder acts as an inhomogeneous friction force, pinning parts of the interface. However, other parts are free to advance, and they try to drive the neighboring parts of the interface as well.

The most general equation describing the motion of a driven interface is the KPZ equation. In a disordered medium, however, more important than the *thermal* noise (always present) is the *quenched* noise generated by the disorder. Thus the thermal noise term $\eta(\mathbf{x}, t)$ is replaced by the quenched noise term $\eta(\mathbf{x}, h)$. The KPZ equation becomes

$$\frac{\partial h}{\partial t} = F + \nu\nabla^2 h + \frac{\lambda}{2}(\nabla h)^2 + \eta(\mathbf{x}, h). \qquad \rightarrow (9.4)$$

We assume that $\eta(\mathbf{x}, h)$ has zero mean, $\langle\eta(\mathbf{x}, h)\rangle = 0$, and correlations of the form

$$\langle\eta(\mathbf{x}, h)\eta(\mathbf{x}', h')\rangle = \delta^d(\mathbf{x} - \mathbf{x}')\Delta(h - h'), \qquad (9.5)$$

where the angular brackets represent the average over the different realizations of randomness. The explicit form of the function $\Delta(h - h')$ will be discussed below.

Note that if we fix the quenched randomness $\eta(\mathbf{x}, h)$ for a given system, the evolution of the interface is *deterministic* – i.e., we always obtain the same final interface if we start our simulation from the same initial conditions. In contrast, the KPZ equation is *stochastic*;

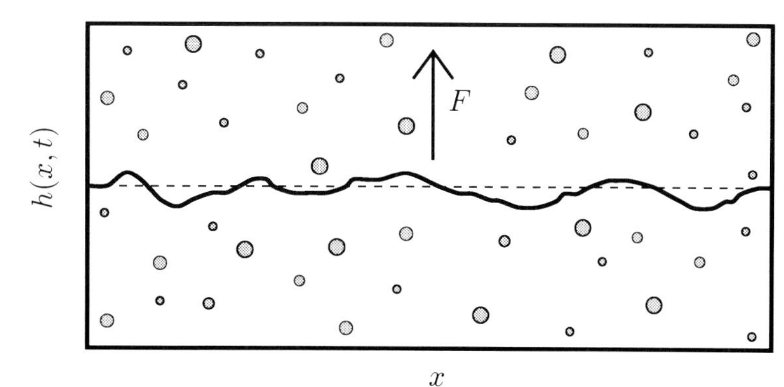

Figure 9.2 Schematic representation of an interface in a random environment. The shaded circles represent randomly distributed pinning centers, whose position and pinning strength is independent of time (quenched).

restarting the interface from the same initial conditions results in a different final configuration.†

There are three regimes predicted by the growth equation (9.4) as indicated in Fig. 9.3.

(a) *Pinned phase* – In the absence of an external field, the interface moves until it finds the closest configuration where the energy has a local minimum, whereupon it becomes pinned. If we add a *small* driving force F, the interface tends to move in the direction of F, but will eventually become pinned by impurities.

(b) *Critical moving phase* – If we increase the driving force, the interface overcomes the pinning force of the impurities at a critical force F_c, and begins to move with a finite velocity. In the immediate vicinity of F_c, the velocity follows (9.3). The motion of the interface just above the threshold is not uniform. At a given moment the interface consists of pinned and unpinned regions. Once the combined effect of the driving and elastic forces overcomes the pinning forces in a particular region, the interface 'jumps' ahead, but is eventually stopped again by an another region of strong pinning sites. Thus the interface exhibits a slow, smooth motion interspersed with jumps.

We denote by ξ the *correlation length*, corresponding to the

† The motion of the interface is deterministic so long as we can ignore thermal fluctuations, i.e., so long as we are in the *zero temperature* limit. At finite temperatures, we must add thermal noise $\eta(\mathbf{x}, t)$ to the growth equation (9.4).

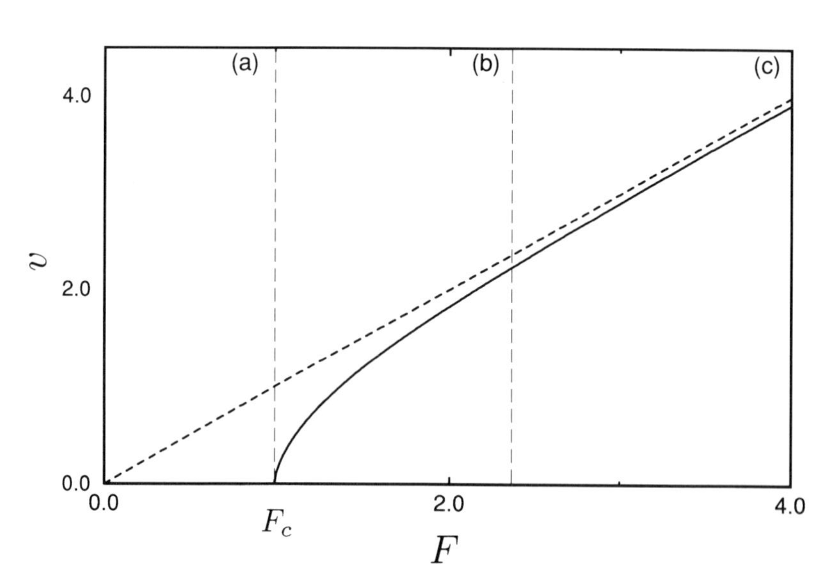

Figure 9.3 The velocity of the driven interface as a function of the driving force F. We can distinguish the three main regimes (a)–(c) on the figure. (a) For $F < F_c$ the interface velocity is zero, being pinned by the quenched randomness. (b) Near the depinning transition, for $F > F_c$, the velocity varies according to (9.3). (c) Finally, for very large driving forces, $F \gg F_c$, the velocity of the interface is proportional to the driving force. The velocity and force units are arbitrary.

characteristic size of the pinned domains. As we approach the critical force $F \to F_c$, the size of the pinned regions diverges as

$$\xi \sim (F - F_c)^{-\nu}, \tag{9.6}$$

where ν is called the *correlation length exponent.*

(c) *Large velocity regime* – If $F \gg F_c$, the interface feels a rapidly fluctuating noise, so the velocity increases linearly with F. In this regime the motion can be described using the KPZ equation, motion-generated thermal fluctuations washing out the effect of quenched randomness. The correlation length becomes equal to the lattice spacing in the models.

9.3 Scaling arguments

Let us consider the continuum growth equation (9.4), with the noise of the general form (9.5). Here $\Delta(u) = \Delta(-u)$, and $\Delta(u)$ is a monotonically decreasing function of u for $u > 0$ and decays rapidly to zero over a finite distance a. A special case of this correlation function is $\Delta(u) = \delta(u)$, which corresponds† to $a = 0$.

In this section we describe the scaling properties of a driven interface [56, 127, 336, 339, 462].

Depinning transition – A simple scaling argument suggests that there is an important length scale in the problem. Let us consider a domain of size ℓ that blocks the motion of the interface (see Fig. 9.4). For this portion, we can rewrite the equation of motion (9.4) in terms

† This is called random field disorder, discussed in §26.4.1

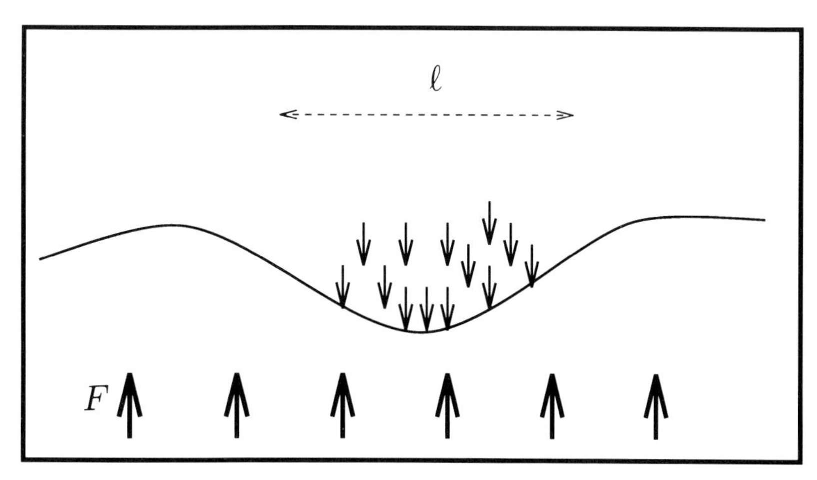

Figure 9.4 Schematic illustration of the pinning of an interface portion of size ℓ. The local impurities (small arrows) oppose the motion of the interface, acting against the external driving force F and the elastic forces generated by the local curvature of the interface.

of its essential length scales as

$$\ell^d \nu \ell^{-2} h + \ell^d F - [\Delta(0)]^{1/2} \ell^{d/2} = 0, \tag{9.7}$$

where we have for now neglected the KPZ nonlinearity. The first term comes from the Laplacian, the second is the driving force, and the last is the contribution from the noise, which has a negative sign because we assume that it opposes the motion of the interface. As a first approximation, we assume that the interface is pinned if the Laplacian term – the local curvature – is weaker than the pinning force. Thus $a\nu\ell^{d-2} \ll [\Delta(0)]^{1/2}\ell^{d/2}$ provides the characteristic length scale

$$\ell_c \sim \left(\frac{\nu^2 a^2}{\Delta(0)} \right)^{1/(4-d)}. \tag{9.8}$$

For $d < 4$, the smoothing effect of the elastic term dominates for length scales $\ell \ll \ell_c$, while for $\ell \gg \ell_c$ the interface wanders, taking advantage of the low energy configurations in the disorder. Thus $d = 4$ is the critical dimension of the model.

The maximum pinning force can be obtained from (9.7), equating the driving force with the pinning effect of the disorder, i.e., $F_c \sim [\Delta(0)/\ell_c^d]^{1/2}$, from which, using (9.8), we obtain

$$F_c \sim [\Delta(0)]^{2/(4-d)} (a\nu)^{-d/(4-d)}. \tag{9.9}$$

The critical force (9.9) represents the maximum force at which the interface remains pinned and is *independent* of the size of the segment. For forces larger than F_c, the interface moves with a finite steady velocity.

Moving interface – There are four critical exponents characterizing the interface at the depinning transition: the velocity exponent θ, the correlation exponent ν, the roughness exponent α, and the dynamic exponent z. We now discuss a scaling relation valid in the vicinity of the depinning transition that connects the four exponents. The motion of the interface close to the threshold is composed of jumps of size ξ. Such a jump moves the interface forward by ξ^α over a time period ξ^z. Thus the velocity of the interface is

$$v \sim \frac{\xi^\alpha}{\xi^z} \sim (F - F_c)^{\nu(z-\alpha)}. \tag{9.10}$$

Comparing (9.10) with (9.3), we obtain for the velocity exponent

$$\theta = (z - \alpha)\nu. \qquad \rightarrow (9.11)$$

This scaling relation reduces the number of independent exponents to three.

9.4 Thermal noise

The results discussed in the previous sections were obtained in the zero-temperature limit, where thermal noise is negligible. However, thermal noise is always present in experiments, so we now discuss its effect on the motion of interfaces in a medium with quenched randomness. At finite temperature, we generalize (9.4) to include both quenched noise $\eta(\mathbf{x}, h)$ and thermal noise $\eta(\mathbf{x}, t)$,

$$\frac{\partial h}{\partial t} = F + \nu \nabla^2 h + \frac{\lambda}{2}(\nabla h)^2 + \eta(\mathbf{x}, h) + \eta(\mathbf{x}, t). \tag{9.12}$$

Here $\eta(\mathbf{x}, t)$ obeys (4.8) and (4.9).

The thermal noise affects the motion of the interface close to and below the depinning transition. At finite temperature the transition from an unpinned to a pinned interface is somewhat rounded; the interface moves with a nonzero average velocity even for $F < F_c$, which can be understood by looking more closely at the nature of thermal fluctuations. Let us consider that the interface is trapped in a local potential minimum, generated by the quenched disorder. In order to advance further, it must pass a potential barrier ΔV. At zero temperature the interface remains in the local minimum forever, its velocity being zero. However, at finite temperature, the interface can pass the barrier with a probability $P \sim \int_{\Delta V}^{\infty} dE \exp(-E/k_B T)$. A schematic illustration of the resulting velocity-driving force diagram at finite temperatures is shown in Fig. 9.5.

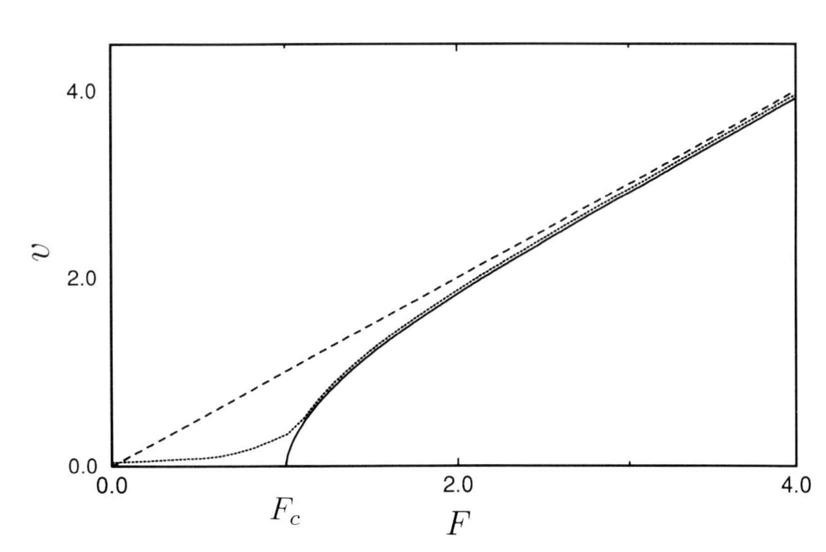

Figure 9.5 Dependence of the velocity on the driving force at nonzero temperature (middle curve), compared with idealized behavior at $T = 0$ (bottom curve). Thermal fluctuations help the interface advance even at forces lower than F_c.

9.5 Discussion

We have seen that quenched noise, generated by the randomness of the substrate, has a nontrivial effect on the motion and morphology of an interface. The interface is pinned by disorder if the driving force is small, but for forces just above the critical threshold force F_c the interface advances by jumping from one pinning path to another. We paid special attention to the critical regime, near the depinning transition, where new exponents θ and ν must be defined to provide a complete description of the system. It is the goal of the next chapter to show how to calculate these exponents for various models, and to discuss the basic ingredients determining the universality class of the system.

The influence of the quenched noise on a moving manifold and the interplay between the quenched and thermal noise has been the subject of intense study in two related areas of physics: charge density waves [154], and the motion of a vortex line in a disordered superconductor [46]. Many results described in this chapter have their counterparts in problems related to these two fields.

Suggested further reading:

[13, 339]

Exercises:

9.1 In Chapter 5 we were able to solve the EW equation exactly. Can we do the same for Eq. (9.4) with $\lambda = 0$?

9.2 The RG method presented in Chapter 7 and Appendix B allowed us to obtain the scaling exponents for the KPZ equation exactly for $d = 1$. Can we use this same method to obtain the scaling exponents for (9.4)? Why not? Find the point at which the RG calculation begins to encounter difficulties.

9.3 How does the correlation length ξ depend on the the velocity of the interface in the vicinity of the depinning transition?

10 Quenched noise

The goal of understanding the effect of quenched noise on interface morphology has motivated a large number of numerical studies. Several models have been developed, both for understanding specific phenomena such as fluid flow or domain growth, and also for the efficient determination of the scaling exponents. In general, it is agreed that quenched noise produces anomalous roughening, with a roughness exponent larger than the values predicted by the KPZ or EW equations.

We can partition the numerical efforts in two distinct classes, according to the *morphology* of the interface.

(i) There is a family of models that neglects overhangs on the interface, by proposing a growth rule that produces self-affine interfaces [31, 44, 58, 108, 162, 195, 204, 267, 272, 359, 409, 411, 412, 443, 457, 461]. In some of these models, the scaling exponents can be obtained *exactly* by mapping the interface at the depinning transition onto a directed percolation problem.

(ii) There is a second family of models that allows overhangs, often leading to self-similar interfaces [42, 78, 79, 80, 140, 235, 293, 294, 295, 350, 351, 427]. In some cases, the interface generated by these models can be mapped onto a site percolation problem, and the scaling exponents can be estimated by exploiting this mapping.

Moreover, models leading to self-affine interfaces can also be classified in two main universality classes. As we show, the two classes have different scaling exponents, which can be obtained using numerical and analytical methods. In this chapter we discuss the details of these two universality classes, and the models described by them.

10.1 Universality classes

In the previous chapter, we used the KPZ equation with quenched noise (9.4) to characterize interface roughening in a disordered medium. In the light of our knowledge regarding roughening with thermal noise, the natural question arises: are nonlinear terms, such as the KPZ nonlinearity $\lambda(\nabla h)^2$, relevant in the presence of quenched noise? In this section we discuss recent numerical and analytical evidence that indicates that the quenched noise problem in many aspects is analogous to the case of thermal noise. The various discrete models fall into *two* distinct universality classes [16, 280], both universality classes being described by Eq. (9.4). For the first universality class the coefficient $\lambda = 0$ (or $\lambda \to 0$ as $f \to 0$), while for the second λ diverges at the depinning transition,

$$\lambda \sim f^{-\phi} \qquad [\phi > 0]. \tag{10.1}$$

One can obtain evidence concerning the existence of the nonlinear terms for a given model by measuring the tilt-dependent velocity, which follows a parabola if λ is nonzero (see §A.2). To illustrate the difference between the two universality classes, in Fig. 10.1 we show the tilt-dependent velocity for two different models that will be discussed in detail in the following sections: (a) the directed percolation depinning (DPD) model (§10.2) and (b) the random field Ising model (RFIM) (§10.3.2). As Fig. 10.1 shows, for both models the velocity follows a parabola. However, as we approach the depinning transition, the curvature changes differently for the two models: for the DPD model the curvature increases, indicating an increasing λ, in accord with (10.1), while for the RFIM the curvature decreases, indicating $\lambda \to 0$.

Let us consider first the DPD model and assume that λ indeed diverges as predicted by (10.1). From the fact that the velocity of the untilted interface follows (A.13), and using (9.3), we obtain

$$v(m, f) \sim f^{\theta} + f^{-\phi} m^2, \tag{10.2}$$

which indicates that $v(m, f)/f^{\theta}$ is a function of $m^2/f^{\theta+\phi}$ only. Thus the velocity is described by the scaling law

$$v(m, f) = f^{\theta} g\left(\frac{m^2}{f^{\theta+\phi}}\right), \tag{10.3}$$

where $g(x) \sim$ const for $x \ll 1$, and $g(x) \sim x^{\theta/(\theta+\phi)}$ for $x \gg 1$. In Fig. 10.2, we replot the velocities of Fig. 10.1 in the form suggested by (10.3). We obtain a remarkable data collapse, confirming the correctness of assumption (10.1) that λ diverges at the depinning transition. The same scaling law (10.3) applies to the RFIM as well, but with a negative ϕ. Indeed, as shown in Fig. 10.2, we obtain a good rescaling of the velocity curves using $\phi = -0.70 \pm 0.11$.

The measurement of λ proves to be an efficient tool for grouping different models of quenched noise into two distinct universality classes.

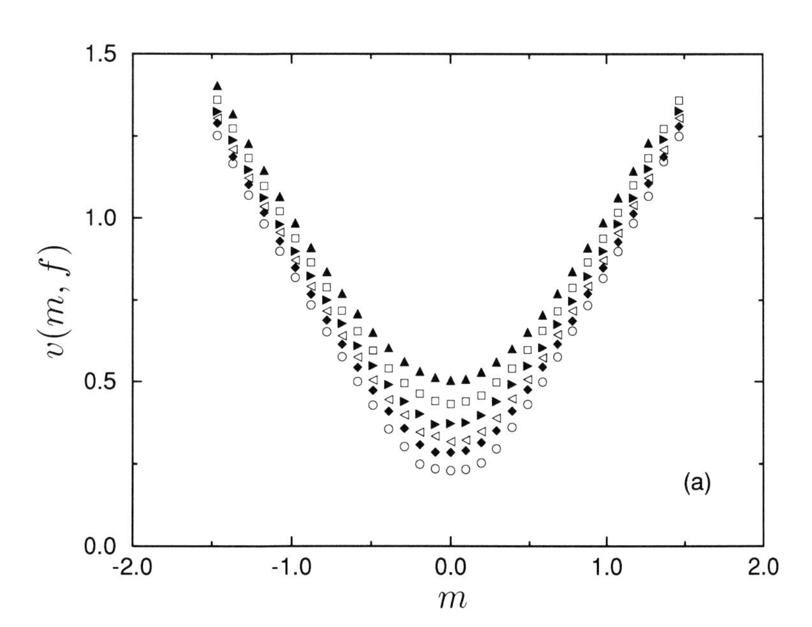

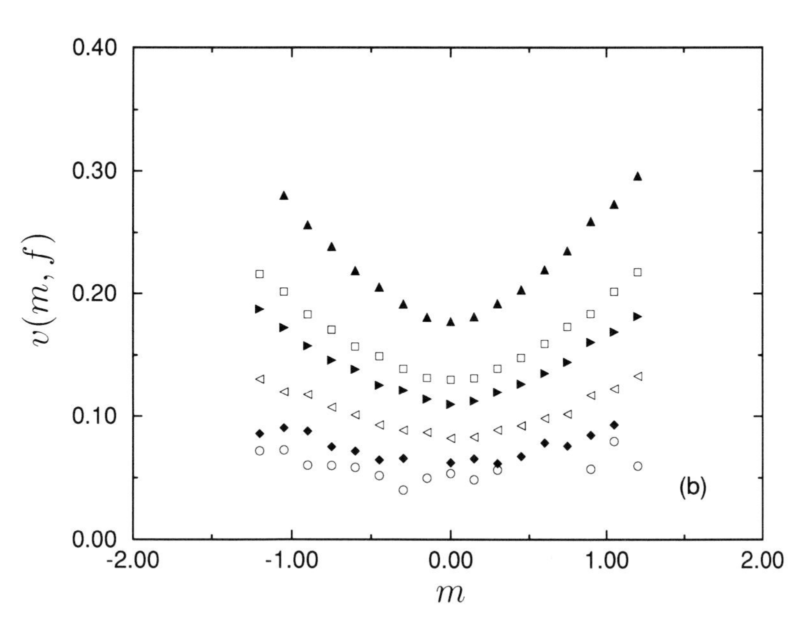

Figure 10.1 (a) Dependence on the tilt m of the average velocity, in the one dimensional DPD model. Data for different forces f are indicated by different symbols, ranging from 0.015 (bottom curve) to 0.072 (top curve); the system size is 256. (b) Tilt-dependent velocities for several forces for the two-dimensional RFIM. f varies from 0.014 (bottom curve) to 0.143 (top curve); the system size is 40×40. (After [16]).

(i) A number of models [16, 58, 63, 359, 443] have a divergent coefficient λ of the nonlinear term, so for these models the nonlinear term is relevant. Hence in order to understand the properties of the system at the depinning transition, we must consider Eq. (9.4).

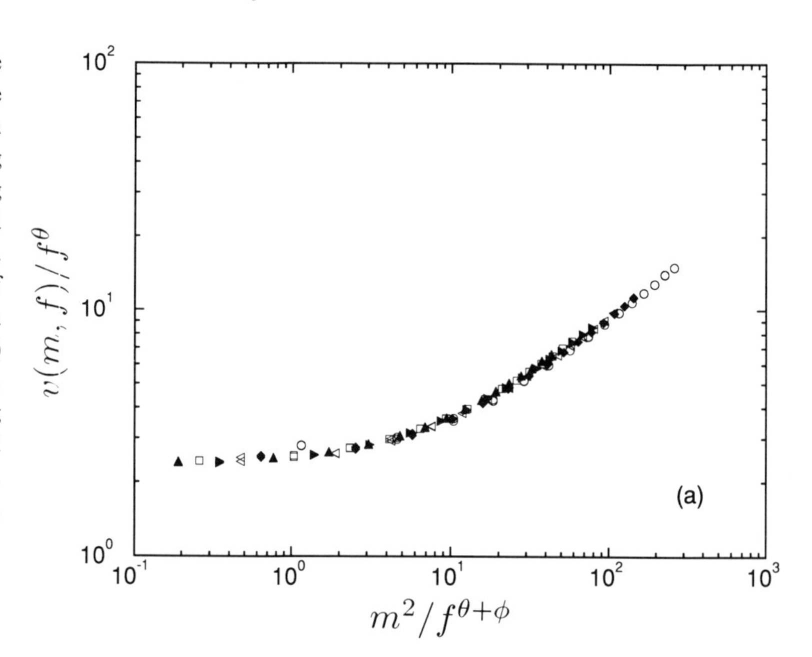

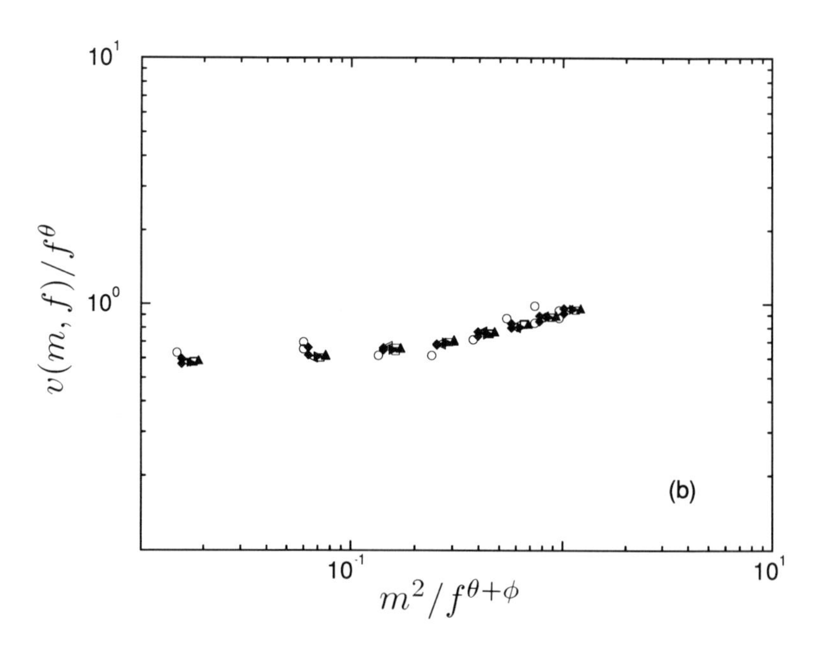

Figure 10.2 (a) The rescaling of the velocities shown in Fig. 10.1(a) according to Eq. (10.3), using $\theta = 0.59 \pm 0.12$ and $\phi = 0.55 \pm 0.12$. (b) The rescaling of the velocities shown in Fig. 10.1(b) according to Eq. (10.3), using $\theta = 0.60 \pm 0.11$ and $\phi = -0.70 \pm 0.11$. (After [16]).

We will shortly see that all these models predict $\alpha \approx 0.63$, obtained from mapping onto directed percolation.

(ii) Other models have either $\lambda = 0$ or $\lambda \to 0$ as $f \to 0$ [108, 202, 203, 235, 272, 280, 350, 351], so these models are correctly described by (9.4) with $\lambda = 0$ in the vicinity of the depinning transition.

It was recently suggested that the origin of the two universality classes observed in the numerical simulations is the anisotropy of the disordered medium [442].

The calculations indicate that the nonlinearities may have two different origins. First, we obtain the nonlinear term $(\nabla h)^2$ with a coefficient λ proportional to the velocity $v = \partial h/\partial t$. Such a nonlinear term has a *kinematic* origin, and vanishes as the velocity goes to zero – exactly the behavior observed for the RFIM. However, for an anisotropic system for which the quenched noise has different correlations in the directions perpendicular and parallel to the growth direction, even if we start without a nonlinear term, λm^2 will be generated by the disorder. Numerical studies show that the nonlinear term diverges. However, if the system is isotropic ($\Delta_h = \Delta_x$), such a nonlinearity will not be generated. In this case the only nonlinearity in the growth equation is kinematically generated, and vanishes at the depinning transition.

There are three additional results provided in [442]: (i) The tilted interface belongs to a third universality class, which has highly anisotropic scaling exponents that can be calculated exactly. We shall not discuss this third universality class further, since in most natural phenomena the interfaces have a zero average tilt. (ii) For the first universality class (λ nonzero) by tilting the interface, the critical force will change as $F_c(m) - F_c(0) \sim -|m|^{1/\nu(1-\alpha)}$, where $F_c(m)$ is the threshold force for the interface with average slope m. (iii) The new exponent ϕ defined by (10.1) can be related to the other scaling exponents by the scaling relation $\phi = \nu(2 - \alpha - z)$.

10.2 Pinning by directed percolation

In this section, we discuss the directed percolation depinning (DPD) model, which has a diverging nonlinear term. The scaling exponents can be obtained exactly by mapping the depinning problem onto a variant of percolation termed *directed percolation.* The connection between surface growth with quenched disorder and directed percolation was originally proposed by two independent studies [58, 443]. A number of seemingly different models also belong to the same universality class [63, 162, 272, 275, 354, 409, 410, 444].

10.2.1 Directed Percolation Depinning (DPD) model

The model is defined as follows: on a square lattice of edge L (with periodic boundary conditions), we *block* a fraction p of the cells to correspond to the quenched disorder. Blocked cells will try to stop the growth, while the interface is free to advance on unblocked cells. At $t = 0$, the 'interface' is the bold horizontal line shown in Fig. 10.3(a). At $t = 1$ we randomly choose a cell (labeled X in Fig. 10.3(b)) from among the unblocked cells that are nearest-neighbors to the interface. We 'wet' cell X and *any cells that are below it in the same column*. This process is then iterated. For example, Fig. 10.3(c) shows that at $t = 2$ we choose a second unblocked cell, cell Y, to wet, while Fig. 10.3(d) shows that at $t = 3$ we wet cell Z *and also cell Z′ below it*.

10.2.2 Scaling properties

For p below a critical threshold $p_c = p_c(L)$ the interface propagates without stopping, while for p above p_c the interface is pinned by the blocked cells. Figure 10.4 displays a typical interface after the growth has stopped; this occurs just as the surface meets a spanning path of blocked cells connected through nearest and next-nearest neighbors.

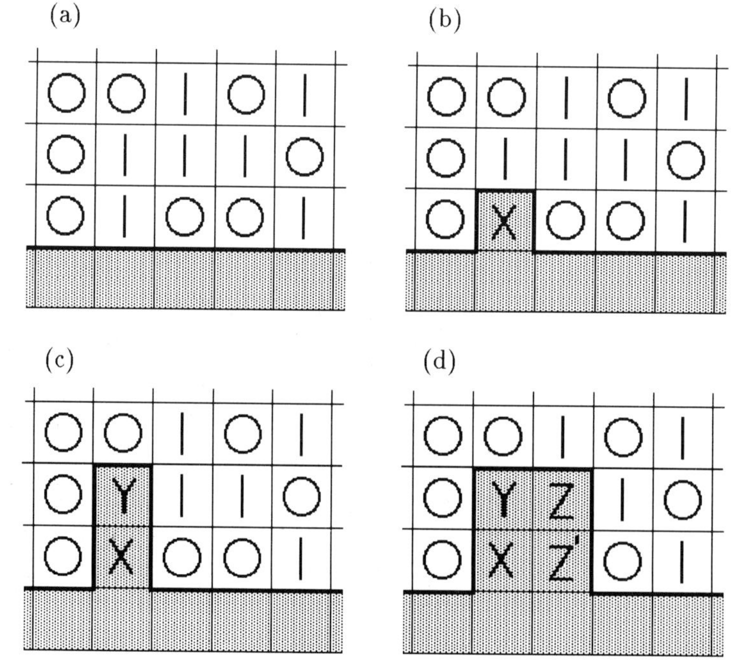

Figure 10.3 The DPD (directed percolation depinning) model for interface growth with erosion of overhangs. 'Wet' cells are shaded, while dry cells are randomly blocked with probability p (indicated by O) or unblocked with probability $(1-p)$ (indicated by |). The interface between wet and dry cells is shown by a heavy line. (a) $t = 0$, (b) $t = 1$, (c) $t = 2$ and (d) $t = 3$. (After [58]).

The definition of the model precludes this path having overhangs, and thus it is equivalent to a spanning path on a square lattice that goes from left to right, is able to turn up and down, but can never turn left. Such a directed path is, in fact, a path on a directed percolation cluster. Indeed, one can show that the pinned interface in the model corresponds to the hull of a directed percolation cluster.

When the probability of blocked cells p is above p_c, the growth is halted by a spanning path of a directed percolation cluster. Such a directed path is characterized by a correlation length parallel to the interface, $\xi_{\parallel}$, and one perpendicular to it, $\xi_{\perp}$. Their meaning is the following: blocking a fraction p of the sites on a lattice, the *blocked* sites form directed paths whose average length is $\xi_{\parallel}$ and whose width is $\xi_{\perp}$. The two correlation lengths diverge in the vicinity of p_c,

$$\xi_{\parallel} \sim |p - p_c|^{-\nu_{\parallel}} \qquad \xi_{\perp} \sim |p - p_c|^{-\nu_{\perp}}. \tag{10.4}$$

From calculations on directed percolation, it is known that $\nu_{\parallel} \approx 1.733$ and $\nu_{\perp} \approx 1.097$ [64, 233].

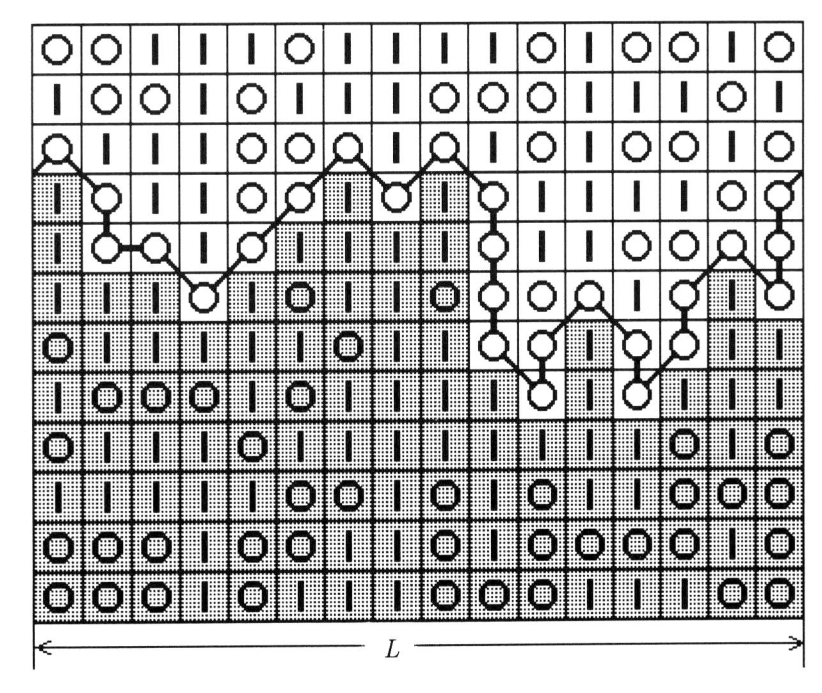

Figure 10.4 Shown as a bold line is a *spanning* path formed by connected nearest-neighbor and next-nearest-neighbor blocked cells which pin the interface. Note that various *nonspanning* clusters of blocked cells (found inside the wet region) are not sufficient to pin the interface. (After [58]).

10.2.3 Pinned interfaces

For $p \ll p_c$, $\xi_\parallel \ll L$. The directed paths *locally* pin the interface, but the interface eventually moves with a nonzero average velocity by advancing in between the blocked paths. Complete pinning appears when $\xi_\parallel$ becomes equal to the system size L. The width, w, of such a path is of the order of $\xi_\perp$. Hence using (10.4) and (2.5) we obtain the scaling of the width

$$w \sim \xi_\perp \sim |p - p_c|^{-\nu_\perp} \sim \xi_\parallel^{\nu_\perp/\nu_\parallel} \sim L^{\nu_\perp/\nu_\parallel} = L^\alpha, \tag{10.5}$$

from which we obtain for the roughness exponent

$$\alpha = \frac{\nu_\perp}{\nu_\parallel} = 0.633 \pm 0.001. \qquad \rightarrow (10.6)$$

This result is confirmed by numerical studies on the scaling of the pinned interface. The critical probability p_c must coincide with the critical probability of the underlying directed percolation problem, giving $p_c \approx 0.47$.

It is straightforward to connect the qualitative picture of an interface moving in a medium with quenched disorder with the concrete behavior displayed by the model. The driving force F corresponds to the probability $1 - p$, and F_c to $1 - p_c$. For $p < p_c$ the interface moves with a finite velocity, being locally pinned by the noise over domains of characteristic horizontal size $\xi_\parallel$. Thus $\xi_\parallel$ can be identified with ξ in the interface language [see (9.6)], and $\nu_\parallel$ with ν.

At this stage, we have obtained *exact* values for two of the four exponents (α and ν). Next we estimate the interface velocity. To do this we must define the *time* in the model. Usually in growth models, the unit time corresponds to L growth attempts. In the model, not all growth attempts result in growth: if we choose a given site on the interface, and attempt to grow it in a direction where a blocked site is, the interface cannot grow. Thus the number of sites adjacent to the interface which may grow is proportional to the number of free sites neighboring the interface. The number of *blocked* sites neighboring the interface is proportional to $\xi_\parallel$, thus one can assume that the number of free sites is proportional to $1/\xi_\parallel$. If the interface manages to jump through a pinning path, it advances over a vertical length of order $\xi_\perp$, the width of the blocking directed path. Thus we obtain for the interface velocity near p_c

$$v(p) \sim \xi_\perp/\xi_\parallel \sim |p - p_c|^{\nu_\parallel - \nu_\perp}, \tag{10.7}$$

from which the velocity exponent (9.3) is

$$\theta = \nu_\parallel - \nu_\perp = 0.636. \qquad \rightarrow (10.8)$$

A simple argument provides the value of the dynamic exponent z. The time necessary to depin a domain of size $\xi_\parallel$ is linear in t, so $\xi_\parallel \sim t$. Hence $z = 1$, according to (2.6). Table 10.1 provides a summary of the calculated exponents, from which we can check that the exponents verify the scaling relation (9.11).

10.2.4 Moving interface

In the previous discussion we focused on the *pinned* interface ($p = p_c$). However, in general one is interested in the dynamical properties of the roughening process, which can be studied by looking at the *moving* interface ($p < p_c$). Numerical studies on the moving interface give $\alpha = 0.70 \pm 0.05$, and $\beta = 0.70 \pm 0.05$ [58]. Scaling arguments, supported by extensive numerical simulations, show that the dynamic exponent is equal to the fractal dimension of the shortest path, d_{min}, of isotropic percolation [13, 161]. These two values indicate that $z = \alpha/\beta = 1$. The scaling relation $\alpha + z = 2$ is not fulfilled. However, this relation is not expected to hold for quenched noise since the coefficient λ of the non-linear term should renormalize.

The mapping to directed percolation exists only for one-dimensional interfaces, since the higher dimensional generalization of directed percolation considers directed paths embedded in higher dimensional spaces. The higher dimensional version of the model is of general interest, in particular because many experimentally relevant interfaces are two-dimensional. The problem can be still mapped onto a percolation problem, that of 'percolation of directed surfaces' [59]. The model has been studied up to $d = 4$, the relevant exponents being listed in Table 10.1.

Tang and Leschhorn studied an equivalent SOS-type model with quenched disorder. In their case, the disorder is a random number between zero and one, quenched into the lattice [443]. An external driving force controls the depinning transition and the velocity of the interface. They demonstrate that pinning occurs when a directed percolation path, which the interface cannot pass, spans the entire system. The rule defining the directed percolation path in this model differs from that of the previous one, which does not affect the value of the scaling exponents, but only the value of the critical probability p_c. Having the same characteristic exponents, the two models belong to the same universality class.

Table 10.1 *Numerical estimates of the scaling exponents for interface roughening with quenched noise. The second column denotes the universality class of the model, I denotes models with isotropic noise, and DP denotes models belonging to the directed percolation universality class. Note that the exponents for the DPD models refer to the pinned interface. The same models give* $\alpha \approx 0.7$ *for a moving interface. (*) In ([228]) a crossover has been observed from* $\alpha = 0.75$ *to* 1. *(**)* $\theta = 0.22$ *has been determined from the advance of the mean height. The increase in the fluid volume leads to* $\theta = 0.41$.

Model	UC	d	α	β	z	ν	θ	Reference
DPD	DP	1	0.633	0.633	1	1.733	0.636	[443]
DPD	DP	1	0.633	0.633	1	1.733	0.636	[58]
DPD	DP	1	0.63	-	-	-	-	[409]
Resistor	DP	1	0.63	0.63	1.01	1.73	-	[63]
Parisi	DP	1	-	0.75	-	-	-	[359]
RFIM-SS	I	1	1	-	1.16	1.33	0.22**	[351]
MCR	I	1	0.81	-	-	1.30	-	[293]
String	I	1	0.97	-	-	1.05	0.24	[108]
Automaton	I	1	1.25	0.88	1.42	-	0.25	[272]
Eq. (9.4)	I	1	0.75*	-	-	-	-	[228]
Eq. (9.4)	I	1	1.25	0.88	-	-	-	[273]
Eq. (9.4)	I	1	0.72	0.61	1.16	-	0.50	[88]
DPD	DP	2	0.48	0.41	1.16	1.06	-	[31, 61]
RFIM-SA	I	2	0.67	-	1.9	0.75	-	[379]
RFIM-SS	I	2	1	-	-	0.9	-	[203]
Automaton	I	2	0.75	0.47	1.58	-	0.65	[272]
DPD	DP	3	0.38	0.28	1.32	-	-	[63]
Automaton	I	3	0.35	0.2	1.75	-	0.84	[273]
DPD	DP	4	0.27	0.18	1.50	-	-	[63]

10.2.5 Self-Organized Depinning (SOD) model

A variant of the DPD model with *global* updating called the self-organized depinning (SOD) model has been recently introduced – specifically, the interface grows on the site at which the pinning force is the smallest [162, 409, 411]. Although we might expect that the non-local character of the updating rule could introduce a new universality class, it has been argued [162, 273, 275, 444] that the same roughness exponent is obtained as for the DPD model. The temporal behavior of the fluctuations is nontrivial, leading to multi-affine behavior (see Chapter 24). The origin of the time-dependent interfacial advance has been discussed in terms of 'avalanches' by several authors [31, 61, 63, 275, 354, 409, 411] (see Fig. 10.5). Such burst-like growth is fairly widespread [139, 431].

The main difference between the original DPD model and the global updating variant is that the latter does not require a tunable driving force. Instead, the growth rule self-tunes the interface such that it is always at the critical point F_c – in this sense, the variant exemplifies ideas of self-organized criticality. The scaling properties of this variant of the DPD model coincide with those of the DPD model at $F = F_c$ [162, 275, 354].

10.3 Isotropic growth models

In light of our previous discussion, we shall call *isotropic growth models* those for which the nonlinear term either vanishes with the velocity,

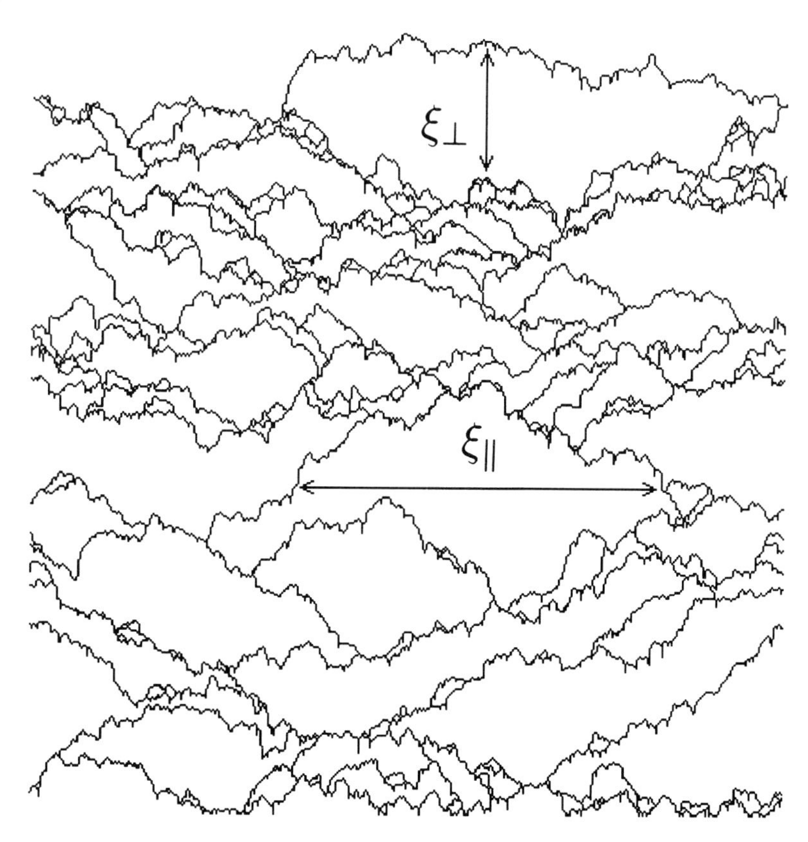

Figure 10.5 Illustration of interface propagation via avalanches in the DPD model. Close to p_c the interface is pinned by a directed percolation path. Removing a randomly chosen site from the pinning path, the interface advances further, being pinned at a later position by another spanning path. The difference between two consecutive pinned configurations defines an *avalanche*. The size of the avalanches can be characterized by the two lengths $\xi_\parallel$ and $\xi_\perp$ defined on the figure. (After [31]).

or is identically zero for any driving force. At the depinning transition, these models are described by (9.4) with $\lambda = 0$.

10.3.1 RG results

Due to the absence of the nonlinear term, (9.4) can be studied using RG methods [336, 339]. The calculation indicates the existence of an additional scaling relation

$$\nu = \frac{1}{2-\alpha}, \qquad \rightarrow (10.9)$$

obtained from the fact that the static response function does not renormalize. Equations (9.11) and (10.9) leave two exponents undetermined, α and z. An expansion of (9.4) in the variable $\epsilon \equiv 4 - d$ predicts

$$\alpha = \frac{\epsilon}{3}, \qquad z = 2 - \frac{2}{9}\epsilon. \qquad \rightarrow (10.10)$$

How reliable are these RG predictions? They result from an expansion in ϵ, so they can provide trustworthy results only for sufficiently small ϵ – in the vicinity of the critical dimension $d_c = 4$. Moreover, the derivation assumes that there are no overhangs in the interface. Overhangs usually can be neglected if $\alpha < 1$. But for one dimension ($d = 1$), (10.10) predicts $\alpha = 1$. Thus the effect of the overhangs is marginal.

10.3.2 Random field Ising model

Next, we discuss the RFIM which, as noted above, is described by the isotropic theory [202, 203, 234]. The RFIM is defined by spins with possible values $s_i = \pm 1$ that are placed on sites i of a lattice, interacting with Hamiltonian

$$\mathscr{H} \equiv -J \sum_{\langle i,j \rangle} s_i s_j - \sum_i (h_i + H) s_i. \qquad \rightarrow (10.11)$$

Here J is the exchange energy between the ferromagnetically-coupled nearest-neighbor spins, $\sum_{\langle i,j \rangle}$ denotes the summation over all nearest-neighbor pairs $\langle i, j \rangle$, $\sum_i$ is the sum over all spins in the system, H is the external uniform field, and h_i is the quenched local field, corresponding to the quenched disorder. In the simplest version of the model we take $J = 1$, and allow h_i to be uniformly distributed over the interval $[-\Delta, \Delta]$, so Δ characterizes the effective strength of the disorder.

A typical simulation starts from a straight interface, the spins under the surface being 'flipped' ($s = +1$), while all the others are 'unflipped'

($s = -1$). To eliminate the effects of nucleating $s = +1$ domains in the unflipped bulk, only spins on the interface are allowed to flip. A (−) spin flips if it is adjacent to the interface, and by flipping lowers the total energy of the system. At zero temperature, once a spin flips, it never returns to its original value.

The role of the driving force is played by the external field H. At zero field ($H = 0$), the interface moves in an arbitrary direction until the system finds a local energy minima, and then it stops. Increasing H makes the $s = -1$ phase unstable, so the $s = +1$ phase will try to advance by moving the interface at the expense of the (−) phase. For small H, the impurities will pin the interface, but for $H > H_c$ the interface becomes depinned, moving with a finite velocity v. The critical field H_c depends on the strength of the disorder Δ.

For the two-dimensional RFIM there are two morphologically-different regimes (see Fig. 10.6), depending on Δ. For $\Delta < 1.0$ the interface is faceted (F), while for $\Delta > 1.0$, it is self-similar (SS).

By definition, the models studied in the previous sections have generated *single-valued* interfaces, i.e., no overhangs have been allowed. In contrast, the RFIM interface forms overhangs (see Fig. 10.7). At short length scales the scaling of the interface width is affected by these overhangs, inducing a large intrinsic width.

In the SS regime the interface can be mapped onto site percolation: the stability of a spin is determined only by the local field $H + h_i$. All spins which have large enough local field are flipped, and

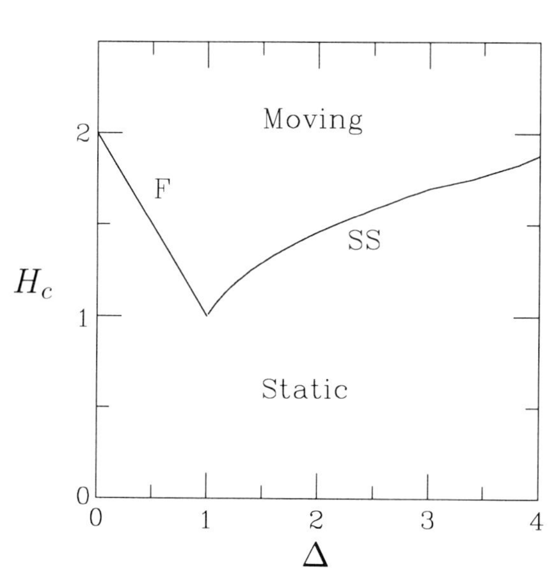

Figure 10.6 Phase diagram for the two dimensional RFIM on a square lattice. The continuous curve indicates the critical magnetic field separating the static and moving states. SS denotes the self-similar regime, while F corresponds to a faceted interface morphology (After [202]).

the interface is the perimeter of the generated percolation cluster [202, 203, 234, 350, 351]. The fractal dimension of the interface coincides with the fractal dimension of the percolation cluster, $d_f = 1.89 \pm 0.01$ [424].

For length scales shorter than ξ, the correlation length of the percolation problem, the interface is pinned by the percolation cluster, while for larger length scales it is free to move. Thus the correlation length ξ for the moving interface is the correlation length of the percolation problem, giving for the correlation exponent $\nu = 4/3$. For $\ell < \xi$ the interface is *self-similar*.† For $\ell > \xi$, the effect of the quenched noise no longer dominates the morphology and there is a crossover to $\alpha = 0.5$.

A simple argument [351] using a mapping of the SS interface onto the percolation problem provides the velocity exponent $\theta = \nu(d_{\rm min} + 1 - d_f)$. Here $d_{\rm min}$ is the fractal dimension of the shortest path across a percolation cluster [176], and d_f is the fractal dimension of the incipient infinite cluster appearing at the percolation threshold. The predicted value $\theta = 0.41 \pm 0.03$ for $d = 1$ is in good agreement with the numerical result $\theta = 0.42 \pm 0.05$. However, it is different from the value $\theta = 1/3$, predicted by the RG calculations. This discrepancy is not surprising, since the RG was performed by assuming that the interface is self-affine and not self-similar.

So far we have focused on the two-dimensional RFIM, which generates a one-dimensional interface. One peculiar aspect of the two-dimensional RFIM is that it does not generate a self-affine interface.

† Since the interface is not self-affine, but self-similar, we have $\alpha = 1$ (See §3.3).

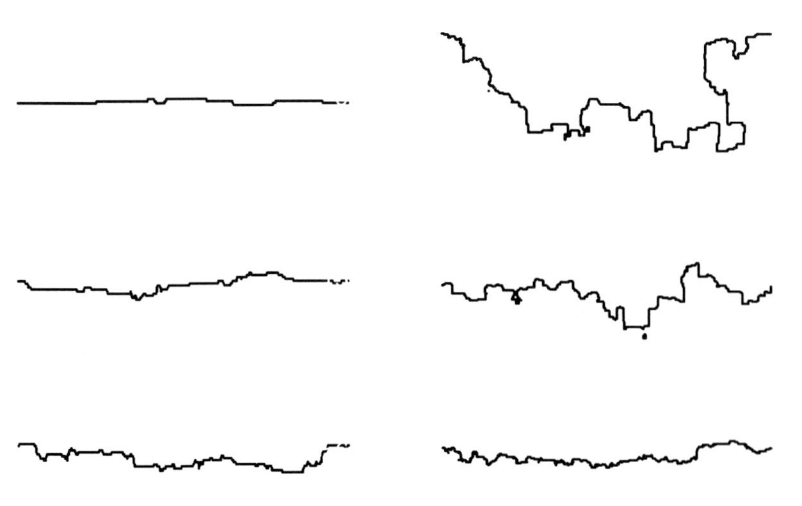

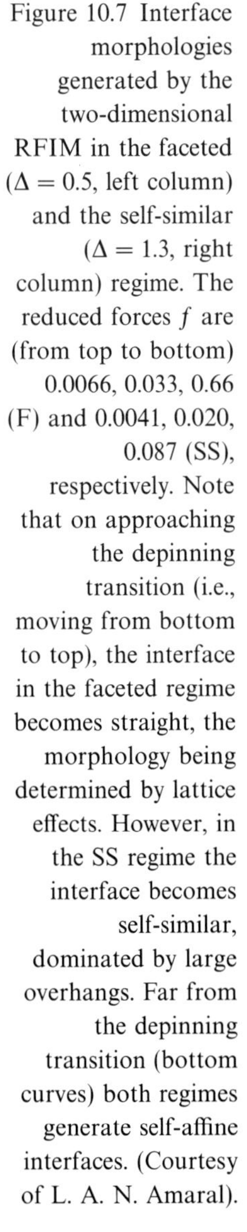

Figure 10.7 Interface morphologies generated by the two-dimensional RFIM in the faceted ($\Delta = 0.5$, left column) and the self-similar ($\Delta = 1.3$, right column) regime. The reduced forces f are (from top to bottom) 0.0066, 0.033, 0.66 (F) and 0.0041, 0.020, 0.087 (SS), respectively. Note that on approaching the depinning transition (i.e., moving from bottom to top), the interface in the faceted regime becomes straight, the morphology being determined by lattice effects. However, in the SS regime the interface becomes self-similar, dominated by large overhangs. Far from the depinning transition (bottom curves) both regimes generate self-affine interfaces. (Courtesy of L. A. N. Amaral).

However, for the two-dimensional interface generated by the three-dimensional RFIM there is a F regime ($\Delta < 2.4$), a SS regime ($\Delta > 3.4$), and also a self-affine (SA) regime in between ($2.4 < \Delta < 3.4$). Numerical simulations in the self-affine regime give $\alpha = 0.67 \pm 0.03$, in good agreement with the prediction (10.10) [203, 379].

The first isotropic interface model for which the depinning transition was identified was proposed by Martys, Cieplak and Robbins (MCR) [79, 80, 293]. The two-dimensional model considers fluid moving among a set of randomly-placed discs (Fig. 10.8).

The morphology and the velocity of the interface are dominated by capillary effects, the relevant control parameters being the contact angle of the invading fluid and the fluid pressure. The resulting scaling exponents are provided in Table 10.1. The critical fluid pressure at which the depinning transition occurs is independent of the interface orientation, implying that the model belongs to the same universality class as the RFIM.

10.4 Discussion

Motivated by the interest in understanding the effects of quenched noise, a number of studies estimated the exponents from the integration of the growth equations [87, 88, 108, 228, 273, 274]. Indeed, direct integration of the KPZ equation has provided a useful approach for checking the predictions of RG calculations, and for determining scaling exponents for $d > 1$. Also, direct integration of Eq. (9.4) serves

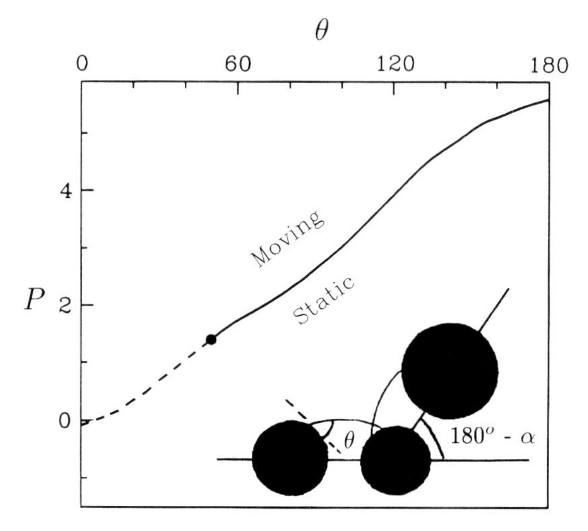

Figure 10.8 The morphology phase diagram predicted by the MCR model. Shown is the dependence of the critical pressure P upon the static contact angle θ of the invading fluid, which varies from 180° to 0° as the fluid changes from nonwetting to wetting. The solid and dashed lines indicate SS and SA growth, respectively. The inset shows the arcs between two successive pairs of beads on an interface, where the invading fluid is *below* the interface. (After [293]).

as a test for the scaling picture discussed in the previous sections, as well as providing accurate values for the exponents (Table 10.1).

The most important result of this chapter is that quenched noise *changes* the scaling exponents. The various models can be grouped into two main universality classes. Some models, including the RFIM, are described by (9.4) with $\lambda = 0$; the exponents for this universality class can be obtained using RG methods. Other models, including the DPD model, have relevant nonlinear terms, whose coefficient λ diverges as we approach the depinning transition; these models are expected to be described by (9.4).

The understanding of roughening dominated by quenched noise is especially important for interpreting the results of various experiments, for which it is generally believed that the disorder of the substrate determines the morphology of the interface. These experiments are discussed in detail in the following chapter.

Suggested further reading:

[13, 31, 146, 339, 379]

Exercises:

10.1 Consider the following modification of the DPD (directed percolation depinning) model: on every site of the lattice we fix a number $p = ah + r$, where r is a random variable uniformly distributed between zero and one. Every site for which $p > 1$ is a blocked site. Let the interface advance from $h = 0$. The increasing density of random sites will stop the interface after a critical height, h_c, is reached. Calculate h_c as a function of a. How is the interface width affected by the gradient in the disorder? How does the interface width depend on a? Generalize the Family-Vicsek scaling relation (2.8) to include the gradient in the disorder.

10.2 Using the definition of avalanches given in Fig. 10.5, calculate the exponent χ_v that describes the divergence of the characteristic avalanche volume, $V \sim |p - p_c|^{-\chi_v}$, for the d-dimensional DPD model. The volume is the number of sites removed by the interface between two consecutive pinned configurations.

11 Experiments

The increasing interest of researchers in the basic properties of growth processes has provided the initiative for a number of experimental studies designed to check the applicability of various theoretical ideas to experimental systems. While many experimental studies have been inspired by the KPZ theory, most have failed to provide support for the KPZ prediction that $\alpha = 1/2$. Instead, most data suggest that $\alpha > 1/2$. These experimental results initiated a closer look at the theory, and led to the discovery that quenched noise affects the scaling exponents in unexpected ways. In this chapter, we discuss some of these key experiments, including fluid-flow experiments, paper wetting, propagation of burning fronts, growth of bacterial colonies and paper tearing. Atom deposition in molecular beam epitaxy, which is one important class of experiments on kinetic roughening of interfaces, will be discussed in Chapter 12. The new theoretical ideas needed to understand the effect of atomic diffusion on the roughening process will be developed at that time.

11.1 Fluid flow in a porous medium

Two-phase fluid flow experiments have long been used to study various growth phenomena. The Hele–Shaw cell, well-known from studies on growth instabilities in viscous fingering [126, 283, 346, 456], has proved to be a useful experimental setup for the study of the growth of self-affine interfaces. A typical setup used in these experiments (Fig. 11.1) is a thin horizontal Hele–Shaw cell made from two transparent plates. A gap is set between the two plates and filled with glass beads of average diameter a. The random packing of the glass beads creates a model porous medium. Fluid (water or glycerol) is injected into the cell from one end. The motion of the interface is recorded using a video camera, the image of which is then digitized and analyzed with a computer.

The first experimental evidence for the existence of the depinning transition for interface motion was provided by Stokes, Kushnick and Robbins [427], who measured the velocity of the fluid interface moving in a porous medium as a function of the pressure drop P driving that motion. They found that the interface is pinned for pressures below a threshold pressure P_c and moves with a finite velocity for $P > P_c$, in agreement with what one would expect from the discussion in the previous chapters.

Next, we review three experiments on the propagation of a wetting fluid in a porous medium. For convenience we refer to them as REDG [384], HFV [181] and HKW [166].

The actual morphology of the interface can be explained as a result of the competition between the nonlocal fluid pressure P_f, and the local random capillary forces. The fluid pressure obeys the Laplace equation

$$\nabla^2 P_f = 0, \tag{11.1}$$

which *smooths* the interface, while the capillary forces serve to *roughen* the interface. The local fluctuations in the capillary pressure P_{cap} are generated by the randomness in the packing of the glass beads. Since the position of the beads does not change in time, their random packing acts as *quenched* noise and affects the evolution of the interface. The interplay between the wetting angle of the fluid and the local geometry of the bead positions may cause a local pinning of the interface. The competition between the smoothing effect of the nonlocal fluid pressure and the local capillary forces determines the final roughness of the interface.

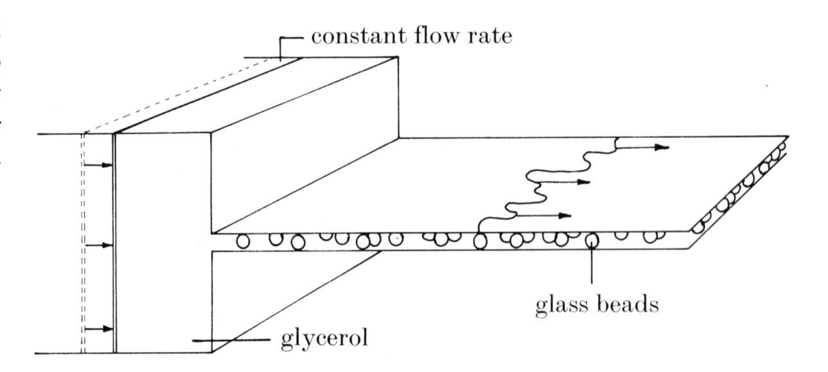

Figure 11.1 Typical experimental setup used in fluid-flow experiments (After [483]).

Table 11.1 *The scaling exponents as measured in different experimental investigations.*

Experiment	α	β	Reference
Fluid flow	0.73	-	[384]
	0.81	0.65	[181]
	0.65–0.91	-	[166]
Paper wetting	0.63 (d=1)	-	[58]
	0.50 (d=2)		[59]
	0.62–0.78 (d=1)	0.3–0.4	[122]
Bacteria growth	0.78	-	[459]
Burning fronts	0.71	-	[504]

A relevant control parameter is the capillary number

$$C_a \equiv \frac{v\mu_w a^2}{k\sigma}, \tag{11.2}$$

where v is the average velocity of the interface, μ_w the dynamic viscosity of the wetting fluid, k the hydraulic permeability, and σ the fluid-air interfacial tension. From Darcy's law, the average pressure gradient is $\langle|\nabla P_f|\rangle = v\mu_w/k$. Since $P_{\text{cap}} \sim \sigma/a$, we have

$$C_a = a\frac{\langle|\nabla P_f|\rangle}{P_{\text{cap}}}. \tag{11.3}$$

Thus C_a is a measure of the relative importance of the smoothing effect of the fluid pressure and the roughening generated by the local capillary forces. Since the capillary number is proportional to the interface velocity, (11.2) shows that roughening due to the local capillary forces becomes increasingly important as the average velocity of the interface goes to zero.

The measured values of the two exponents of primary interest, α and β, are summarized in Table 11.1. The spread calls into question the universality of the growth mechanism. Although the origin of the different values is not fully understood, there are some indications from simulations and experiments that the reported discrepancies arise not only from experimental uncertainties, but the experimentally-accessible exponents may have small systematic variations. In the following, we present some arguments to account for these apparent discrepancies.

The random packing of the beads generates quenched disorder. As discussed in Chapter 10, in the presence of quenched noise there exists a correlation length ξ quantifying the length scale below which the quenched noise dominates the roughening process. For length scales larger than ξ, thermal fluctuations induced by the motion of the interface determine the roughness. Thus we expect the correlation function to show a crossover from anomalous roughening with an exponent $\alpha > 1/2$ to a region with $\alpha = 1/2$. The existence of this crossover is supported by numerical simulations on the DPD, RFIM, and MCR models [58, 293, 350, 379]. Experimental support for the existence of the correlation length ξ is provided by the HFV measurements reproduced in Fig. 11.2, which display a crossover from $\alpha \approx 0.81$ to the KPZ prediction $\alpha \approx 0.49$ in the scaling of the height–height correlation function.

From the discussion in Chapter 10, we expect that changing the driving force should change the correlation length ξ. However, in fluid-flow experiments the driving force is not well defined. The only well-controlled quantity is the average velocity v of the interface. According to (9.3) and (9.6), we have

$$\xi \sim v^{-\nu/\theta}. \tag{11.4}$$

If the correlation length is not known, there is a chance that one fits data mainly in the crossover region, measuring erroneous roughness exponents. The same arguments suggests that the effective roughness

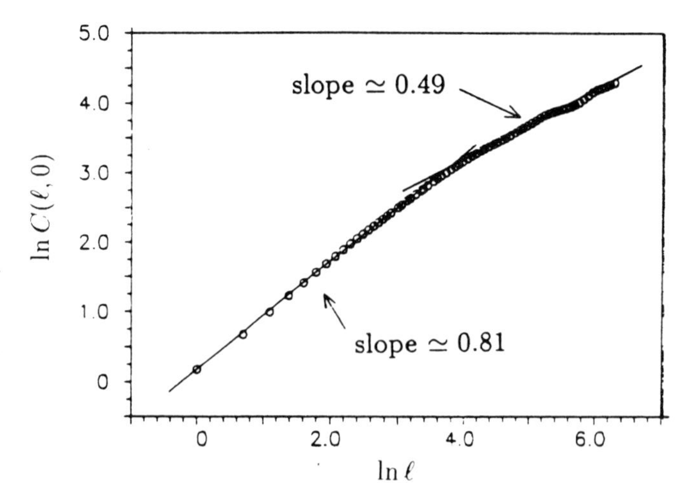

Figure 11.2 Spatial scaling of the height–height correlation function in fluid-displacement experiments. The slope of the straight line fitting the short length scale part of the curve is $\alpha \approx 0.81$, while the long length scale behavior is fitted by a line with exponent $\alpha \approx 0.49$. (After [181]).

exponent decreases as the velocity or capillary number increases (and the correlation length decreases according to (11.4)). Indeed, the HKW experiments varied the capillary number C_a between 10^{-5} and 10^{-2}, and found a systematic decrease in the effective roughness exponent α as C_a increases (see Fig. 11.3).

There is only one reported measurement of the value of the exponent β characterizing the dynamics of the roughening process [181]. The scaling of the height–height correlation function with *time* is consistent with $\beta = 0.65$. This value implies the presence of anomalous roughening, β being larger than 1/3, the KPZ prediction, or 1/4, the EW prediction, and 1/2, the RD prediction.

11.2 Paper wetting

In this section we describe a set of simple experiments that does not require sophisticated equipment. The measured exponents are robust, and are in good agreement with the theoretical predictions of simple

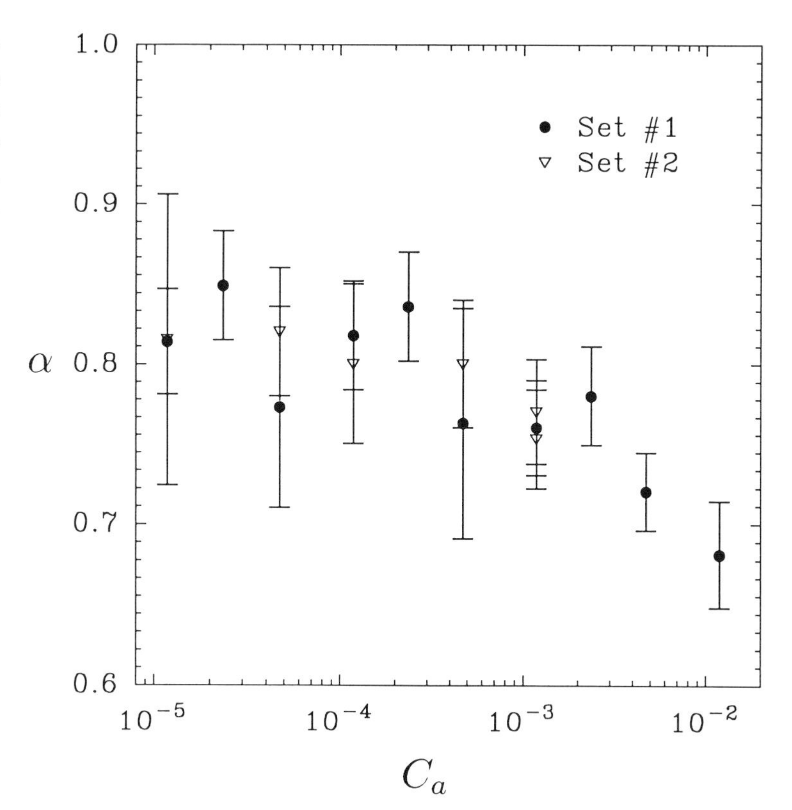

Figure 11.3 The variation of the effective roughness exponent α with the capillary number. (After [166]).

discrete models. Thus they can be used as hands-on demonstrations to test the basic ideas of interface roughening [14, 15, 31, 58, 59].

The idea underlying these experiments stems from the simple observation that a drop of ink on a sheet of paper spreads and forms a patch with a rough interface between the wet and dry areas. The circular form of the patch is not optimal for interface analysis. Luckily a simple modification of the experiment can generate the desired strip geometry: one can simply dip a strip of paper into a basin filled with a suspension of ink (Fig. 1.2). The suspension is absorbed into the paper, creating a rough interface. One allows the interface to rise up the paper *until it stops*, and finds no further change in either the height or shape of the interface.† After drying, this rough interface is digitized (Fig. 1.2).

The roughness exponent of the interface is calculated by averaging over different samples. The height–height correlation function defined in Appendix A scales as $C(\ell,0) \sim \ell^{\alpha}$, with

$$\alpha = 0.63 \pm 0.04 \tag{11.5}$$

(see Fig. 11.4). No difference in the value of α is observed between fronts propagating in an upward direction and fronts propagating downward, so *gravity* plays a negligible role relative to other driving forces.

† The most important difference between paper wetting and fluid-flow in porous media discussed in the previous section is that in paper wetting the *pinned* interface is studied in contrast to the moving fluid front in fluid-flow experiments.

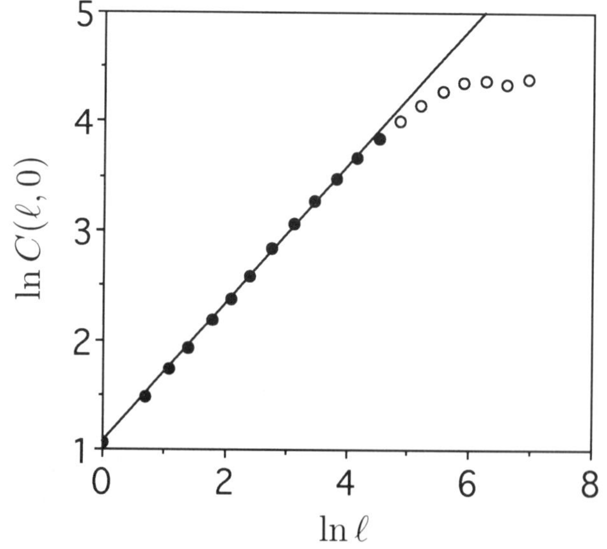

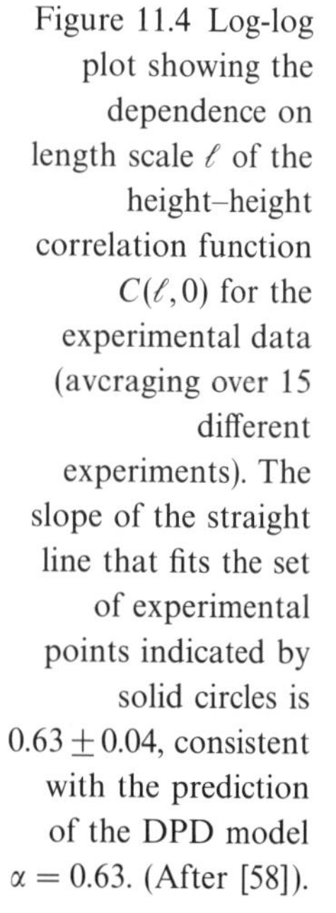

Figure 11.4 Log-log plot showing the dependence on length scale ℓ of the height–height correlation function $C(\ell,0)$ for the experimental data (averaging over 15 different experiments). The slope of the straight line that fits the set of experimental points indicated by solid circles is 0.63 ± 0.04, consistent with the prediction of the DPD model $\alpha = 0.63$. (After [58]).

The experimental results can be interpreted in the framework of the DPD model (see §10.2). The properties of the pinned interface can be understood using the mapping to directed percolation (see Eq. (10.6)), which predicts the roughness exponent $\alpha = 0.63$, in good agreement with the experimental result (11.5).

In the model, the mapping to directed percolation exists only when the concentration of the disorder is critical – i.e., only at the critical point $p = p_c$. Thus the natural question arises: What makes the wet interface critical – i.e., what is the physical mechanism that stops the interface *exactly* at the critical state?

At microscopic length scales, paper is an extremely disordered substance, formed by long fibers that are randomly distributed and randomly connected (see Fig. 11.5). The wetting fluid propagates mainly due to capillary forces. The random nature of the fiber network and the particles in the suspension constantly obstruct the fluid flow. As we depart from the water source, evaporation decreases the fluid pressure, making it increasingly difficult for the fluid to overcome the microscopic 'obstructions.' At the critical height, the fluid pressure balances the effect of the pinning obstacles and the fluid stops propagating.

Thus the presence of evaporation decreases the fluid pressure, which in turn increases the pinning effect of the inhomogeneities in the paper – ultimately leading to the stopping of the interface. Although

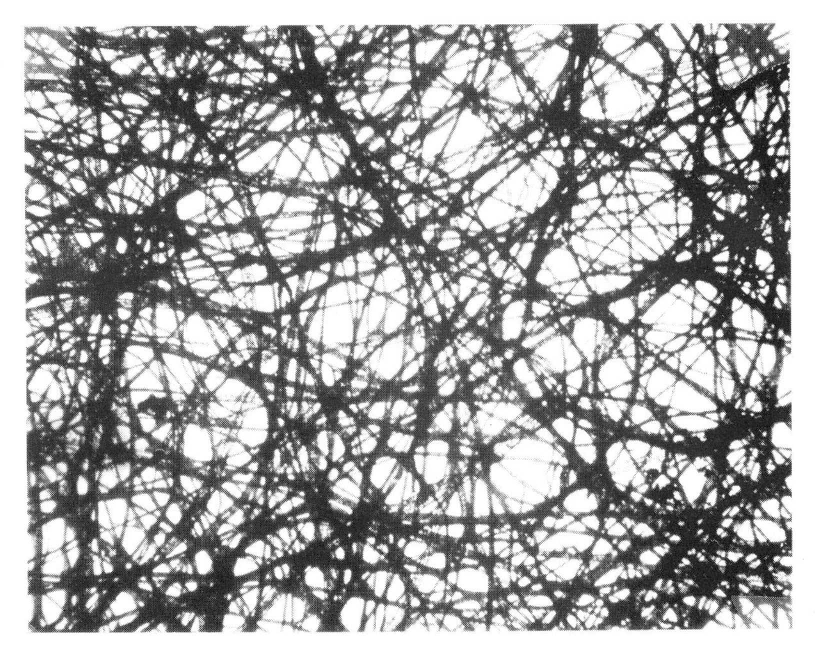

Figure 11.5 Magnified picture of paper (magnification 20). At this length scale we note the extremely inhomogeneous structure of the paper. (After [504]).

the actual disorder in the paper is *not* height-dependent, its *effect* in pinning the fluid interface is increasing with height due to the decrease in the fluid pressure. To include this effect, we can modify the DPD model by considering an increasing 'effective' density of pinning sites as the distance between the interface and reservoir increases – i.e., considering a *gradient* ∇p in the density of disorder [45, 149]. The existence of a gradient in the effective disorder generates a critical height at which the interface stops, and changes the width of the pinned interface [13–15]. Experimental results obtained on the variation of h_c and w with the evaporation rate can be understood quantitatively by studying the ∇p dependence of these quantities in the discrete model [15].

All fluid flow and wetting experiments discussed so far were carried out for two-dimensional systems, leading to one-dimensional interfaces. Recently Buldyrev *et al.* performed a set of three-dimensional experiments, with two-dimensional interfaces [59]. These experiments are the only available higher dimensional experimental results related to growth with quenched noise, and thus serve as a test system for the higher dimensional generalization of the proposed models and theories.

In the experiments, a highly porous, sponge-like material was used (the material is called 'oasis' and is commonly used by florists). A brick of square horizontal section of size 7×7 cm^2 was placed inside a large beaker, and an approximately constant increase (5 cm per hour) of the level of invading fluid (Bingo ink) was maintained by a constant flow of ink via a siphon. The ink was allowed to propagate upward for an hour, after which the brick was longitudinally sliced and the interface between wet and dry regions scanned.

The data are consistent with a roughness exponent $\alpha = 0.5 \pm 0.05$. This value is larger than the roughness exponent $\alpha \cong 0.4$ predicted by the KPZ equation for $2+1$ dimensions (see Table 8.2), indicating the relevance of the quenched noise in generating anomalous roughening. Indeed, simulations on the two-dimensional generalization of the DPD model predicts $\alpha = 0.49$, in good agreement with the experimental result (see Table 10.1).

11.3 Propagation of burning fronts

Another natural phenomenon leading to self-affine interfaces is the propagation of a burning front. If we ignite a sheet of paper with a match, the burning front spreads radially from the ignition point. If we start the burning process from a straight line, controlled burning can be used as a model experiment to study interface propagation and roughening.

In the experiments of Zhang *et al.* [504] lightweight paper (91 g/m^2, with thickness $\sim$ 45 μm) was used. The lightness helps keep heat production to a minimum. Flames and extreme heat production can produce uncontrollable air circulation, leading to the generation of unwanted long-range forces. Thus the goal is to burn flamelessly, with a relatively slow burning speed. To ensure uniform propagation, the paper was treated with KNO_3, an oxidation aid. Depending on the concentration of KNO_3 used, it was possible to control the mean velocity of the burning front, reaching velocities of the order 5.5–8.2 mm/s.

A 46 cm wide and 110 cm long paper was ignited using a straight electric wire, and the developed burning front photographed and digitized. A typical burning front is shown in Fig. 1.3. After digitization it was possible to calculate the interface width $w(l)$, whose scaling provides the roughness exponent

$$\alpha = 0.71 \pm 0.05. \tag{11.6}$$

As noted by Zhang *et al.*, the inhomogeneous nature of the paper (see Fig. 11.5) is what determines the nonuniform propagation of the burning front. There are local irregularities in porosity and chemical composition that are independent of time. These directly affect the local speed of the burning process. Thus, as in the case of imbibition experiments, the quenched nature of irregularities is likely to generate the anomalously large exponent value. Indeed, the value of α coincides with the values obtained in DPD model with propagating interfaces (see §10.2).

11.4 Growth of bacterial colonies

The growth of bacterial colonies has been investigated by both biologists and physicists [39, 40, 138, 296, 299, 300, 396, 460] with the purpose of understanding the cooperative motion leading to the experimentally-observed complex structures. Motivated by the interest in self-affine interfaces, Vicsek *et al.* [459] have studied experimentally the surface properties of the growing *Escherichia coli* and *Bacillus subtilis* colonies. The bacteria were inoculated on 150 mm diameter plates of a thick (5mm) layer of nutrient rich agar. The inoculations were carried out along a straight line, so that the bacterial growth area would configure itself into a strip geometry, thereby facilitating study of the interface properties of the growing colonies (see Fig. 11.6).

In the presence of the nutrient, the bacteria can multiply such that within a time scale of several (3–6) days one can observe a growing colony with a random interface. A representative colony obtained experimentally is shown in Fig. 11.7, together with a digitized picture. Using the digitized picture, it is possible to study the scaling of the interface width. The scaling is consistent with a roughness exponent $\alpha = 0.78$, much larger than the value $\alpha = 0.5$ obtained for the Eden model.†

In the case of fluid-flow experiments, we argued that the anomalous exponents are due to the quenched randomness generated by the random packing of the glass beads. But for bacterial growth the existence and the effect of such a randomness is not evident. It is possible that the inhomogeneities of the agar substrate act as a source

† Murray Eden originally introduced his model, which is now called the Eden model, to study the growth of cell colonies (see Chapter 8).

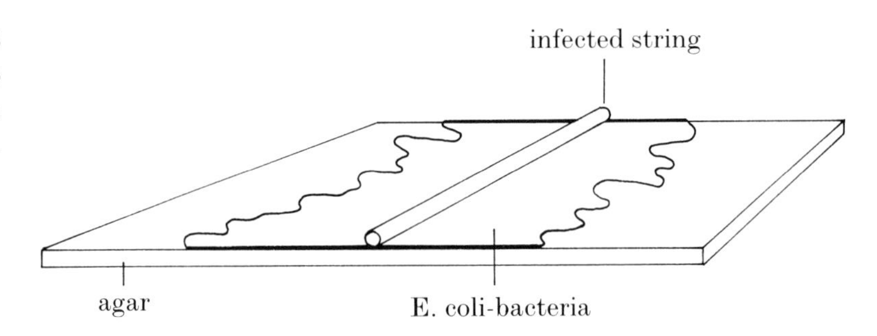

Figure 11.6 Experimental setup for bacterial growth (After [483]).

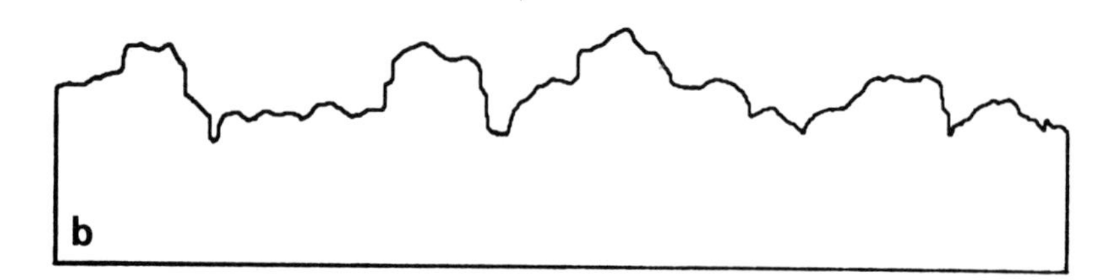

Figure 11.7 (a) A representative colony of *Escherichia coli* taken on the fourth day after inoculation. (b) The digitized picture of the colony shown in (a). (After [459]).

of quenched randomness, but one can think of a variety of alternative explanations as well. One is that the bacteria are mobile in the colony, so nonlocal forces might influence the growth front. The nutrient may diffuse as well, leading to possible coupling between the interface growth and nutrient diffusion, which in some limits might also lead to larger exponents [26, 117].

11.5 Rupture lines in paper sheets

Another easily reproducible experiment leading to a self-affine surface is related to paper rupture. Fractal fracture is a problem, actively investigated in recent years, which describes the fractured material in terms of self-similarity concepts [100, 175]. Recently it has been proposed that the anisotropy of the fracture process might lead to self-affine, rather than self-similar, interfaces [47, 100, 284, 322].

Recently, Kertész *et al.* have investigated a seemingly simple problem: the mechanism of paper fracture, and its characterization using scaling ideas [222]. Although fairly reliable analysis can be carried out by simply tearing paper [141], in the actual experiments an Instron tensile testing machine has been used to control the speed (2 mm/min). Different types of papers of size 300 $\times$ 450 mm were used, and a notch of 1 mm in the middle of one free edge was cut into the paper in order to initiate the rupture propagation.

Figure 11.8 shows a typical rupture line at two different magnifications. The height–height correlation function can be measured using the digitized interface and the scaling is shown in Fig. 11.9. The large scaling region is remarkable, spanning three orders of magnitude. The scaling exponent α for the individual measurements scatters between 0.63 and 0.72.

How do we interpret these results? Rupture seems to have nothing in common with interface growth (except that there is a rough interface) so why did we include it in this book? A very interesting in-

Figure 11.8 Rupture line at two different magnifications. The figure on the right is a three-fold magnification of a portion of the figure on the left. (After [222]).

terpretation of the experimental results has been proposed by Kertész *et al.* [222], whereby paper rupture can be understood in terms of directed polymers in a random medium, which in turn can be mapped onto the KPZ equation (see §26.1).

The relation of the rupture problem to directed polymers can be understood if we study a classic model for paper rupture, introduced in 1964 [454]. Paper is an inhomogeneous material, with local variations in density, composition, etc. The model assumes that flaws are located in the low density regions. The path across the strip for which the 'energy' sum over the local values of the paper density ρ_i, $\mathscr{E} \equiv \sum \rho_i$, is minimal, represents a weak path where most of stress can be released during rupture. Thus most likely the rupture line follows this minimal density path.

Having knowledge about directed polymers, it is not difficult to observe the connection between this path and a directed polymer in random medium: we have a lattice of randomly distributed densities, and we select the directed path with the lowest overall density, exactly as directed polymers select the path with the lowest energy. In one dimension, the mapping to the KPZ equation (see §26.1) gives the roughness exponent $\alpha = 2/3$, in good agreement with the results obtained from paper rupture. According to this model, paper rupture is a global optimization problem: the slow accumulation of stress selects the path over which the rupture line proceeds as being that path needing the smallest tension.

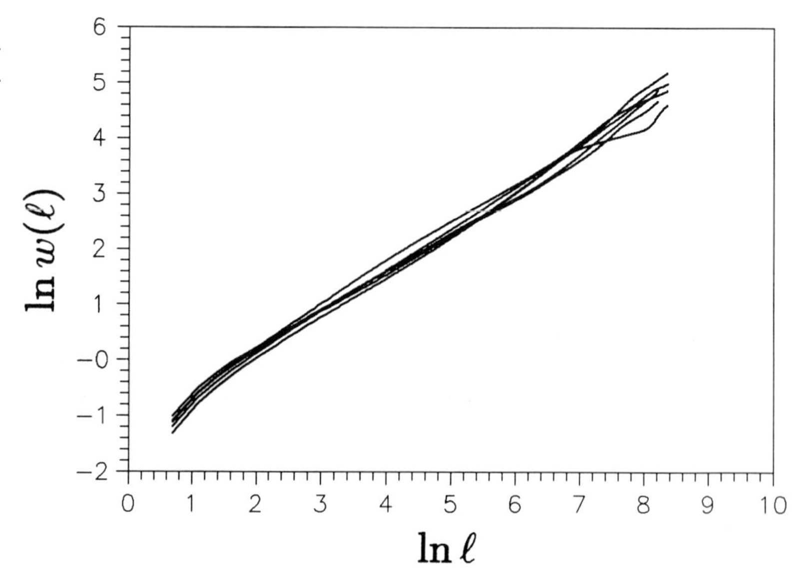

Figure 11.9 Scaling of the width of rupture lines of paper. The different lines correspond to different experiments. (After [222]).

11.6 Discussion

As can be seen from Table 11.1, which summarizes the exponents found in various experimental investigations, the agreement among different experiments is far from perfect. In previous sections, we discussed some of the reasons for the discrepancies, such as the uncontrollable nature of the scaling region and the crossover length scale. The experiments were carried out before an understanding of quenched-noise dominated roughening had emerged. At that time no theoretical knowledge was available concerning the importance of the depinning transition and the limited scaling region around it. Hopefully the theoretical advances achieved in understanding the basic phenomenology will initiate further experiments for checking the theoretical results.

Although the exponents sometimes do not quantitatively agree with one another, all experiments do support rather well the scaling picture developed above. One important question not addressed by the experiments, however, is the universality class to which these growth processes belong. We presented evidence in the previous chapter that there are two universality classes, depending on the relevance of the nonlinear term $\lambda(\nabla h)^2$. Experimental measurements of the value of λ would seem to be possible, and might shed light on the relevant universality class describing these natural processes.

Suggested further reading:

[13, 379, 490]

Exercises:

11.1 Perform the paper towel imbibition experiment described in the text. Digitize the interface and measure its roughness exponent.

11.2 Perform the paper tearing experiment and measure the roughness exponent of the interface.

11.3 Choose one of the experiments mentioned in this chapter and read the related literature. How strong is the evidence presented in this literature for scaling and for the scaling exponents? Can you suggest ways of altering the setup of your chosen experiment so that it provides an even stronger test for scaling?

11.4 Design an experiment that allows us to measure λ in the fluid flow experiments, and thus to determine the relevant universality class.

PART 4 Molecular beam epitaxy

12 Basic phenomena of MBE

12.1 Introduction

The discovery that scaling laws and continuum theories are applicable to molecular beam epitaxy (MBE) has generated increasing interest among both experimentalists and theorists [208, 256]. The closer study of these deposition processes reveals the decisive role played by surface diffusion of the deposited particles. From the experimental point of view, these studies re-focus attention on a neglected aspect of MBE growth processes: roughening of a growing interface.

There are two complementary approaches to crystal growth:

(a) *Atomistic* approaches, in which the position of every atom is well defined. Our knowledge of the behavior of individual atoms has increased due to the high resolution of scanning tunneling microscopy (STM). STM is capable of identifying not only the structure of the lattice, but the positions of the individual atoms as well. First principles calculations provide insight into the energetics of atomic motion on solid surfaces. Based on this detailed information, modeling of different growth processes on the atomic level is becoming a widely used tool to gain deeper insight on the collective nature of atomic motion and deposition processes.

(b) *Continuum* approaches view the interface on a *coarse-grained* scale, in which every property is averaged over a small volume containing many atoms. Neglecting the discrete nature of the growth process, continuum theories attempt to capture the essential mechanisms determining the growth morphology. Their predictive power is limited to length scales larger than the typical interatom distance, providing information on the collective nature of the growth process, such as the variation in the interface roughness or correlation functions.

In this and the following chapters we discuss both approaches. Microscopic considerations, such as the mechanism of atomic diffusion, desorption, or deposition processes, will be used to construct realistic models of crystal growth. We must understand the basic symmetries and conservation laws affecting the growth in order to derive the appropriate continuum equations describing the roughening of the interface.

The models to be discussed here are motivated by MBE, but they stand as independent problems in statistical mechanics as well. The KPZ equation describes *nonconservative* growth processes – i.e., the basic mechanisms contributing to the formation of the interface need not conserve the number of particles. In contrast, surface diffusion is a relaxation process that *conserves* the number of particles on the interface. The introduction of the conservation laws into the growth equations will lead us to consider altogether new dynamical processes, similar to the way conservation laws change the dynamic properties of Ising systems [179].

Finally, it is important to understand how experimental observations relate to the predictions of the continuum theories. High temperature growth, where the diffusion process enables good quality ('flat') multilayers to be obtained, is the most experimentally-investigated case. Under these conditions, the interface roughness oscillates in time, as observed using reflection high energy electron diffraction (RHEED). Unfortunately, the continuum theories fail to reproduce this oscillatory behavior [463]. When the temperature is lowered, however, the interface begins to grow in a three-dimensional growth mode, and acquires nonzero roughness. This is the regime where continuum theories become applicable.†

Until recently, roughening of growing interfaces was a largely neglected subject from the experimental side, in spite of the great technological importance of understanding the basic laws governing growth. One reason for this neglect is that the experimental goal is to grow *smooth* interfaces, needed in technological and industrial applications. Rough semiconductor or metal interfaces have poor contact properties and are practically unusable for electronic devices. Thus most experimental investigations focus on the study of surfaces with good

† Even in the high-temperature oscillatory regime, the long-time behavior follows dynamic scaling. The oscillatory behavior is, in fact, a finite-size transient (with the size depending on temperature); continuum equations describe only the asymptotic behavior, not transients.

contact properties, avoiding as much as possible the roughening process. However, in many cases roughening cannot be avoided, even in materials with promising technological potential. Understanding the roughening process might help map the experimental and technological limits, hopefully leading to the discovery of methods to control the interface roughness.

Motivated by the advances in modeling and predicting the roughening processes, a number of experimental groups recently have turned their attention to checking the theoretical predictions. In Chapter 16 we describe some experiments that confront theoretical results with the reality of the laboratory.

12.2 Microscopic processes on crystal surfaces

Before beginning a detailed study of the possible growth equations describing the growth of crystal interfaces by atom deposition , we review the relevant microscopic processes taking place on the crystal interface (see Fig. 12.1). The morphology of the interface is determined by the interplay between *deposition*, *desorption*, and *surface diffusion*.

12.2.1 Deposition

An atom from the vapor arrives at a random position on the interface, forms bonds with the surface atoms, and sticks. This process is termed *deposition* (see Fig. 12.1). Crystals grow by atomic deposition:

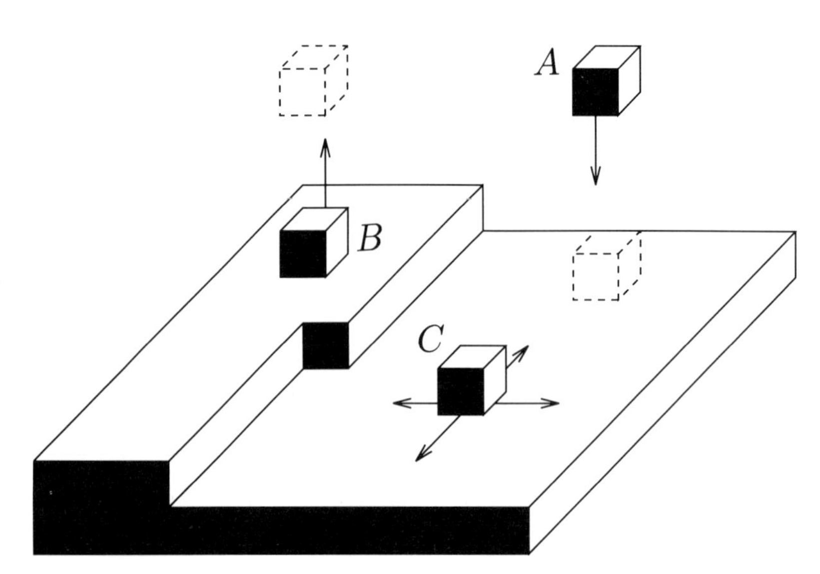

Figure 12.1 Elementary processes on a crystal surface. (A) Atoms arrive on the surface, where they are deposited. (B) An atom on the surface may *desorb*, leaving the surface of the crystal. (C) Atoms *diffuse* in random directions on the surface.

atoms arrive on the surface from a 'reservoir,' the nature of which – the atomic source – depends on the technique used to grow the crystal. MBE experiments are usually performed under ultra-high vacuum conditions (pressures smaller than 10^{-10} torr) to minimize uncontrollable deposition of impurities such as H_2, CO_2, CO, or H_2O. Typically, the deposition material is thermally evaporated from a source, forming a beam of neutral atoms or molecules that have thermal velocities. This beam (or vapor) is directed to the growing substrate.

12.2.2 Desorption

An effect competing with deposition is *desorption* – some atoms deposited on the surface leave the interface (see Fig. 12.1). The desorption probability depends on how strongly the atom is bonded to the crystal surface. When an atom is deposited on a surface, it forms bonds that must be broken before desorption can occur. The strength of the bonds depends on (i) the type of atom and (ii) the local geometry of the surface where the atom sticks. Bonding energies differ from solid to solid. The strength of a bond is expressed in terms of the amount of energy needed to break it.

The usual procedure for measuring the lifetime of a deposited molecule or atom is to measure the average time τ spent by the atom on the surface from deposition to desorption. For example, for Ga on GaAs(111)† substrate over the temperature range 860–960 K, τ was found to obey the Arrhenius law [18]

$$\tau = \tau_0 \exp\left(\frac{E_D}{kT}\right). \tag{12.1}$$

Here $\tau_0 \approx 10^{-14}$s, and $E_D \approx 2.5$ eV is the characteristic desorption energy to release an atom from the surface.

Desorption is negligible for many materials under typical MBE conditions. In order to estimate the relevance of desorption, one must compare the characteristic desorption time τ with the time scale set by the deposition process. A deposition rate of 10^{16} atoms cm^{-2} s^{-1}

† The numbers in the parenthesis after the name of the material are the Miller indices, and refer to the orientation of the interface with respect to the unit crystal cell. The unit cell is described by three lattice vectors $\mathbf{a}_1$, $\mathbf{a}_2$ and $\mathbf{a}_3$. Any plane related to the crystal may be described by the points or intercepts at which it cuts the three unit cell axes. These intercepts may be expressed in terms of the length of the corresponding lattice vectors α_1, α_2 and α_3. The Miller indices are found by taking the reciprocals of α_i and dividing each by any common factor c: $h_1 = 1/(\alpha_1 c)$, $h_2 = 1/(\alpha_2 c)$ and $h_3 = 1/(\alpha_3 c)$. These indices can always be reduced to integer form, and the usual notation is $(h_1 h_2 h_3)$.

corresponds to approximately 10 atoms per second per lattice site. At 850 K, (12.1) predicts that the lifetime of the deposited molecule is ≈ 2 s, so, on average, for every 20 deposited atoms one is desorbed. For a deposition rate of 10^{11} atoms cm^{-2} s^{-1} (which corresponds to $\approx 10^{-4}$ atoms s^{-1} site^{-1}), the desorption rate is orders of magnitudes larger than the deposition rate, so the interface does not grow but rather evaporates!

According to (12.1), the desorption rate is sensitive to changes in the substrate temperature. At $T \leq 750$ K the sticking coefficient of Ga is effectively unity. Desorption practically stops and only the deposition rate determines the growth velocity of the interface.

12.2.3 Surface diffusion

We have studied various deposition models in previous chapters. In BD, atoms stick to the first point on the surface with which they come in contact. In RD with surface relaxation, atoms 'search' for the *lowest height position*. In contrast to these processes, in MBE the deposited atoms diffuse on the crystal surface, searching for the *energetically most favorable position*.† The diffusion length can be quite large, and depends on the temperature and the binding energies.

From a microscopic point of view, surface diffusion is an activated process. The discrete positions of the atoms are determined by the crystal lattice. For an atom on the surface to diffuse to the next lattice position, it must overcome the lattice potential existing between two neighboring positions. This excess energy required for diffusion, E_0, is

† Strictly speaking, the deposited atoms randomly encounter surface sites that bind them more or less well. Less well bound atoms break loose and continue to diffuse.

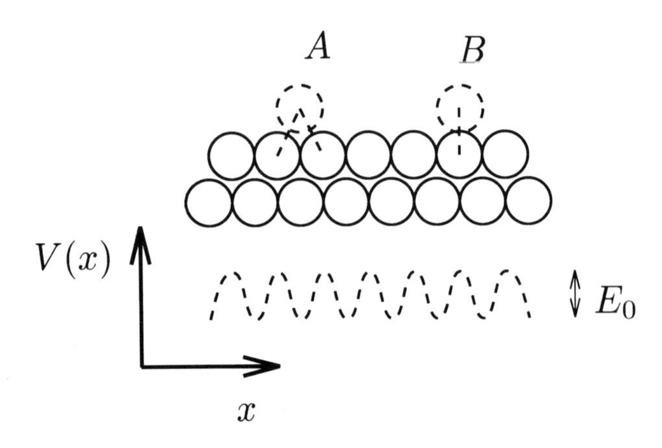

Figure 12.2 Schematic illustration of the lattice potential and its microscopic origin. Atom A forms two bonds (dotted lines) with the surface atoms, and it is in an energetically favorable position, corresponding to a minimum of the lattice potential $V(x)$. In order to diffuse, it must move into an intermediate position (such as shown in B) where there is only one strong bond. This is an energetically unstable configuration, corresponding to maximum in the lattice potential.

the microscopic origin of the lattice potential schematically illustrated in Fig. 12.2.

The magnitude of the diffusion barrier depends both on the substrate orientation and the nature of the diffusing atom. For example, a W atom on a W(110) surface must overcome a diffusion barrier of 0.87–0.92 eV while a Pt atom on the same surface must overcome a barrier of only 0.61–0.67 eV [451]. The fact that $E_0 \approx 0.53 - 0.76$ eV for a W atom on a W(221) surface – much smaller than it would be on a W(110) surface – illustrates how diffusion depends on the miscut direction.

The average number of jumps in a unit time interval has an exponential temperature dependence, given by the Arrhenius law

$$N = \omega_D \exp\left(-\frac{E_0}{k_B T}\right), \tag{12.2}$$

where ω_D is the frequency of small atomic oscillations (Debye frequency). According to (12.2), if a W atom on the W(110) surface

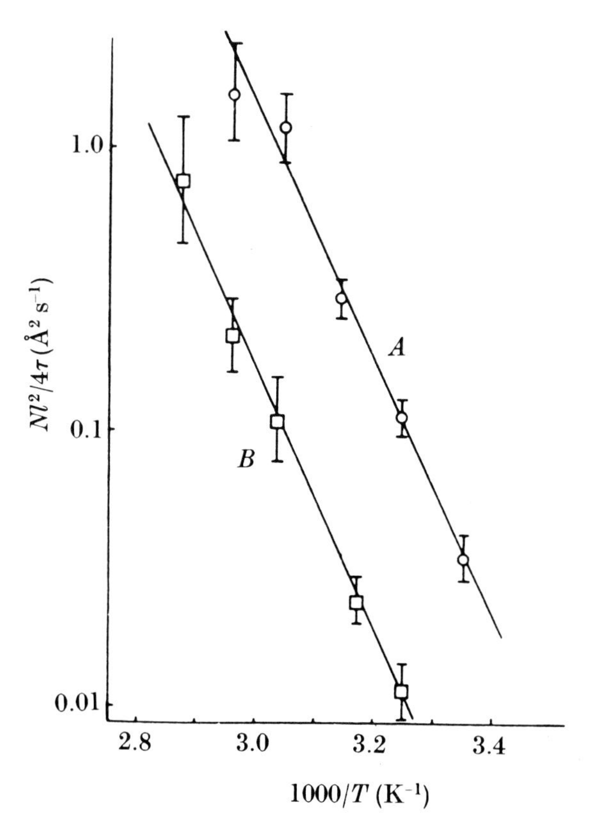

Figure 12.3 Arrhenius plot for single atom (A) and diatomic (B) diffusion of W on a W(110) substrate (After [451]).

travels an average distance ℓ at room temperature (approximately 300K) in a given amount of time, in the same time interval it will travel $\sim 10^3$ times farther at 400 K, and 10^7 times less far at 200 K. This simple calculation illustrates how sensitive atomic diffusion is to changes in substrate temperature.

Measuring the number of jumps (i.e., the average distance traveled by an atom on the surface) is a standard experimental technique used for determining the activation energy E_0 needed for diffusion. The results of a such a measurement are shown in Fig. 12.3, and confirm the validity of the Arrhenius law (12.2). For how long does an atom diffuse on the surface? Is there any mechanism that will stop this diffusion process? An atom continues to diffuse so long as it is on the surface, but its diffusivity can decrease sharply if it meets another atom, an edge of an island, or a step. As an example, Fig. 12.3 shows the number of diffusive steps during the diffusion of both atomic W and diatomic W. We see that the diffusion probability of a diatomic cluster is an order of magnitude smaller than that of a single W atom.

A second example is the situation in which an atom sticks to the edge of an island (Fig. 12.4). 'Sticking' means that atom A forms a bond with an atom on the edge of the island. The probability that the atom breaks away from the edge (i.e., of moves one step in the x direction) is $P \sim \exp[-(E_0 + E_N)/kT]$, because in order to move it must overcome the lattice potential for diffusion E_0, and must break the bond formed with the atom on the edge of the island. If we assume the bonding energy to be $E_N \approx 1$ eV, then this probability at 600 K is 10^8 times smaller than the characteristic diffusion probability of an atom on the surface – so once an atom sticks to the edge of an island it will remain there for a long time. It will take even longer for the atom B in Fig. 12.4 to move in the x direction, since it has two neighbors, and hence two bonds to break. Thus when an atom finds the edge of a step or island, in most cases it sticks nearly irreversibly. The breakaway probability depends exponentially on the temperature: at low T, sticking is almost irreversible, while increasing the substrate temperature allows the bonds to break more easily.

Depending on the atomic structure and the direction we cut the crystal, surfaces may be highly anisotropic. The clearest indicator of this anisotropy is seen from studies of surface diffusion: atoms may readily diffuse in a given direction, and strongly resist diffusion in other directions. A technologically-important example is the diffusion of Si atoms on a Si(001) surface. Neighboring atoms form dimer bonds, thereby generating long valleys that can be seen with a STM (see Fig. 1.6). A Si atom placed on this surface has a large diffusion constant along the valleys and a very small diffusion constant perpendicular to them. This can be explained by the difference in the activation energy for diffusion: along the dimer rows $E_{\parallel} \approx 0.6 - 0.7$ eV, while perpendicular to them $E_{\perp} = 1$ eV [325, 54]. This difference can account for the experimental fact that Si atoms diffuse 1000 times faster along the rows than transverse to them.

Although anisotropy is important in understanding particular materials, in many growth models it is neglected. Nevertheless, anisotropy may drastically change the scaling behavior (see §25.2).

12.2.4 Diffusion bias

When a diffusing atom meets an island where the interface height increases by one atom, it exhibits a sticking preference for atoms located on the edge of the step. But what happens if the atom diffuses *on* the island and approaches its edge from above? Will it jump off the island? There is an additional potential barrier at the edge of the island that the atom must pass in order to jump off (Fig. 12.5), so the probability of its being reflected is higher than the probability of its jumping off. This additional potential barrier is sometimes called the *Schwoebel barrier* [113, 393, 394], and the reflection of the diffusing atoms on the edge of the islands is termed *diffusion bias*.

The microscopic origin of this potential barrier can be understood using a simple microscopic model. For this we count the existing bonds an atom has before it jumps off. The dotted atom in Fig. 12.5 must break its bond with the atom on its left before jumping down the step, but there is no atom on its right to help the diffusion process in that direction. The resulting asymmetry in the lattice potential generates a higher probability of moving to the left than to the right. There have

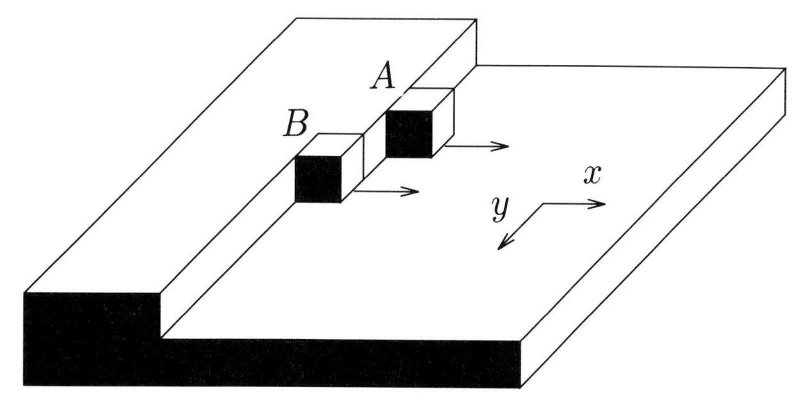

Figure 12.4 Atoms may stick to the edge of an island or step, forming additional bonds. This reduces their diffusion: to move in the x direction, atom A must break an additional bond it formed with the atom on the perimeter of the island. Even less probable is the breakaway of atom B, which must break two bonds to diffuse. However, atoms A and B may diffuse along the edge of the step (in the y direction) which is less expensive energetically.

been extensive experimental investigations regarding the magnitude of the Schwoebel barrier for different materials [113, 451, 471, 472, 473]. For example, for W atoms moving on a W(100) surface, the additional diffusion barrier is $\Delta E_b \approx 0.215$ eV at 370 K.

12.3 Discussion

Before turning to the detailed investigation of the continuum theories in the next chapter, we summarize the overall picture concerning basic processes occurring on a crystal surface.† Consider the deposition of an arbitrary material X:

(i) At high temperatures, the diffusion length is very large and atoms can potentially find terraces or steps where they can stick. Under these conditions it is possible for the interface to grow in a layer-by-layer mode, the roughness not exceeding roughly one atomic thickness.

(ii) Lowering the temperature decreases the diffusion length. Before

† In this book we neglect the effect of the *stress* on the atomic motion and interface morphology. It is known that in some conditions stress has a decisive role in shaping the interface morphology, but there is as yet no understanding of how to include the effects of stress into the continuum growth equations. [355]

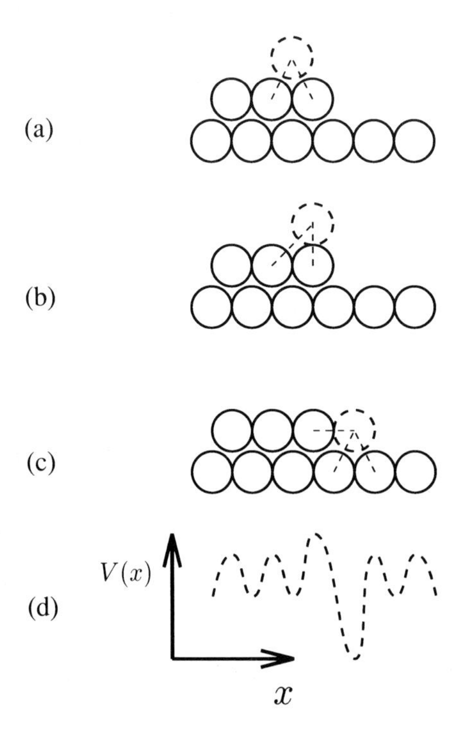

Figure 12.5 Cross-section of a monoatomic step in a surface and the lattice potential associated with the diffusion of an atom over the step. Two major modifications of the potential may occur near the step. (a) The dotted atom may diffuse to the left or to the right. (b) Diffusing to the right, in the intermediate position the atom has two bonds, but there is a missing bond to the right. The absence of this bond (which, if present, would help the atom in its diffusion process) generates a further increase in the lattice potential, as shown in (d). In position (c) the atom forms three bonds with the neighbors, corresponding to a deep minimum in the lattice potential.

finding the edge of the existing island, atoms will meet and nucleate new islands on the surface. Moreover, islands will nucleate on the top of the existing islands as well. This regime is called three-dimensional growth, or 'islanding.' The surface diffusion is still relevant, but the interface becomes rough.

(iii) Lowering the temperature still further, the diffusion length becomes very short, shorter than the lattice spacing. In this limit only deposition determines the growth, and the material becomes amorphous with a rough surface.

There are three main effects determining the final interface morphology of a growing interface: deposition, desorption, and surface diffusion. Their relative importance depends on the microscopic properties of the interface, such as the magnitude of the bonding energies and diffusion barriers. These parameters cannot be modified experimentally, except by changing the substrate. The experimentally-controllable parameters are the deposition rate and temperature.† By tuning these parameters a rich variety of morphologies can be achieved, from layer-by-layer growth – which results in an essentially smooth interface – to rough self-affine surfaces.

The previous examples convince us that the surface energetics are the most important microscopic phenomena, and determine the macroscopic parameters. Of all the possible atomic motions that are allowed, the energetic conditions select a few that are indeed relevant. Even at very low temperatures, there is a nonzero probability of surface diffusion of the deposited atoms, but the diffusion rate is so small that, on experimentally-available time scales, diffusion cannot be observed. Tuning the relevance of different effects is used in experimental investigations to separate various desired qualities, and thereby control the obtained interface morphology.

Suggested further reading:

[256, 257, 484, 492, 502]

Exercises:

12.1 Consider a Si atom diffusing on an Si surface. The diffusion barrier along the rows is 0.67 eV, and perpendicular to the rows is 1 eV [325]. (a) Calculate the relation between the distance traveled by the atom after a time t in the two different directions. (b) Denote by $P(x, y, t)$ the probability of finding the particle at

† We emphasize that all the processes (except deposition) are activated, so it is only E/kT that is important. Hence changing E (i.e., the surface) should be equivalent to changing T.

moment t in position (x, y), if the particle is released from $(0, 0)$ at moment $t = 0$. What is the shape of the curve representing the locus of equal probability, $P(x, y, t) = \text{const}$, and how does it depend on time and temperature?

12.2 Why is the diffusion probability of a diatomic W much smaller than that of a monoatomic W? Estimate the activation energy for both monoatomic and diatomic W from Fig. 12.3.

13 Linear theory of MBE

Continuum theories are able to respond to the challenge posed by the variety of competing effects by predicting different morphologies and different universality classes as various effects dominate the growth process. Even for a system with fixed experimental parameters, interesting crossovers are expected as a function of the deposition time, or the number of deposited layers.

In this chapter, we discuss how these continuum theories incorporate the effects of deposition, desorption, and surface diffusion, which we have seen are important to take into account if we want to understand MBE. We focus on linear terms here, possible nonlinear contributions being left for the next chapter. We shall see that a number of essential phenomena may be successfully described by the linear theory.

13.1 Surface diffusion

The physical mechanism that distinguishes MBE from previously-discussed growth processes is the surface diffusion of the deposited particles. In this section, we discuss possible ways of including surface diffusion in the growth equations. For the moment, we assume that desorption is negligible.

Let us consider the simplest scenario: atoms are deposited on a surface, whereupon they diffuse. The goal is to find a continuum equation of the form

$$\frac{\partial h}{\partial t} = F(h, \mathbf{x}, t) \tag{13.1}$$

that describes the variations in the interface height $h(\mathbf{x}, t)$.†

† A close look at the growth process tells us that the symmetries (i–iv) of Section 5.2 are all present. We want isotropy in the plane (translational and rotational invariance)

Surface diffusion implies that the deposited particles generate a macroscopic current $\mathbf{j}(\mathbf{x}, t)$, which is a vector parallel to the average surface direction. If we neglect desorption, the diffusing particles can never leave the surface. Thus the local changes in the surface height are the result of the nonzero currents along the surface. Since the total number of particles remains unchanged during the diffusion process, the current must obey the continuity equation

$$\frac{\partial h}{\partial t} = -\nabla \cdot \mathbf{j}(\mathbf{x}, t). \tag{13.2}$$

On the other hand, the surface current is driven by the differences in the *local chemical potential* $\mu(\mathbf{x}, t)$

$$\mathbf{j}(\mathbf{x}, t) \propto -\nabla \mu(\mathbf{x}, t). \tag{13.3}$$

To determine the local chemical potential, we may use either (a) physical reasoning or (b) symmetry arguments.

(a) Surface diffusion is an activated process. The motion of an atom does not depend on the local height of the interface, but only on the number of bonds that must be broken for diffusion to take place. The number of bonds a particle may form increases with the local *curvature* of the interface at that point (see Fig. 13.1). If the local radius of curvature R is positive, the atom has a large number of neighbors, and moving away from the site will be difficult. In contrast, if R is negative, the atom has a few neighbors, and is able to diffuse easily. The simplest assumption is that the chemical potential controlling the diffusion probability is proportional to $-1/R$, which in turn is proportional to $\nabla^2 h(\mathbf{x}, t)$. Hence

$$\mu(\mathbf{x}, t) \propto -\nabla^2 h(\mathbf{x}, t). \tag{13.4}$$

(b) Symmetry arguments can lead us to the same result (13.4). The diffusion probability of an atom should not depend explicitly on the height h of the interface, since it does not depend on where we define the origin. Similarly, if the chemical potential were to depend on a power of (∇h), then the sticking probability would be different if our system of coordinates were tilted. Thus we are left with (13.4) as the lowest order form for the chemical potential.

and translational invariance along h, since the growth process normally does not depend on the choice of the origin of h. These symmetries exclude explicit h and x dependences of F, leaving combinations ∂h and ∂x to form the growth equation.

Combining (13.2)–(13.4), we obtain the equation describing relaxation by surface diffusion [173, 333]

$$\frac{\partial h}{\partial t} = -K\nabla^4 h. \qquad \rightarrow (13.5)$$

Equation (13.5) is *deterministic*. However, during deposition there is an inherent *randomness* in the system coming from the fluctuations in the intensity of the atomic beam. $F(\mathbf{x}, t)$ describes the incoming atomic flux, which is defined as the number of particles arriving on the unit surface (or per lattice site) in a unit time. At large length scales the beam is homogeneous with an average intensity F, but there are local random fluctuations $\eta(\mathbf{x}, t) \equiv \delta F(\mathbf{x}, t)$, uncorrelated in space and time. We can include deposition in the growth equation (13.5), by considering the deposition flux to be made up by the average flux F and the noise η with

$$\langle \eta(\mathbf{x}, t)\rangle = 0 \qquad (13.6)$$

and correlations described by

$$\langle \eta(\mathbf{x}, t)\eta(\mathbf{x}', t')\rangle = 2D\delta(\mathbf{x} - \mathbf{x}')\delta(t - t'). \qquad (13.7)$$

Since the relative fluctuations should not change upon changing the flux F, $\delta F/F \cong$ const. Hence we can estimate the magnitude of D [463]

$$\delta F \sim F \sim D^{1/2}. \qquad (13.8)$$

Thus the stochastic growth equation incorporating surface diffusion

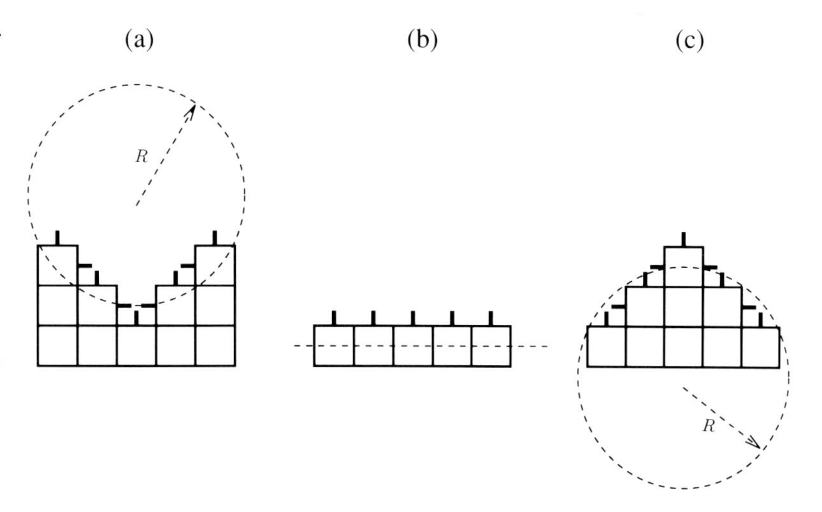

Figure 13.1 Schematic illustration of three possible local geometries. (a) A local 'valley' on the interface is an energetically-favorable place for atoms to stick, since an atom landing in the free site in the center of the valley has three neighbors with which to form bonds. Since such a site is an energetically-favorable position, the deposited atom remains for a long time. The local Laplacian (being the inverse of the radius R of the circle 'fitting' the interface locally) is positive at that site, corresponding to a negative chemical potential. (b) The chemical potential of an atom on a flat interface is zero, since $R = \infty$. (c) It is energetically unfavorable for an atom to stay at the center of this interface configuration, which has a positive chemical potential, and hence a negative value of the local Laplacian.

and deposition has the form

$$\frac{\partial h}{\partial t} = -K\nabla^4 h + F + \eta(\mathbf{x}, t). \qquad \rightarrow (13.9)$$

This variant of (13.5) was introduced independently by Wolf and Villain [488], and Das Sarma and Tamborenea [99], and played a leading role in developing our understanding of MBE.

The growth equation (13.9) is a stochastic differential equation, similar in form to the EW equation. The difference comes in the linear term: in the EW equation we have $\nu\nabla^2 h$, while in (13.5) we have $-K\nabla^4 h$. This difference affects the value of the scaling exponents. However, the two equations are similar in the sense that they both provide descriptions of the growth process at a coarse-grained length scale – (13.9) describes the dynamics and morphology of an interface at large length scales (larger than the typical interatomic distance) and for large times (after the deposition of several monolayers).

13.2 Solving the diffusive growth equation

Due to the linear character of the growth equation (13.9), we can solve it exactly – using Fourier transformation. The analysis is parallel to that used for the EW equation in §5.4.2. However, since we are interested primarily in the scaling properties of the system, we can use simple scaling arguments to obtain the exponents.

Rescaling the growth equation (13.9) with $F = 0$ and multiplying both sides by $b^{z-\alpha}$ (see §5.4.1), we find

$$\frac{\partial h}{\partial t} = -Kb^{z-4}\nabla^4 h + b^{(z-d-2\alpha)/2}\eta. \qquad (13.10)$$

From the requirement of scale invariance, we obtain the exponents

$$\alpha = \frac{4-d}{2}, \qquad \beta = \frac{4-d}{8}, \qquad z = 4. \qquad \rightarrow (13.11)$$

Note that the roughness exponent is large,† with $\alpha = 3/2, 1$ for $d = 1, 2$, respectively, a 'first warning' concerning the possible relevance of nonlinear terms since nonlinear terms cannot be neglected if $\alpha > 1$.

13.3 Growth with desorption

Since the purpose of the continuum growth equations is to describe the asymptotic dynamics of the system, we also must study the relevance of

† A large roughness exponent is consistent with the picture that surface diffusion, which *microscopically* smooths the interface (acts to increase the coordination number of the particles), at *large* length scales generates a rough surface.

the desorption process. Neglecting for a moment surface diffusion, we may assume that the deposition-desorption dominated growth process is governed by the difference between the average chemical potential in the vapor $\bar{\mu}$ and the local chemical potential on the surface $\mu(\mathbf{x}, t)$ [463]

$$\frac{\partial h}{\partial t} = -B\,[\mu(\mathbf{x}, t) - \bar{\mu}]\,. \tag{13.12}$$

Using (13.4), the growth equation incorporating desorption becomes

$$\frac{\partial h}{\partial t} = \nu\nabla^2 h + B\bar{\mu}. \tag{13.13}$$

To (13.13), we must add the flux term describing deposition, obtaining

$$\frac{\partial h}{\partial t} = \nu\nabla^2 h + B\bar{\mu} + F + \eta, \tag{13.14}$$

which is essentially the EW equation.

If we include surface diffusion as well, we obtain

$$\frac{\partial h}{\partial t} = \nu\nabla^2 h - K\nabla^4 h + B\bar{\mu} + F + \eta. \tag{13.15}$$

This equation contains all physically-relevant *linear* terms describing the growth of interfaces by MBE. The competition between the linear terms $\nabla^2 h$ and $\nabla^4 h$ generates a characteristic horizontal length scale

$$L_1 = \left(\frac{K}{\nu}\right)^{1/2}. \tag{13.16}$$

Rescaling the two linear terms using $\mathbf{x} \to b\mathbf{x}$ and $h \to b^{\alpha}h$, we obtain $\nu b^{\alpha-2}\nabla^2 h$ and $K b^{\alpha-4}\nabla^4 h$. Thus for short length scales ($b \to 0$) the diffusive term determines the scaling behavior, while for large length scales ($b \to \infty$) the desorption-generated Laplacian governs the scaling.

For $L \ll L_1$, the $K\nabla^4 h$ term determines the scaling behavior, and the exponents are given by (13.11). However, for $L \gg L_1$, the desorption-generated $\nu\nabla^2 h$ term controls the scaling, and the exponents are given by (5.16) (see Fig. 13.2).

To summarize, the linear growth equation (13.15) supports the existence of two distinct phases, each with different morphologies and scaling exponents α and β. They are delimited by the crossover length L_1. In general, we are interested in the *asymptotic* scaling behavior. In the presence of desorption, the asymptotic scaling is described by the EW equation. However, scaling exponents consistent with it can be seen only for length scales larger than L_1, and L_1 can be very large in certain experimental situations.

From an experimental point of view, it is important to determine which regime is appropriate under any given set of external conditions. Hence we must be able to estimate the magnitude of L_1.

According to (13.16), L_1 depends on ν and K. The value of ν can be determined from numerical simulations [250], and for some models it may be possible to calculate L_1 [470]. In general one expects that ν vanishes if desorption is irrelevant, a fact confirmed by analytic calculations.

The origin of K is surface diffusion. Since both diffusion and desorption are activated processes, one expects that both depend exponentially on the inverse temperature. In particular, K should vary as $\exp(-E_0/kT)$, where E_0 is the activation energy for diffusion. Thus L_1 should depend on the temperature, and it is expected to diverge if desorption becomes irrelevant ($\nu \to 0$).

13.4 Discussion

The continuum theories in the linear approximation discussed in this chapter provide a convincing example of the complexity of the problem

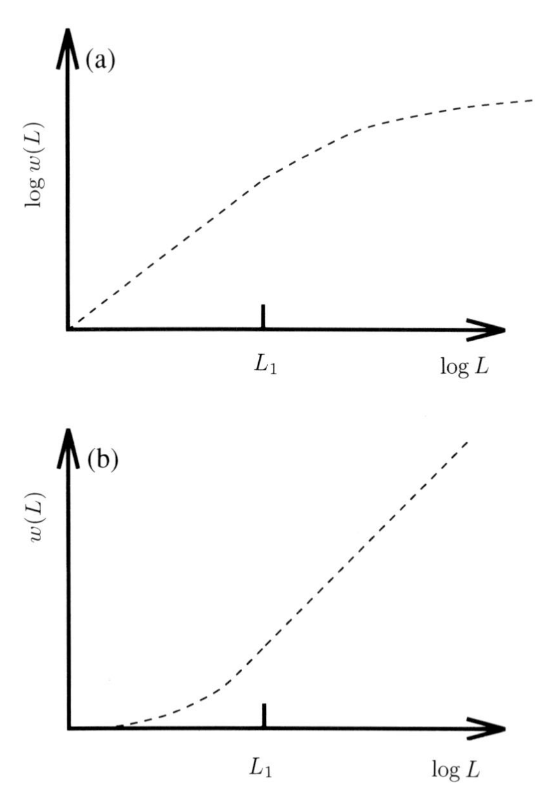

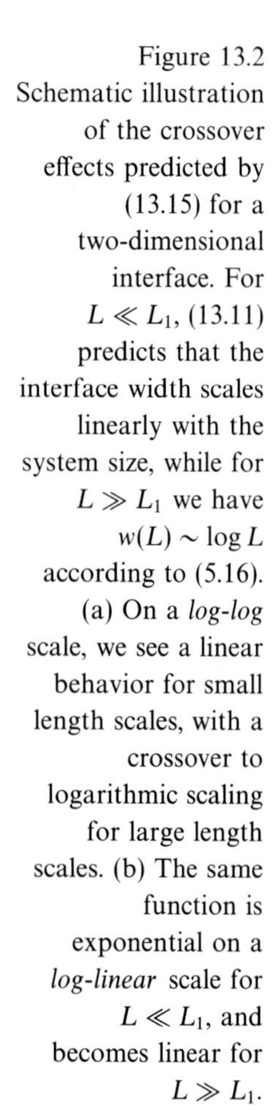
Figure 13.2 Schematic illustration of the crossover effects predicted by (13.15) for a two-dimensional interface. For $L \ll L_1$, (13.11) predicts that the interface width scales linearly with the system size, while for $L \gg L_1$ we have $w(L) \sim \log L$ according to (5.16). (a) On a *log-log* scale, we see a linear behavior for small length scales, with a crossover to logarithmic scaling for large length scales. (b) The same function is exponential on a *log-linear* scale for $L \ll L_1$, and becomes linear for $L \gg L_1$.

Table 13.1 *Summary of the two growth equations discussed in this chapter. The two equations are particular cases of (13.15), and they are relevant on different length scales.*

Length Scale	Equation	α	β	z
$L \ll L_1$	$\partial h/\partial t = -K\nabla^4 h + \eta(\mathbf{x}, t).$	$\frac{4-d}{2}$	$\frac{4-d}{8}$	4
$L \gg L_1$	$\partial h/\partial t = \nu\nabla^2 h + \eta(\mathbf{x}, t).$	$\frac{2-d}{2}$	$\frac{2-d}{4}$	2

we must deal with in order to describe MBE. There are two major effects, each corresponding to a different term in the growth equation (13.15). The two main relaxation mechanisms are generated by surface diffusion ($-K\nabla^4 h$) and desorption ($\nu\nabla^2 h$). The competition between these two effects generates the characteristic length scale L_1 of (13.16), which delimits two scaling regimes with different scaling exponents.

Thus the linear growth equation (13.15) leads to two distinct phases, each with different morphologies and scaling exponents α, β and z (Table 13.1). The relation between the system size L and the crossover length L_1 determines the observable scaling behavior.

We emphasize that the asymptotic behavior of (13.15) is controlled by the EW equation, and all other phases and exponents are observable only over finite scaling intervals, bounded by L_1. However, under typical MBE conditions desorption is negligible, so ν is small, and L_1 is very large. In this case the main experimentally-observable relaxation mechanism may be surface diffusion.

Finally, we note that since the growth equations discussed in this chapter are linear, we can generate them from properly-defined Hamiltonians (see Appendix C).

Suggested further reading:

[94, 448, 463, 488]

Exercises:

13.1 Calculate the exponents in (13.11) by Fourier transforming the growth equation (13.9) and obtaining an exact solution for the time-dependent width.

13.2 Show that the deterministic equation (13.5) smooths an initially rough interface.

14 Nonlinear theory for MBE

Our experience with the EW and the KPZ equations has already taught us an important lesson: *nonlinear* terms may be present for certain growth models and, if present, may control the scaling behavior. For the problem of MBE dominated by surface diffusion, there is no reason to ignore the possibility of nonlinear terms, so we must carefully examine when nonlinear terms are relevant and explore the resulting scaling behavior.

In order to address the relevance of certain nonlinear terms that can affect the growth equation (13.9), it will be necessary to use the dynamic RG method, as we did with the KPZ equation. We shall see that the dynamic RG method leads to exact exponents in any dimension, in contrast to the case of the KPZ equation where results are found only for $d = 1$.

14.1 Surface diffusion: Nonlinear effects

If the relevant relaxation process is surface diffusion, then the growth equation must satisfy the continuity equation (13.2). If the surface current is driven by the gradient of the local chemical potential (13.4), we argued that the diffusive growth process in the *linear* approximation is described by the growth equation (13.9).

For $d = 1, 2$, Eq. (13.9) predicts $\alpha = 3/2, 1$ respectively. However, continuum growth equations are valid in the 'small slope' approximation, $|\nabla h| \ll 1$, which means that the local slopes are small at every stage of the growth process. With increasing system size, the local slopes scale as $\nabla h \sim L^{\alpha-1}$, so if $\alpha > 1$, nonlinear terms become relevant for large length scales.

14.1.1 Continuum equations

In Chapters 5 and 6, we used symmetry arguments to construct the EW and KPZ equations, and we also discussed the relevance of various linear and nonlinear terms. We can use the same approach to construct the relevant nonlinear equation for MBE. To this end, we add the possible nonlinear terms to (13.9), and eliminate those that do not obey the basic symmetry requirements of the growth process. Our goal is to obtain an equation that obeys translational invariance both in the growth direction ($h \to h+\delta_h$) and in the direction perpendicular to growth ($\mathbf{x} \to \mathbf{x}+\delta_x$), and has rotational symmetry ($\mathbf{x} \to -\mathbf{x}$), thereby excluding odd-order derivatives in $\mathbf{x}$.

Any nonlinear term of the diffusion-dominated growth process must also obey the continuity equation (13.2) – i.e., the allowed terms must conserve the number of particles on the surface. This condition eliminates, e.g., the KPZ nonlinearity $(\nabla h)^2$, which cannot be written as a gradient of a current.† Thus the generalization of the linear equation (13.15) has the form

$$\frac{\partial h}{\partial t} = -K\nabla^4 h + \lambda_1 \nabla^2 (\nabla h)^2 + \lambda_2 \nabla \cdot (\nabla h)^3 + F + \eta(\mathbf{x}, t), \qquad (14.1)$$

where we keep terms up to the fourth order in ∂_x [258].

Before examining (14.1) in detail, we consider the nonlinear term $\nabla \cdot (\nabla h)^3$. According to (13.2), this term corresponds to a current $\mathbf{j} = -(\nabla h)^3$, which cannot be written as a gradient of a chemical potential. Its effect is similar to that of the linear term $\nabla^2 h$ [compare Fig. 5.4 with Figs. 14.1(c) and 14.1(f)]. Hence it has been argued that $\nabla \cdot (\nabla h)^3$ acts as a higher-order correction to the surface tension (see Appendix C). Thus far, no physical mechanism generating the λ_2 term has been discovered [258], so we shall not discuss it further. Nevertheless, the λ_2 term is more relevant than the λ_1 term: it will likely determine the scaling behavior whenever physical arguments indicate the presence of the λ_2 term.

† Symmetry and conservation arguments allow the linear term $\nabla^2 h$. However, we must exclude it, as it predicts a gravity-like chemical potential $\mu \sim h$, which is not physical in the absence of desorption or for partial sticking. While gravity does not play a role in MBE, other effects, e.g., the existence of a Schwoebel barrier and diffusion bias, lead to a $\nabla^2 h$ term. These effects are discussed in detail in Chapter 20.

14.1.2 Dynamic RG analysis

Suppose we neglect the flux term F, which does not affect the scaling exponents,† and the 'physically unmotivated' $\nabla \cdot (\nabla h)^3$ term. Then the growth equation (14.1) becomes

$$\frac{\partial h}{\partial t} = -K\nabla^4 h + \lambda_1 \nabla^2 (\nabla h)^2 + \eta(\mathbf{x}, t), \qquad \rightarrow (14.2)$$

which is the lowest order nonlinear equation exhibiting conserved dynamics with nonconserved noise [258, 463]. If $\lambda_1 = 0$, we recover the linear equation (13.9), with the scaling exponents given by (13.11). Rescaling our system using $\mathbf{x} \rightarrow b\mathbf{x}$, $h \rightarrow b^\alpha h$ and $t \rightarrow b^z t$, we obtain

$$\frac{\partial h}{\partial t} = -Kb^{z-4}\nabla^4 h + \lambda_1 b^{\alpha+z-4}\nabla^2(\nabla h)^2 + b^{(z-d-2\alpha)/2}\eta. \qquad (14.3)$$

If we compare the scaling factor of the linear term with that of the nonlinear term, we see that the nonlinear term determines the scaling behavior. In order to obtain the scaling exponents, we must study the

† In fact we can remove F by viewing the interface from a co-moving reference frame, $h \rightarrow h - Ft$.

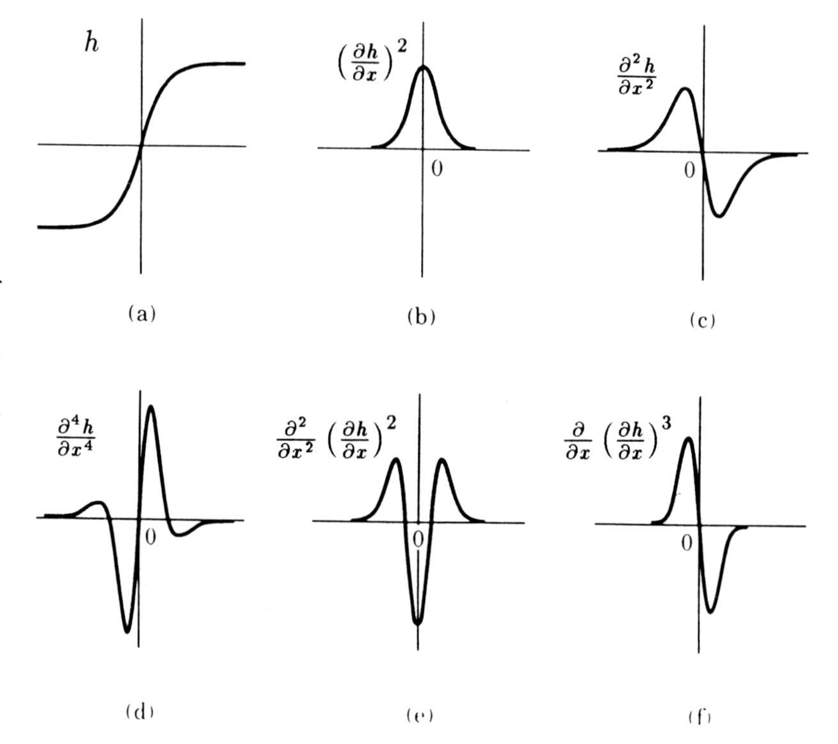

Figure 14.1 Geometrical interpretation of the different linear and nonlinear terms appearing in the continuum growth equations. In (a) we show a segment of the surface. (b)–(f) show the effect of the various terms on this segment. (After [258]).

effects of the nonlinear terms using the dynamic RG (see Appendix B). We find the flow equations [258]

$$\frac{dK}{dl} = K\left[z - 4 + K_d \frac{\lambda_1^2 D}{K^3}\,\frac{6-d}{4d}\right], \tag{14.4}$$

$$\frac{dD}{dl} = D[z - 2\alpha - d], \tag{14.5}$$

$$\frac{d\lambda_1}{dl} = \lambda_1[z + \alpha - 4]. \tag{14.6}$$

Here $K_d \equiv S_d/(2\pi)^d$, and S_d is the surface area of the d-dimensional unit sphere. The coupling constant $g_f^2 = \lambda_1^2 D/K^3$ flows under rescaling as

$$\frac{dg_f}{dl} = \frac{4-d}{2} g_f + K_d \frac{3(d-6)}{8d} g_f^3, \tag{14.7}$$

indicating that for this model the critical dimension is $d_c = 4$.

Because we have two scaling relations, we can determine the exponents exactly. The first one is the result of the non-renormalization of the noise D, obtained by setting to zero the lhs of (14.5)

$$z - 2\alpha - d = 0. \tag{14.8}$$

This scaling relation is exact, and is valid for any equation that combines conserved dynamics and nonconserved noise. The second scaling relation can be obtained from the non-renormalization of the nonlinear term† λ_1, obtained by equating to zero the lhs of (14.6)

$$z + \alpha - 4 = 0. \tag{14.9}$$

We can use the two scaling relations, (14.8) and (14.9), to obtain the scaling exponents

$$\alpha = \frac{4-d}{3} \qquad \beta = \frac{4-d}{8+d} \qquad z = \frac{8+d}{3}. \qquad \rightarrow (14.10)$$

For MBE, the physically-interesting dimension is $d = 2$, and we obtain $\alpha = 2/3$, $\beta = 1/5$ and $z = 10/3$. All three values differ from the predictions (13.11) of the linear theory. In particular, the roughness exponent is smaller than one, so the problem arising from the large roughness exponent has been resolved by the presence of the nonlinear term.

† For the KPZ equation, the non-renormalization of the nonlinear term is the result of Galilean invariance (see §6.4). Equation (14.2) is invariant under the transformation [258, 432] $h \rightarrow h + \epsilon \cdot \mathbf{x}$, and $\mathbf{x} \rightarrow \mathbf{x} - \lambda_1 t \epsilon \nabla^2$, where ϵ is a constant vector. The existence of this additional symmetry implies that (14.9) is exact, and higher loop calculations will not modify it. An explicit two-loop calculation verifies the exactness of Eq. (14.9) [97].

Since the nonlinear term λ_1 and the noise D are not renormalized, the exponents α and z can be obtained using simple scaling arguments as well (considering that the nonlinear term and the noise determines the scaling of the system) [266, 463]. Note that D is renormalized in the case of the KPZ equation, and scaling arguments fail in that case.

Equation (14.2) is the simplest nonlinear growth equation describing growth and relaxation by surface diffusion. When we investigate in the next chapter the various discrete models proposed to describe MBE, we will see that (14.2) correctly describes the scaling behavior of a number of atomistic models.

14.2 Growth with desorption

As noted in §13.3, desorption from the surface induces the linear term $\nabla^2 h$, which is allowed since the desorption process need not obey the continuity equation (13.2). Thus if we wish to construct the most general growth equation, we can relax the conservation requirement imposed in the previous section. The most general nonlinear growth equation with nonconservative dynamics is the KPZ equation. Therefore if the relevant relaxation mechanisms are desorption and deposition, the growth of the interface is described by the KPZ equation (6.4). However, we expect surface diffusion to play a role as well at moderately low temperatures. Including the relaxation terms generated by surface diffusion and desorption, we obtain the nonlinear generalization of equation (13.15) describing the roughening of the interface in the presence of adsorption,

$$\frac{\partial h}{\partial t} = \nu\nabla^2 h + \frac{\lambda}{2}(\nabla h)^2 - K\nabla^4 h + \lambda_1\nabla^2(\nabla h)^2 + F + \eta(\mathbf{x}, t). \qquad (14.11)$$

In the remainder of this section we focus our attention on the implications of this equation.

14.2.1 Microscopic support

Heuristic and symmetry arguments were used to construct (14.11), whose validity we can question for describing the effects of such activated processes as surface diffusion and desorption†. It has recently become possible, however, to actually derive (14.11) starting from a microscopic model with activated diffusion. Indeed, Vvedensky *et al.*

† We can question its validity, even though the viability of the arguments has been demonstrated in the preceding chapters, where we compared the predictions of the continuum theories with the results obtained on discrete models.

have considered a model with deposition, desorption and diffusion [470]. Particles are deposited at a constant rate to form a solid-on-solid surface, and the rate of desorption is given by an Arrhenius activation law of the type (12.1). The activation energy for desorption E_D is a linear combination of the substrate binding energy U, and a contribution E_N from each of the n lateral neighbors the particle has before desorption. Hence

$$E_D = U + nE_N. \qquad (14.12)$$

Surface diffusion was also considered to be an activated process, with the hopping probability given (12.2), where again the activation energy for diffusion depends on the number of bonds the diffusing atom forms with the neighbors. Starting from these microscopic concepts, it is possible to derive a master equation governing the surface dynamics. The continuum limit of the master equation is (14.11) and the coefficients ν, λ, K and λ_1 can be related directly to the microscopic parameters. The results demonstrate that ν and λ arise exclusively from desorption – i.e., if we neglect desorption, $\nu = \lambda = 0$, we end up with the growth equation (14.2). Similarly, surface diffusion contributes only to K and λ_1, providing additional support for the above arguments that surface diffusion excludes the terms $\nabla^2 h$ and $(\nabla h)^2$ from the growth equation.

Without reproducing the details of the lengthy calculations, we shall discuss the results and the consequences of this theoretical undertaking. It is particularly instructive to consider the length scale L_1 controlling the crossover between surface diffusion and desorption. From (13.16),

$$L_1 \sim \left[\frac{D_s}{\omega_D}\exp(U/kT) + \frac{a_0^2}{12}\right]^{1/2}, \qquad (14.13)$$

where a_0 is the lattice constant and D_s is the surface diffusion constant. As we anticipated on physical grounds, when desorption becomes negligible (which can be achieved by either decreasing T or increasing U), (14.13) predicts that L_1 diverges and the scaling is controlled by the diffusive growth process.

If surface diffusion is negligible ($D_s \to 0$), L_1 becomes equal to the lattice spacing a_0. Hence the scaling behavior is described by the KPZ equation.

14.2.2 Scaling regimes

We can classify the possible behavior of the system into two scaling regimes (cf. §13.3), according to the relationship between the system size L and the crossover length L_1.

For $L \ll L_1$, the conservative dynamics determines the scaling behavior. This regime can be described by neglecting desorption. Thus we may set $\nu = \lambda = 0$ in (14.11), and regain (14.2), as discussed in §14.1.

For $L \gg L_1$, the growth is described by the KPZ equation. As discussed in Chapter 6, we have exact exponents for the one-dimensional case but only numerical results for the two-dimensional interface (see Table 8.2). It is important, however, to note that the KPZ equation gives the asymptotic behavior of a growth process when desorption is allowed. Hence, when the system is sufficiently large and the deposition time sufficiently long, the KPZ equation describes the final roughness and dynamics – no matter how important is the diffusion process.

14.3 Discussion

Mathematically, the surface tension $\nabla^2 h$ and the KPZ nonlinearity dominate the terms generated by surface diffusion, and determine the asymptotic scaling. However, as discussed in §13.3, in some experimental setups L_1 can be quite large – comparable with the system size – in which case the surface-diffusion dominated scaling may be the only experimental observable. Because a large number of linear and nonlinear terms contribute to the dynamics of the growing interface, we can expect complex crossover behavior [147]. Table 27.1 summarizes the principal growth equations, together with the predicted exponents, in an effort to help guide the reader among the many universality classes.

In the next chapter we present some discrete models proposed for MBE, and compare their scaling properties with the predictions of the continuum theories discussed in this and in the previous chapter.

Suggested further reading:

[94, 463]

Exercises:

14.1 Find the fixed points in flow equations (14.4)–(14.6) and analyze their stability. Show how they lead to the exponents in (14.10).

14.2 In Eq. (14.1), show that the λ_2 term, if present, determines the scaling behavior.

15 Discrete models for MBE

The elementary processes taking place at crystal surfaces suggest that numerical simulations of discrete models might be helpful in understanding the collective behavior of the atoms during growth. In recent years, a number of models have been proposed to describe the evolving morphology of crystal surfaces.

There are two main sources of interest in motivating the study of simple discrete models. First, there is the hope that the studies on simple models might serve as a guide in understanding the experimentally-observable morphologies and scaling behavior. In this respect, discrete models might serve as an intermediate step between experiments and the continuum theories discussed in previous chapters. The evident advantage of the models is the separability of different secondary effects always present in experiments, thereby offering the possibility of studying the influence of selected mechanisms, such as surface diffusion or desorption, on the growth process.

Second, apart from the connection with MBE, models with surface diffusion represent potentially new universality classes in the family of kinetic growth models, distinct from both the KPZ and EW universality classes. Thus their study poses exciting theoretical questions for statistical mechanics and condensed matter physics.

The models that we discuss in this chapter can be classified into three main categories.

(i) The simplest set of models is motivated by the desire to understand the scaling behavior and the corresponding universality classes. These are intrinsically nonequilibrium models, and encompass random deposition, local relaxation and sticking rules. They capture the essential properties of the roughening process, but do not include the effects of thermal activation of the atoms on the surface. The deposited atoms usually take a few steps on the surface, after which they stick irreversibly. For this reason we call them irreversible growth models. Due to the simplicity of the relaxation rules, these models are the most efficient from numerical point of view, allowing large scale simulations with long running times, and thus the possibility of reaching the asymptotic scaling behavior characteristic of the underlying relaxation mechanism.

(ii) A second, more realistic, approach regarding the resemblance to the typical processes taking place during MBE is to study an activated type of model for which all surface atoms are mobile, with a hopping probability given by the Arrhenius law. These are intrinsically finite-temperature models, with temperature as a tunable parameter. While these models are more realistic from the MBE point of view, they are not too helpful in uncovering the underlying scaling behavior. Due to the time-consuming activation process, these models are limited to small system sizes and short simulation times. In addition it is common to find strong finite size effects, again hampering efforts to determine the scaling exponents. However, combined with the experience gained from the study of irreversible models, models with reversible sticking and diffusion are extremely useful in tracing the effect of activated diffusion on the scaling behavior.

(iii) Third, there are a number of phenomenological models, which do not start from the elementary processes observed during MBE, but rather use activated processes driven by a Hamiltonian. Despite rather little resemblance to MBE, these models can capture some essential properties of the scaling behavior generated by surface diffusion.

We shall not discuss molecular dynamics simulations, which consider the interatom potentials acting between different atoms. Molecular dynamics is extremely useful in understanding the local effects taking place on the surface of the crystals, but is limited by very small sizes and short simulation times.

15.1 Irreversible growth models

The physical dimension for MBE is $d = 2$. However, many model studies have focused on $d = 1$, in part because the existence of new universality classes can usually be demonstrated using $d = 1$, and for the technical reason that simulations on two-dimensional interfaces are extremely time consuming. Indeed, for two distinct universality classes the exponents differ in each dimension, so if we are able to establish the existence of a universality class for a given growth process in a given dimension, and if we are successful in connecting a continuum equation to the respective universality class, then the continuum equation will predict the scaling exponents in other dimensions.

The first nonequilibrium growth models including deposition and surface diffusion were introduced independently by Wolf and Villain [488] and Das Sarma and Tamborenea [99]. The two models are only slightly different as regards the relaxation process. Figure 15.1 shows the relaxation rules for the two models. The models are defined as follows: at every time step a particle is added at a randomly chosen position. Then the particle has three choices: to remain where it was deposited, or to move left or right by one site. The particle will choose one of these three sites based on the following criteria: (a) in the Wolf–Villain model, the particle will choose the site that offers the

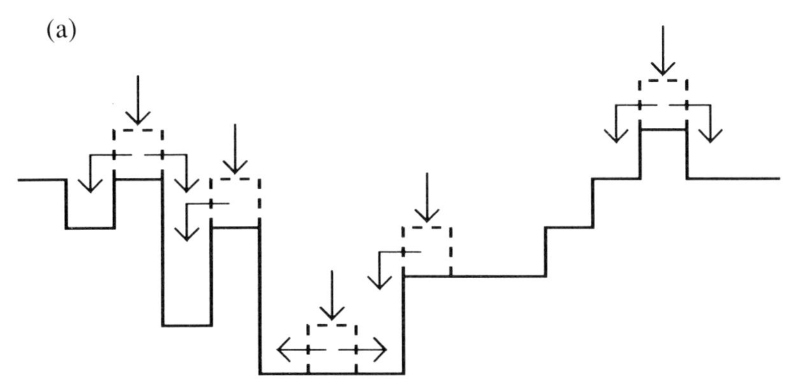

Figure 15.1 The relaxation mechanisms of the (a) Wolf–Villain and (b) Das Sarma–Tamborenea models.

(b)

strongest binding to the surface – i.e., the site that has the *largest* number of neighbors; (b) in the Das Sarma–Tamborenea model the particle moves to the nearest kink site – i.e., it seeks only to *increase* the number of neighbors.

Both relaxation rules contain the essential property of surface relaxation: atoms diffuse so as to maximize their number of neighbors. The irreversible sticking after the relaxation mimics the fact that at low temperatures atoms attached to the edge of an island are likely to remain attached for a long time, due to the strong bonds they form with the neighboring atoms.

One unphysical aspect of the SOS models is related to the formation of very large steps, or large local slopes, as shown in Fig. 15.2. These arise from the fact that in one step atoms can jump as low as is possible to maximize the numbers of neighbors during relaxation. Such phenomena are thought not to occur in MBE.

A key question concerns to which universality class the models belong. We can select some candidates by inspecting the conservation laws incorporated in the model. The random deposition of particles suggests the presence of a nonconservative noise. The relaxation mechanism conserves the number of particles during surface diffusion. Hence we can hope that the linear approximation (13.9) will suffice, without requiring the nonlinear equation (14.2). Calculations of the

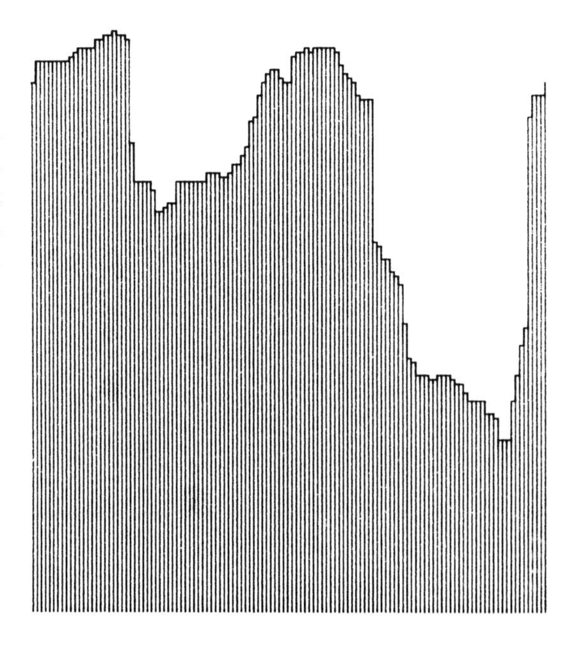

Figure 15.2 A typical interface generated by the Wolf–Villain model. The system size is $L = 120$, and only the upper part of the aggregate is shown. (After [488]).

scaling exponents should give the final verdict between these two growth equations.

The scaling with time of the interface width for different system sizes for the Wolf–Villain model is shown in Fig. 15.3. The fit to the early time dependence gives an exponent $\beta = 0.365 \pm 0.015$, while the scaling of the saturation width with the system size, shown in the inset of Fig. 15.3, is consistent with a roughness exponent $\alpha = 1.4 \pm 0.1$. These exponents are close to $\beta = 3/8$ and $\alpha = 3/2$ predicted by (13.11) for the $d = 1$ linear theory, while they differ from the predictions (14.10) of the nonlinear theory.

Although there is agreement between the predictions of the linear theory and the numerical results, a few remarks must be made. First, the roughness exponent α is larger than one; apparently the large roughness exponent arises from the generation of large local slopes, as we can see from Fig. 15.2. If $\alpha > 1$, we expect the continuum approaches to break down, since these approaches were introduced in the approximation of small local slopes. One expects higher order nonlinear terms to become relevant for such a large scaling exponent.

Second, it has been argued that the Wolf–Villain and related models do not obey the scaling hypothesis (2.8) [12, 96, 374, 391]. The origin of this anomalous behavior is a third exponent, independent of α and z, that describes the evolution of the mean square step height,

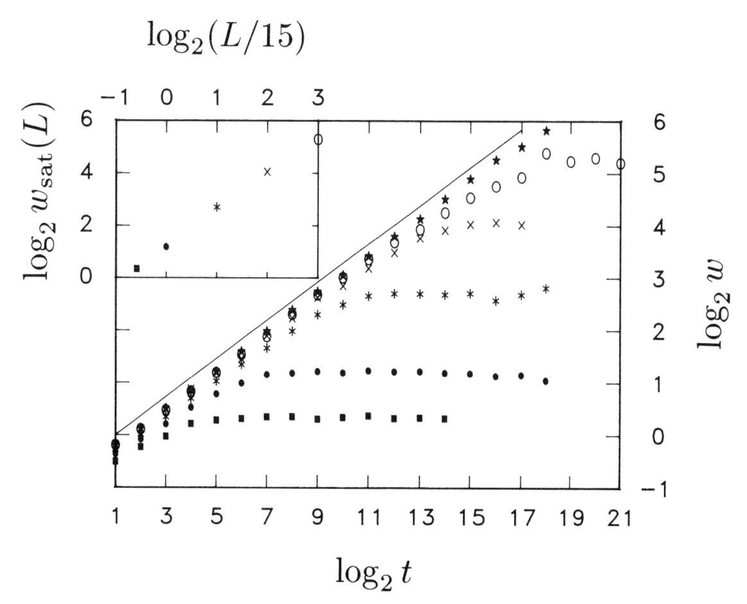

Figure 15.3 Scaling of the interface width with time in the Wolf–Villain model. The system sizes are $L = 10, 15, 30, 60, 120,$ and 450, from the bottom to the top curve, respectively. The solid line has slope $\beta = 0.365$. The inset shows the saturated width as a function of the system size. The solid line has a slope $\alpha = 1.4$. (After [488]).

measured by the nearest neighbor height–height correlation function $C(1,t)$ [cf. Eq. (A.6)]

$$C(1,t) \sim t^{\zeta}, \tag{15.1}$$

with $\zeta = 0.19 \pm 0.01$. This is in fact a time-dependent 'intrinsic width' (see §A.3) affecting the scaling of the interface roughness. It seems that $C(1,t)$ saturates only for system sizes larger than $L = 250$ and $t \gg 10^6$, suggesting that the scaling law (2.8) is valid only in this limit. This possibility has been confirmed in recent large-scale numerical simulations [402]. Moreover, the effect of (15.1) leads to an unexpected multi-affine behavior (see Chapter 24) [245].

The physically most relevant two-dimensional generalization of the nonequilibrium growth models with local surface diffusion was studied by Kotrla *et al.* [241, 403] and Das Sarma and Ghaisas [95]. Kotrla *et al.* found that the two-dimensional generalization of the Wolf–Villain model has exponents $\beta = 0.20 \pm 0.20$ and $\alpha = 0.66 \pm 0.03$, close to the predictions (14.10) of the nonlinear theory (14.2), but different from $\alpha = 1$ and $\beta = 1/4$ predicted by the linear growth equation (13.9). Moreover, Das Sarma and Ghaisas concluded that the Wolf–Villain and Das Sarma–Tamborenea models actually belong to different universality classes; $\beta = 0.192 \pm 0.002$ for the Wolf–Villain model, consistent with the nonlinear theory, while $\beta = 0.237 \pm 0.002$ and $\alpha = 0.95 \pm 0.01$ for the Das Sarma–Tamborenea model, indicating that it belongs to the universality class of the linear growth equation.

What is most interesting is that the Wolf–Villain model gives rise to different universality classes for $d = 1$ and $d = 2$, suggesting that the $d = 1$ model is described by the nonlinear theory as well. There is, however, a very large crossover region preceding the true scaling behavior, which is possible if the bare value of the coupling constant g_f in (14.7) is very small. The unusually large crossover region is also supported by the anomalous scaling effects observed in the one-dimensional Wolf–Villain model [391]. No such long-time crossovers have been found which indicate that the Das Sarma–Tamborenea model deviates from the linear description. A final result concerning the universality class is given by tilt-dependent current measurements, which indicate that the Wolf–Villain model is actually described by the EW equation (see §A.4).

A one-dimensional model belonging to the universality class of the nonlinear growth equation (14.2) has been introduced and studied by Lai and Das Sarma [258]. The model is similar to the Wolf–Villain and Das Sarma–Tamborenea models with the difference that if an atom falls in a kink site, it is allowed to break its two bonds and jump either down *or up* to the nearest kink site with the smallest step height [360]. A kink site is one with two neighbors belonging to the substrate (Fig. 15.4). The scaling of the interface width with time gives $\beta = 0.340 \pm 0.015$ and $\alpha = 1.05 \pm 0.1$, values consistent with the predictions (14.10) for $d = 1$ of the nonlinear theory.

As discussed in §14.2, if nonconservative effects act during the growth process, scaling described by the KPZ equation will eventually dominate the behavior of the system at long enough time and length scales. Such a nonconservative mechanism can be desorption, or the formation of overhangs during growth. Since we imposed the SOS approximation on the growth rules, the previous models all neglect the effects of overhangs and vacancies. Kessler *et al.* [227] and Yan [493] have studied growth models in which the SOS condition is replaced by the ballistic deposition sticking rule. After arriving on the surface, particles diffuse, looking for an energetically-favorable position where they can stick irreversibly. The numerical results indicate that the early time behavior is described either by the EW equation or by surface diffusion, while the asymptotic scaling is described by the KPZ equation. The existence of the crossover from the conservative (diffusion-dominated) to the nonconservative (KPZ) growth confirms the conclusions drawn from (14.11).

15.2 Models with thermal activation

As discussed in Chapter 12, all surface atoms should in principle be mobile. During diffusion the atoms follow the Arrhenius law, the hopping probability being an exponential function of the activation energy divided by the absolute temperature. The activation energy depends on the bonds formed by the diffusion candidate with the neighboring atoms. The time scale is set by the deposition rate. As a function of the temperature, different diffusion processes might be more or less relevant. At high temperatures most of the atoms are

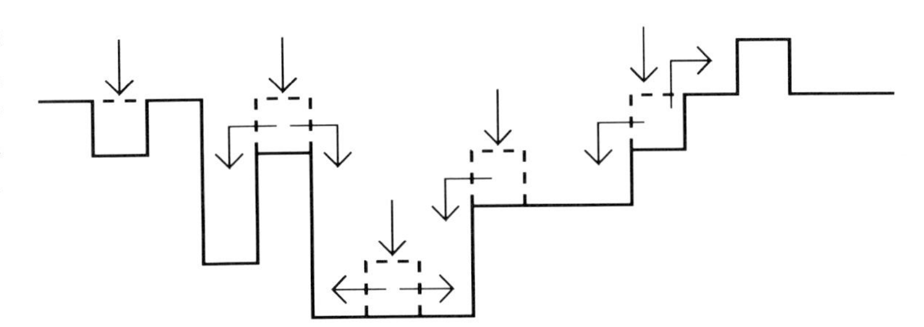

Figure 15.4 The relaxation mechanism of the Lai–Das Sarma model.

highly mobile, while at low temperatures it is difficult to pass even the lattice potential.

Recently a number of models have been studied that incorporate this activated nature of diffusion, both for $d = 1$ [99, 264, 437] and $d = 2$ [81, 98, 205, 357, 405, 477]. In this section we discuss the one-dimensional simulations.

In order to include the effect of the activated surface diffusion, Das Sarma and Tamborenea have studied a model which contains deposition and activated surface diffusion [99]. Atoms are deposited on the surface with a deposition rate $F = 1$ (one atom per site per second). The deposited atoms diffuse with a hopping rate

$$R = R_0 \exp\left(-\frac{E}{kT}\right), \tag{15.2}$$

where the activation energy for hopping is $E = E_0 + nE_N$. Here n is the number of neighbors of the atom, and the parameters were chosen to be $E_0 = 1$ eV and $E_N = 0.3$ eV. Thus $n = 1$ corresponds to the situation when the atom has originally one atom underneath, while with $n = 2$ the atom has an additional nearest neighbor. For $n > 2$, diffusion is forbidden. Thus to improve the effectiveness of the simulation, the model reflects a compromise between the real crystal and the nonequilibrium models: only those atoms that have a maximum of two neighbors are allowed to move.

Due to the activated character of the diffusion rate, the simulation of the model is very time consuming, and saturation occurs only for small system sizes. In Fig. 15.5(a) we show the variation of the measured exponent β as a function of the temperature T. There are three main regimes in temperature:

(i) At low T, surface diffusion can be practically neglected. In this limit atoms stick where they first arrive on the interface, the model corresponding to pure 'random deposition without relaxation.' The RD exponent $\beta = 1/2$ is observed in this limit.

(ii) At intermediate T, surface diffusion becomes relevant, and we find the exponent crossing over to $\beta = 3/8$, a value predicted by the linear theory with surface diffusion (13.9).

(iii) At higher T, a further decrease of the exponent is observed. In this limit, the diffusion length becomes larger than the system size, and the atoms can search the entire system for an energetically favorable position to stick. As a result the interface becomes smooth, the decrease in the exponent being a finite-size effect. The most likely universality class for high T is EW, the no-hopping-up condition of the atoms generating a downward surface current with a gravity-like behavior.

As Fig. 15.5(b) illustrates, essentially the same scaling regimes can be observed if instead of the temperature we vary the deposition rate F. For large F, β is close to the prediction of the random deposition model, while as F decreases, diffusion becomes relevant and $\beta \rightarrow 0$, after a mild inflection at $\beta \approx 0.375$, the prediction of the linear diffusive theory.

It is instructive to examine how the interface morphologies change with temperature (Fig. 15.6). At low temperatures, when surface diffusion is negligible, the interface is very rough. Increasing T enhances

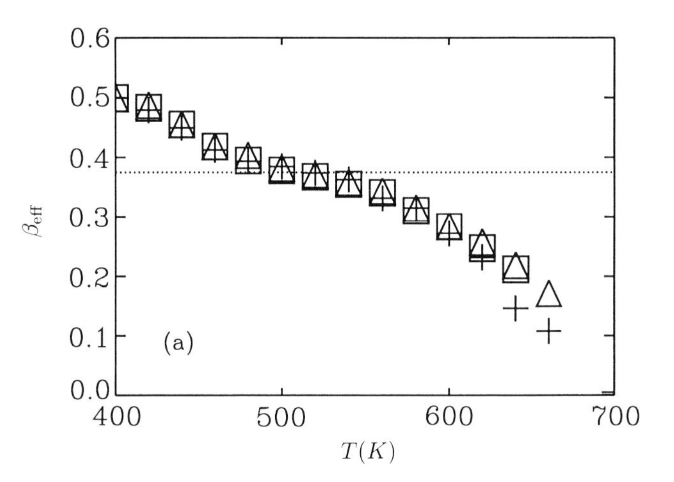

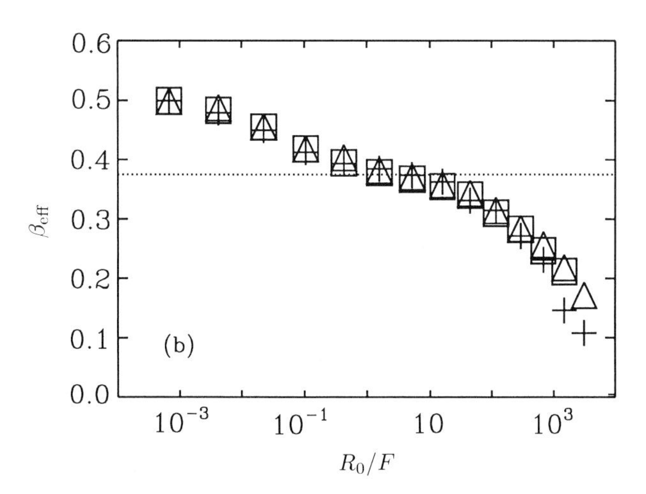

Figure 15.5 The variation of the effective exponent β_{eff} for the activated surface diffusion model and its variation with (a) temperature and (b) the ratio R_0/F. The different symbols correspond to different system sizes: +, $L = 256$; $\triangle$, $L = 512$; and $\square$, $L = 1024$. The dotted line corresponds to $\beta = 0.375$. (After [437]).

surface diffusion, which in turn leads to the smoothing of the interface. Note that the change in the saturated width of the interface may have a different source than the decrease of the roughness exponent α. Even a seemingly rough interface may have a small roughness exponent, since according to (2.5), the roughness exponent characterizes the scaling of the width with the system size. The change in the overall roughness can be attributed to the change with temperature of the *prefactor* of the scaling law (2.5).

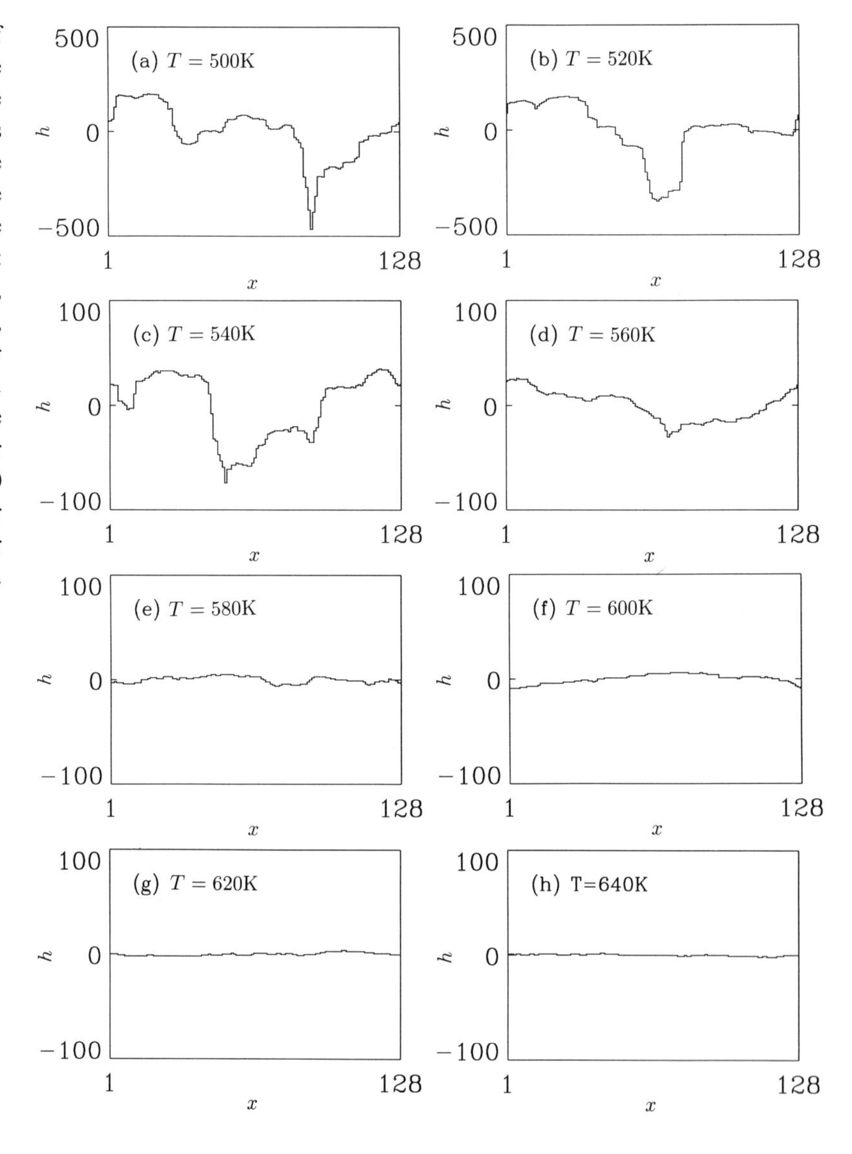

Figure 15.6 Effect of temperature on the interface morphologies generated by the activated surface diffusion model. The temperatures are: $T = 500$, 520, 540, 560, 580, 600, 620, and 640 K, for (a)–(h), respectively. Note that the range of the h axis is larger for $T = 500$ and 520 K than for the other temperatures. (After [437]).

15.3 Hamiltonian models

A class of models approach the understanding of the effects of surface diffusion by focusing on the collective motion of atoms described by the Hamiltonian [250, 397]

$$\mathscr{H} = \sum_i g_1|h_i - h_{i-1}| + g_2(h_i - h_{i-1})^2 + g_4(h_i - h_{i-1})^4 + g_6(h_i - h_{i-1})^6. \quad (15.3)$$

As before, h_i is the (integer) height of the interface at site i. In simulations, a lattice site i is chosen randomly and a particle is deposited on that site with a probability p. With a probability $1 - p$, instead of deposition a diffusive move is attempted. The direction of the diffusion is chosen randomly, and a move from site i to its nearest neighbor j in that direction is accepted with a probability $w_{i \to j} = [1 + \exp(\beta \Delta \mathscr{H}_{i \to j})]^{-1}$. Here $\Delta \mathscr{H}_{i \to j}$ is the energy change associated with the move of the particle.

At a finite deposition rate, a morphological phase transition is observed from a rough interface to a grooved phase†, which seems to be present in the case when g_4 is nonzero.‡ For this phase, the width does not follow the scaling law (2.8), and translational invariance is spontaneously broken. The saturated interface will have one large groove, and the expectation value of the height $\langle h(x, t, L) \rangle = f(x + x_0(t), L)$ obeys in the steady state the scaling law $f(x, L) = L^\zeta g(x/L)$,

† Groove instabilities of different origin have been observed in other MBE models as well [121, 123, 452].

‡ If deposition is absent ($p = 0$), the model reproduces the exponents of the diffusive linear theory with conservative noise, which is discussed in Chapter 21.

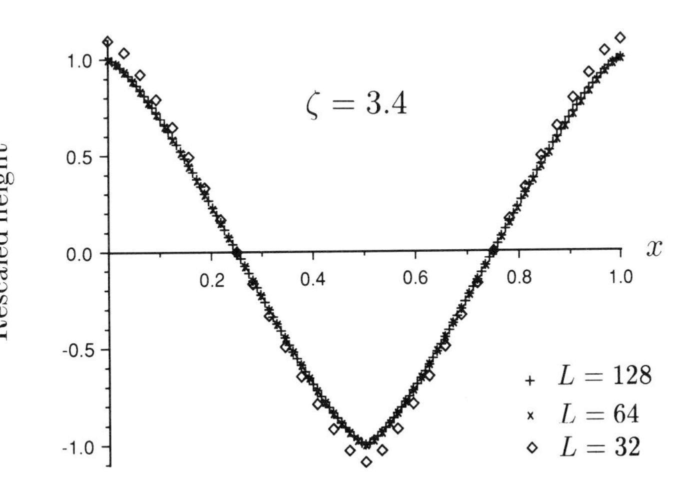

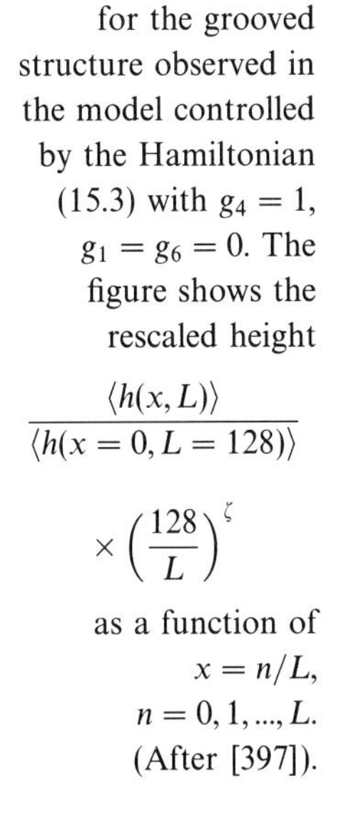

Figure 15.7 Evidence for the grooved structure observed in the model controlled by the Hamiltonian (15.3) with $g_4 = 1$, $g_1 = g_6 = 0$. The figure shows the rescaled height

$$\frac{\langle h(x, L) \rangle}{\langle h(x = 0, L = 128) \rangle} \times \left(\frac{128}{L}\right)^\zeta$$

as a function of $x = n/L$, $n = 0, 1, ..., L$. (After [397]).

as shown in Fig. 15.7. One can estimate a lower bound for the new exponent, $\zeta \geq 3.6$.

The origin of the groove instability can be understood if we consider the tilt dependence of the surface current (see §A.4). Figure 15.8 shows the current for two cases (a) $g_1 \neq 0$ and $g_2 = g_4 = g_6 = 0$, characterized by a negative (downhill) current, consistent with the EW growth equation. (b) $g_1 = g_2 = g_6 = 0$ and $g_4 \neq 0$, for which the current is *uphill*, destabilizing the $m = 0$ interface. Actually, the current is essentially zero up to a critical slope m_c, indicating a nucleation driven transition to the grooved state for $m < m_c$.

15.4 Discussion

The models presented in this section were initiated by the phenomenology of MBE, and incorporate surface diffusion as an essential relaxation mechanism. Unfortunately, the scaling behavior is not as evident as it was in the case of the nonconservative growth processes, due to the rich crossover behavior generated by the competition between the diffusive and the deposition processes. Crossovers are found in most of the models, rendering the determination of universality class somewhat ambiguous. A representative example is the Wolf–Villain model. A simple measurement of the scaling of the interface width gives results in quite good agreement with the *linear* theory, but its two-dimensional variant seems to be described by the *nonlinear* theory (14.2). An even more careful look at the surface currents indicates that

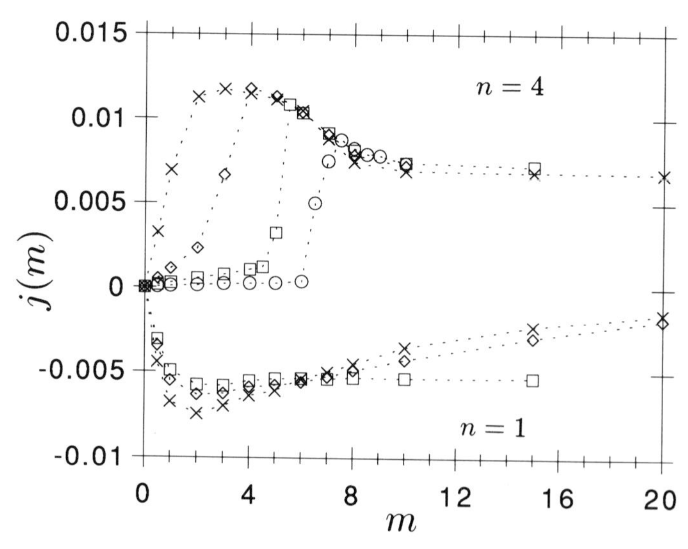

Figure 15.8 Surface diffusion currents j for the one-dimensional SOS deposition model controlled by the Hamiltonian (15.3). The lower curves correspond to $g_2 = g_4 = g_6 = 0$, while the upper to $g_1 = g_2 = g_6 = 0$. The substrate sizes are $L = 16$ ($\times$), $L = 32$ ($\diamond$), $L = 64$ ($\square$) and $L = 128$ ($\circ$). (After [250]).

the asymptotic behavior is described by the EW equation both in one and two dimensions. It is not surprising that more complicated models, such as those including activated diffusion, are also characterized by crossovers.

Generally, it seems that in the hydrodynamic limit most models are described by the EW theory, or if overhangs are present, by the KPZ equation. However, the study of intermediate scaling regimes is important since as we have seen in the case of the Wolf–Villain model – or as we argued for real experimental situations – the asymptotic scaling behavior may never be reached on the available experimental or simulation time scales.

Suggested further reading:

[437, 484]

Exercises:

15.1 Write two computer programs, one simulating the Wolf–Villain model, the other the Das Sarma–Tamborenea model. Compare the scaling of the interface widths obtained for the two models.

15.2 Calculate the surface current of the random deposition model with surface relaxation. What does the result tell us about the growth equation?

16 MBE experiments

As discussed in the previous chapters, we can distinguish the various growth processes based on the concept of universality. Interfaces that belong to the same universality class are described by the same scaling exponents, and they are also described by the same continuum equation.

The universality class is determined by the physical processes taking place at the surface. There are three basic microscopic processes that can play a major role in this respect: deposition, desorption, and surface diffusion. In addition to these, nonlocal effects such as shadowing may play a decisive role in shaping the interface morphology. While deposition must occur, the other microscopic processes may be irrelevant or even absent altogether. For example, in many systems desorption is negligible, while at low temperatures surface diffusion may be negligible.

A number of recent experiments support the existence of kinetic roughening in various deposition processes. It is possible to measure both the roughness exponent α characterizing the interface morphology, and the exponent β quantifying the dynamics of the roughening process. However, the emerging picture is far from complete, and there is no unambiguous support for the various universality classes.

There are a number of reasons for this situation. First, it is only recently that experimental groups have initiated systematic investigations of the various roughening processes. While the results are quite encouraging, more work is needed to obtain a coherent picture. Second, due to the complicated nature of the competing effects discussed in the previous chapters, the interpretation of the data is often not straightforward. In this chapter, we present some of the experimental results. We also discuss the difficulties related to the interpretation of the scaling exponents, and their relation to the universality classes introduced so far.

Early studies on rough interfaces focused on the fractal or self-affine properties of the rough substrates, measuring the fractal dimension or the roughness exponent. These are *static* quantities, characterizing the interface at a given stage of the growth process. In the next section we review these investigations. More recent advances in the theoretical studies related to the *dynamics* of the growing interfaces have initiated a number of studies focusing on the time evolution of the interface roughness, providing information on the dynamic exponents. As we saw in the previous chapters, in many situations the roughness exponent alone is not sufficient to determine the universality class. In this respect, measurements on the dynamic properties of the growth process serve as a major step in comparing the theoretical predictions with experiments. Similarly, comparing the amplitudes of various correlation functions obtained from experiments and models is becoming increasingly necessary to connect experiment and theory.

16.1 Experimental techniques

For mesoscopic surfaces, such as the one generated by fluid flow or bacterial growth, a simple video camera is able to produce sufficient resolution that the scaling properties of the interface can be studied. For MBE, however, we require sophisticated techniques that provide information about the morphology of the interface at *atomic* scales. There are two main methods that are used to measure the roughness of crystal surfaces: (i) diffraction methods and (ii) direct imaging methods. In this section we describe the main elements of these two different techniques.

16.1.1 Diffraction techniques

Until the introduction of STM, diffraction was the main source of microscopic information on surface roughening. In a diffraction measurement a plane wave with wavelength k_1 falls on the interface. If the surface were ideally smooth, one would expect a reflected and a refracted beam according to the Fresnel theory. Roughness reduces the reflected intensity at specular condition and increases the diffuse component.

There are two independent exponents, α and β, that characterize the roughening of an arbitrary interface. To allow measurement of both exponents, the technique must be sensitive to both the static and dynamic properties of the interface.

There are a large number of studies focusing on the characterization of rough interfaces using diffraction methods, both for single interfaces [401, 446, 474, 491, 497, 498] or multilayers [370, 385, 386, 387, 426]. The main result is that diffraction methods are able to separate the dynamical and static properties of interfaces.

Specular reflections, for which the angle of incidence is equal to the angle of reflection, yield information about the dynamical properties of the interface width. Such measurements involve recording the scattered intensity as a function of k_z, the wave vector transfer perpendicular to the surface. Taking these curves for various film thicknesses, one can determine the various surface widths, and thereby the exponent β.

The static properties of the interface can be measured from the diffuse component of the scattering, for which the angle of incidence is different from the angle of reflection. Indeed, the diffuse intensity decays asymptotically as a power of k_z, the exponent depending on α. Thus a log-log plot of the diffuse intensity versus k_z can directly provide an experimental measure of α.

16.1.2 Direct imaging techniques

While AFM, SEM, and TEM can also provide high resolution pictures of the interface, the direct imaging technique currently used in most surface studies is the scanning tunneling microscope (STM). A pin tip – ideally of atomic size – scans the entire interface, providing $h(x, y)$, the height of the interface at the point of coordinates (x, y). Having the height of the interface in every point, the methods discussed in §A.1 can be used directly to determine the roughness exponent of the interface. The resolution provided by STM is of atomic scale: in Fig. 1.6 one can see the individual atoms on the surface. However, there are some disadvantages related to high quality resolution: due to the long scanning time, the size of the interface one can scan is limited. Thus if we are interested in the large scale properties of the interface, it would be too time consuming to obtain such data using STM. Also, as one looks for long wavelength behavior, having lower resolution is a sort of 'coarse-graining' which is more in the spirit of continuum equations.

16.2 Scaling approach for interface roughening

The morphology of experimentally-observable interfaces (see Fig. 1.9) can be conveniently described by fractal concepts. Messier and Yehoda in 1985 suggested that the remarkable similarity at high magnifications of amorphous silicon, germanium, pyrolytic graphite, thick metal films – or the cauliflower (see Fig. 1.10) – can be understood if one assumes that the surfaces are fractals. Hence they proposed that the growth mechanism involved in the formation of some of the thin film morphologies must lead to self-similar morphologies [316].

An early quantitative study on the fractal structure of microscopically rough films has been performed by Pfeifer *et al.* [368, 369]. They used a method developed to measure the fractal dimension of various objects to study the morphology of Ag films [367]. The basic idea is illustrated in Fig. 16.1. A fractal surface is covered by a monolayer of molecules. The smaller the radius of the molecule, the larger the number of molecules adsorbed on the surface. If the surface is fractal, the number of molecules of size R required to cover the surface with a monolayer scales as $N_R \sim R^{-d_f}$, where d_f is the fractal dimension of the interface.

Pfeifer *et al.* used adsorption of N_2 on rough Ag surfaces to determine the fractal dimension of the interface. The number of N_2 molecules changes with the applied pressure, the variation giving information about the fractal dimension of the interface. They find $d_f = 2.3$. Kardar and Indekeu have re-analyzed the experimental data, assuming that the interface is not self-similar, but is self-affine [215, 216]. Their analysis leads to a roughness exponent $\alpha \approx 0.37$, different from the value predicted by $\alpha = 3 - d_f = 0.7$. X-ray reflectivity measurements carried out on Ag films deposited at low temperatures yield an exponent $\alpha \approx 0.46$, in between the two values deduced from adsorption [76]. While the assumption that the interface is self-affine rather than self-similar is correct and is supported by the visual appearance of the interface shown in Fig. 1.9, the situation is still not

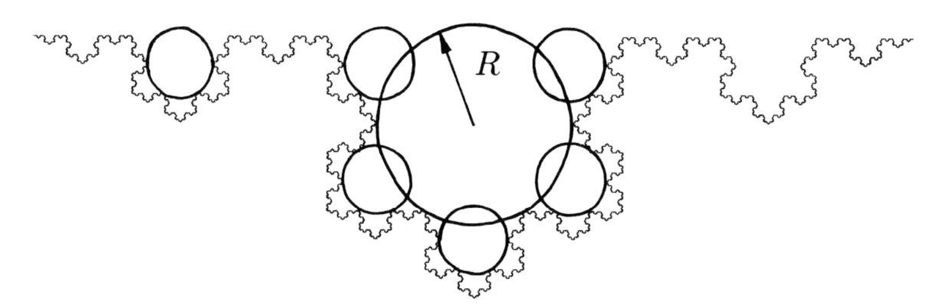

Figure 16.1 The adsorption technique illustrated on the Koch fractal. The surface is covered with balls of radius R. The smaller the radius, the larger the number of balls required to obtain a monolayer coverage. (After [369]).

settled, so the interpretation of the adsorption data remains the topic of a lively debate [358].

Since STM provides the height of the interface at a given point with high resolution, it is suitable for studies on the scaling properties of the interface. Similarly, it can be used to test the accuracy of other methods, such as adsorption. Such a study has been performed by Krim *et al.*, who measured by STM the scaling of the interface roughness for gold films (Fig. 16.2) [242]. They found $\alpha = 0.96 \pm 0.03$, in good agreement with $\alpha = 0.98 \pm 0.01$ determined from adsorption measurements.

16.3 Dynamical properties

As mentioned above, the advances on the theoretical side have motivated a closer look at the microscopic properties of thin film growth. In this section we review two studies, both focusing on roughening by iron deposition at low temperatures. They illustrate how the advances in the experimental techniques lead to the clarification of the various effects involved in the growth process. While the measurement of the exponent β gives a first indication of kinetic roughening, it is not enough to decide about the universality class – one must determine *both* α and β.

The first attempt to follow experimentally the dynamics of the

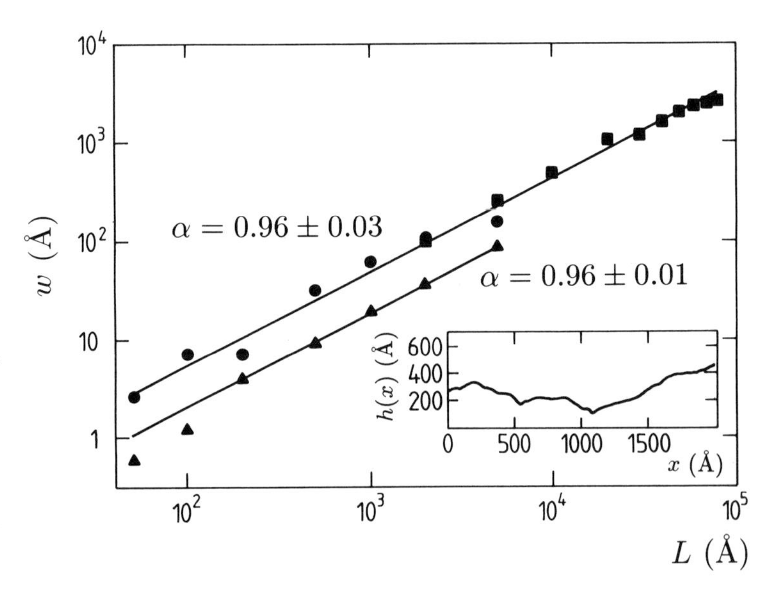

Figure 16.2 Scaling of the interface roughness with the scan size for gold films. The circles and the squares represent data recorded in air by STM, while triangles represent data recorded in vacuum. The slope of the solid line is $\alpha = 0.96$. The inset shows a typical cross section of the interface (After [242]).

roughening process was by Chevrier *et al.*, who studied iron deposition on a Si(111) surface [75]. The experiments were performed near room temperature, where surface diffusion is expected to be limited. The RHEED measurements indicate that iron grows epitaxially on the Si surface. The interface roughness was measured *in situ* using RHEED, which allows one to follow the roughening process in time (as a function of the average thickness of the sample, which is proportional to time if the deposition rate is constant). The experimental results provide a clear indication of kinetic roughening, displaying an increase in the width with time (Fig. 16.3). The results are consistent with an exponent β between 0.22 and 0.3.

What is the origin of this exponent value? The experiments indicate that the film has good crystal quality, excluding void formation and overhangs. Moreover, desorption is expected to be negligible under normal MBE conditions. The absence of void formation and desorption excludes the EW and KPZ terms from the growth equation. This is rather surprising, since the experimentally-obtained exponent is $\beta = 0.25$, close to the value predicted by the KPZ equation.

An alternative interpretation is that surface diffusion controls the scaling regime. While experimentally we expect that surface diffusion is reduced, we saw that in nonequilibrium growth models, where the diffusion length is restricted to one lattice site, even a short diffusion length can lead to the appearance of scaling regimes predicted by the diffusive growth equations. This may be the case here as well, and the

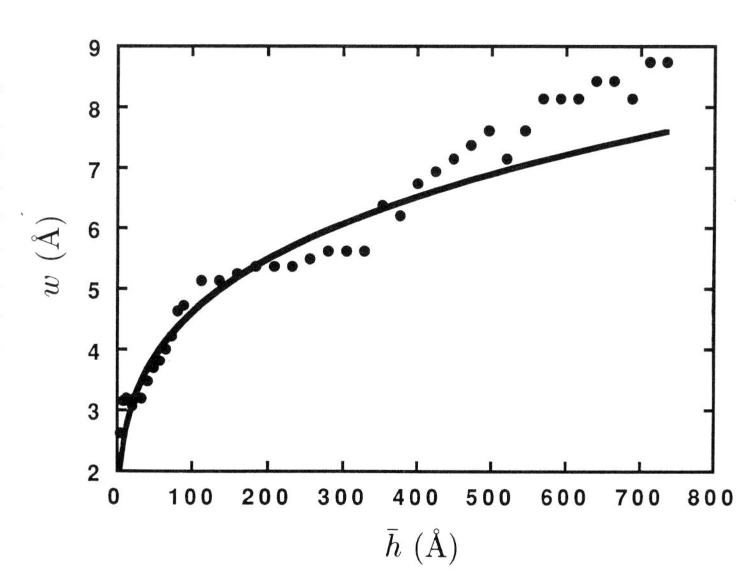

Figure 16.3 Variation of the surface width w with the average thickness $\bar{h}$ of the film for iron deposition on Si(111) surface. The solid line is $w = 1.46\bar{h}^{0.25}$, indicating an exponent $\beta = 0.25$. (After [75]).

exponent predicted by the nonlinear theory (14.2), $\beta = 0.2$, is not too far from the experimentally-observed value.

A final indication of the relevant universality class may come from the simultaneous measurement of the independent scaling exponents. In fact, the measurement of α may help distinguish between the KPZ and the diffusive nonlinear equations (14.2), which predict the values 0.38 and 2/3, respectively.

A major step in this direction was carried out by He *et al.* who, studying iron deposition on Fe(001), were able to simultaneously measure α and β [167, 496, 497]. Iron was deposited on a specially prepared flat buffer layer of Fe(100) at a deposition rate of ~ 2 monolayers/min. Information about the surface morphology was obtained using high-resolution low-energy electron diffraction (HRLEED) [498]. The experiments resulted in $\alpha = 0.79 \pm 0.05$ and $\beta = 0.22 \pm 0.02$.

The value of β is consistent with the result obtained in iron deposition on Si, discussed above. However, the roughness exponent is too high to be consistent with the prediction $\alpha = 0.38$ of the KPZ equation, indicating that surface diffusion might play a decisive role in the experiments. In fact, the exponent of the nonlinear diffusive theory (14.2), $\beta = 0.20$, is in good agreement with the experimental observation, and the $\alpha = 2/3$ prediction of (14.2) is much closer to the experimental result than the KPZ prediction. Although the measurements of He *et al.* are closer to the $\alpha = 2/3$ prediction than the KPZ prediction, the roughness exponent observed is not consistent with either prediction. Similar disagreement exists for the x-ray reflectivity data reported by Thompson *et al.* [446] for Ag films deposited at room temperature onto silicon. These measurements, carried out *in situ* on films ranging from 10 to 150 nm in thickness, yield the values $\alpha = 0.70 \pm 0.10$ and $\beta = 0.26 \pm 0.05$.

16.4 Discussion

In this chapter we have reviewed some of the experimental results regarding the roughening of nonequilibrium interfaces generated by atom deposition. There is a large number of other experiments that we did not discuss in detail, some of them being summarized in Table 16.1. We leave for Chapter 19 the discussion on growth by sputtering and erosion by ion bombardment.

We shall see that in many experiments the Schwoebel barrier plays a key role in shaping the interface morphology. Experimental results on this process are presented in Chapter 20, where we discuss in more detail the effect of diffusion bias. Considerable experimental work has

Table 16.1 *Summary of the scaling exponents reported in various deposition and ion bombardment experiments. (*) Logarithmic.*

System	Method	α	β	Reference
Ag on quartz	Adsorption	0.7 0.37	— —	[368, 369] [215, 216]
Ag evaporation	Adsorption, X-ray	0.46 (80 K)	—	[76]
Au on quartz	X-ray, Adsorbtion	0.95	—	[76]
Ag on Si	X-ray, STM	0.70	0.26	[446]
AlCu	X-ray, STM	0.70	—	[474]
W/Si multilayer	X-ray	$0^{(*)}$	—	[387, 385, 386]
Au on glass	STM	0.35, 0.89	—	[148, 172, 388]
Fe on Fe	HRLEED	0.79	0.22	[167, 496, 497]
Fe on Si	RHEED	—	$0.22 - 0.3$	[75]
InP	STM	—	$0.1 - 0.2$	[86]
CuCl islands on CaF_2	AFM	0.84	—	[447, 449]
Si on Si	STM	0.68	—	[168]
	TEM	—	1.0	[109]
Au deposition	STM	0.96	—	[242]
Cu evaporated on Cu(100)	He atom beam scattering	0.6-1	0.56 (200 K) 0.26 (160 K)	[115]
NbN decorated with AlN	TEM, Multilayer decoration	—	0.27	[321]
Au sputter deposited on Si	X-ray STM	— 0.42	0.40 (300 K) 0.42 (220 K)	[500]
sputter etching with Ar ions of graphite	STM	0.2-0.4	$z = 1.6 - 1.8$	[114]
Fe erosion	STM	0.52		[242]
SiO_2 bombarded with Xe and H	X-ray	—	1	[74, 72]

been done regarding *equilibrium* roughening. Relevant experiments will be discussed in Chapter 18.

The main message of this chapter is that indeed there is experimental evidence that interfaces are rough on certain length scales and the interface profile is self-affine, so scaling theories can be used to characterize the interface morphology. Similarly, experiments indicate that roughening is a dynamic process, described by the scaling laws proposed in the previous sections.

We also note that the identification of the relevant universality class is not a simple task, and usually requires the knowledge of both scaling exponents, α and β. Similarly, the identification of the universality class is hampered by subtle crossover effects, the origin of which is in many cases unknown, due to the limited information we have on the microscopic processes relevant for the given experimental setup.

This chapter leaves us with the conclusion that the experiments performed so far pose intriguing questions that will hopefully be followed up in the near future.

Suggested further reading:

[208, 256]

Exercises:

16.1 Select one of the experiments listed in Table 16.1 and not discussed in the text. Determine which microscopic effects will shape the interface morphology, and determine the most appropriate growth equation.

16.2 In analysing the scaling properties of a rough surface, what are the advantages and disadvantages of the STM method as compared to diffraction methods?

17 Submonolayer deposition

The main focus of our previous discussion of MBE has been the identification of various universality classes. The models we discussed are expected to be valid on a coarse-grained level, at which the exact structure and form of an island does not matter. However, with the perfection of experimental tools, it is possible to observe the interface morphology at the atomic scale – leading to the discovery of rich island morphologies. In this chapter we focus on this early-time morphology, for which the coverage is less than one monolayer; this regime is usually referred to as *submonolayer epitaxy*.

The phenomenology is quite simple. Start with a flat interface, and deposit atoms with a constant flux. The deposited atoms diffuse on the surface until they meet another atom or the edge of an island, whereupon they stick. Thus if at a given moment we would photograph the surface, we would observe a number of clusters – called *islands* – with monomers diffusing between them. What is the typical size and number of the islands? What is their morphology? How do these quantities change with the coverage and with the flux? These are among the questions we address.

17.1 Model

Let us consider in more detail the deposition process outlined above. Consider a perfectly flat crystal surface with no atoms on it. At time zero we begin to deposit atoms with a constant flux F. Atoms arrive on the surface and diffuse (the deposition and diffusion processes take place simultaneously). When one atom meets another, they form an immobile cluster. Similarly, if an atom meets an island, it sticks to its edge, and becomes immobile. If an atom arrives on the top of an existing island, it will continue diffusing, eventually falling down if it reaches the edge of the island – or, due to the effect of the Schwoebel barrier, the atom may not leave the upper terrace.

This model is termed the DDA model, since it includes the three elementary processes of *deposition*, *diffusion* and *aggregation*. There are two competing processes that define the typical time scale: deposition and diffusion. The larger the flux, the less time the deposited atoms will have to diffuse before meeting another atom on the surface. We can understand this by considering the two extremes of the flux. If F is almost zero, the first deposited atom will have a long time to diffuse before the next atom is deposited on the surface. If the flux is large, the deposited atom will soon find another deposited nearby. The diffusion process can be characterized by the diffusion constant D, where $1/D$ is proportional to the typical time between two hops. Due to the competition between deposition and diffusion, one expects that all physical processes depend on the ratio D/F. This means that if we increase the flux and the diffusion constant without changing D/F, we obtain a system similar to the original.

What is the surface morphology for a fixed value of D/F? The answer depends on the coverage, that is the total number of particles deposited on the surface since the beginning of the deposition process. We denote the coverage by Θ, where $\Theta = Ft$. Figure 17.1 shows the morphology of the interface for different values of D/F and for various coverages. For a fixed Θ, on increasing D/F (corresponding to decreasing flux, for example) we note that the typical distance between the islands increases. For a fixed F, on increasing coverage, we see that the density of islands increases as Θ increases. Our goal is to understand quantitatively the variation of the island size and number as a function of these two 'control parameters,' D/F and Θ.

17.2 Scaling theory

17.2.1 Island distribution

The fact that for different fluxes and coverages the observed island morphologies are somewhat similar suggests that scaling laws may be useful in quantitatively characterizing the model. We therefore discuss the basic elements of a scaling theory describing the island formation and distribution [35].

We denote by ρ_n the density of a set of islands that contain exactly n particles, and ρ_1 the density of the monomers. Denoting by ρ the total island density excluding monomers, we have

$$\rho \equiv \sum_{n\geq 2} \rho_n. \tag{17.1}$$

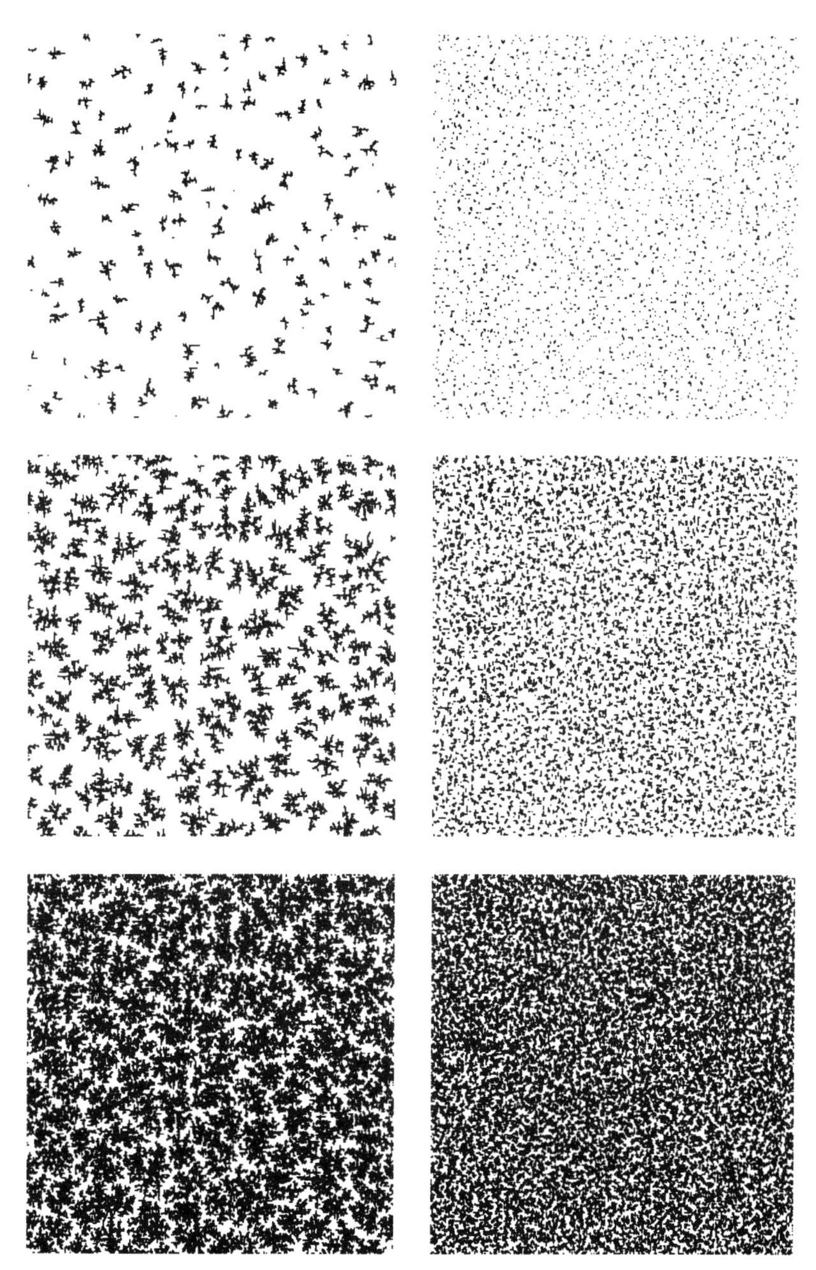

Figure 17.1 Island morphologies for different coverages and fluxes. The left column corresponds to $F = 10^{-7}$, and the right column to $F = 10^{-3}$. The coverages are: 5% (top), 20% (middle), and 50% (bottom). (Courtesy of P. Jensen).

The average island size is then

$$N \equiv \frac{\sum_{n\geq 2} n\rho_n}{\sum_{n\geq 2} \rho_n} = \frac{\Theta - \rho_1}{\rho}. \tag{17.2}$$

We shall assume that there is only one typical length scale in the problem, and shall take this to be the average size of a cluster, N.† Thus in general we can assume that the density depends on n only through the ratio n/N,

$$\rho_n = G(\Theta, N) f\left(\frac{n}{N}\right), \tag{17.3}$$

where f is called a *scaling function.* For the total coverage, we have $\Theta = \sum_{n\geq 1} n\rho_n = G(\Theta, N)N^2 \int du\, uf(u)$, which gives $G(\Theta, N) \sim \Theta/N^2$. Thus from (17.3), we obtain the scaling law

$$\rho_n = \frac{\Theta}{N^2} f\left(\frac{n}{N}\right). \tag{17.4}$$

17.2.2 Island formation and diffusion

At this point we require an argument to connect the typical length scale in the system (for instance, the typical distance between the islands) to the flux. To make progress, we use the properties of random walks on a plane [145, 465]. Denote by τ_1 the lifetime of a monomer in the DDA model. Then the monomer density is given by $\rho_1 = F\tau_1$, since only 'alive' monomers contribute. The distance traveled by the randomly moving monomer during its lifetime is $\ell_d^2 \sim D\tau_1$, so

$$\rho_1 \sim \ell_d^2 \frac{F}{D}. \tag{17.5}$$

Before a monomer is absorbed, it visits ℓ_d^2 sites. Hence the probability of being absorbed is $\rho_1 \ell_d^2$.

The rate of pair formation, $\mathscr{R}$, is proportional to ρ_1^2

$$\mathscr{R} \sim D\rho_1^2. \tag{17.6}$$

We can calculate the same quantity by considering that in the area ℓ_d^2 covered by the monomer during the time necessary to complete a full layer, there should be only one dimer formed, since there is only one island within the diffusion distance ℓ_d. The lifetime of the island to cover the surface is roughly $1/F$. Thus the rate of pair formation is

$$\mathscr{R} \sim F/\ell_d^2. \tag{17.7}$$

† We may as well consider the average distance between the center of two islands, or the typical diffusion length of a monomer. These quantities, although not equal to each other, scale in the same way.

Equations (17.5)–(17.7) lead to the characteristic length scale [145, 428, 429, 465, 464]

$$\ell_d \sim \left(\frac{D}{F}\right)^{\psi_d}, \tag{17.8}$$

with $\psi_d = 1/6$.

How reliable is the result (17.8)? Indeed, there are a number of corrections that must be considered.

(i) As visual inspection of Fig. 17.1 suggests, the islands in the DDA model are fractal. Their fractal dimension coincides with that of DLA, i.e. $d_f \approx 1.7$. We can include the fractality of the islands in the previous arguments by considering that there is only one pair formation in the area ℓ_d^2 during the time necessary to cover the area $\ell_d^{d_f}$, and this time is of order $\ell_d^{d_f-2}/F$. Thus we have $\mathcal{R} \sim F/\ell_d^{d_f}$, instead of (17.7), leading to

$$\psi_d = \frac{1}{4+d_f}, \tag{17.9}$$

which reduces to (17.8) for compact clusters, with $d_f = 2$.

(ii) In deriving (17.6), we assumed that the monomer visits ℓ_d^2 distinct sites in time τ. In fact, a particle undergoing Brownian motion visits a much smaller number of distinct sites during its motion, because there are sites that are visited more than once. This is a well known problem in statistical mechanics, and it is found that the number of sites visited is logarithmic in the diffusion time [327], leading to a logarithmic correction to (17.8). Hence (17.8) must be replaced by

$$\ell_d^6 \ln(\ell_d^2) \sim \frac{D}{F}. \tag{17.10}$$

(iii) In deriving (17.8), we assumed that dimers are stable, and islands do not decay during the deposition process. We also assume that islands do not diffuse. A more careful analysis, including, e.g., cluster diffusion, leads to [465]

$$\psi_d = \frac{n^*}{2+4n^*}, \tag{17.11}$$

where n^* is the size of the largest cluster that is considered to be mobile. For the DDA model, in which all clusters except monomers are immobile, we have $n^* = 1$ and (17.11) reduces to (17.8). However, if all clusters are mobile, i.e., $n^* \to \infty$, we obtain an exponent $1/4$.

17.3 Rate equations

For an ensemble of clusters that grow and decay, we can construct a series of rate equations that describe the time evolution of the cluster size distribution [10, 35, 408, 440, 455, 465]. Here we show how to apply these equations to study the DDA model. In general, we can write that the variations in time of the island and monomer densities are due to various coalescence processes, and have the general form

$$\frac{d\rho}{dt} = D\rho_1^2, \tag{17.12}$$

$$\frac{d\rho_1}{dt} = F - 2D\rho_1^2 - D\rho\rho_1. \tag{17.13}$$

Equation (17.12) states that the number of islands increases due to the fact that new islands are formed by monomer coalescence. Thus the growth rate is proportional to the probability that a monomer meets another monomer, which is $D\rho_1^2$. Equation (17.13) describes the variation in the number of monomers. The monomer density is fueled continuously by the deposition process, incorporated in the first term of the equation. However, the number decreases, due to dimer formation, with a rate proportional to ρ_1^2. The factor of two accounts for the fact that two monomers form a dimer. The second mechanism leading to a decay in ρ_1 is the capture of the monomers by islands. Thus the rate is proportional to the island density *and* monomer density.

In order to study these equations, it is convenient to write them in dimensionless units. To this end, we notice that D and F define typical length and time scales for monomer motion

$$\ell_1 = \left(\frac{D}{F}\right)^{1/4}, \qquad t_1 = \frac{1}{(DF)^{1/2}}. \tag{17.14}$$

Rewriting the rate equations using the dimensionless variables $\tilde{t} \equiv t/t_1$, $\tilde{\rho} \equiv \rho\ell_1^2$, and $\tilde{\rho}_1 \equiv \rho_1\ell_1^2$, we obtain

$$\frac{d\tilde{\rho}}{d\tilde{t}} = \tilde{\rho}_1^2 \tag{17.15}$$

$$\frac{d\tilde{\rho}_1}{d\tilde{t}} = 1 - 2\tilde{\rho}_1^2 - \tilde{\rho}\tilde{\rho}_1. \tag{17.16}$$

While there is no exact solution to these equations, we can obtain the scaling behavior using physically-realistic approximations:

(i) For sufficiently short times such that $\tilde{t} \ll 1$, the density of both monomers and islands is small, so the last two terms in (17.16) can be neglected compared with unity. This gives $\tilde{\rho}_1 \sim \tilde{t}$, which written in the original variables gives

$$\rho_1 \sim Ft. \tag{17.17}$$

From (17.15) we have $\tilde{\rho} \sim \tilde{t}^3$, which translates into

$$\rho \sim F^2 D t^3. \tag{17.18}$$

(ii) At larger times, the number of monomers decreases, the density of the islands being much larger. This is supported by (17.17) and (17.18), which show that ρ increases much faster than ρ_1. Thus we can take $\rho_1 < \rho$, and the second term in (17.16) can be neglected. These terms are relevant when they become comparable with the first term, i.e. when $\tilde{\rho}\tilde{\rho}_1 \approx 1$, so $\tilde{\rho}_1 \sim 1/\tilde{\rho}$. Inserting into (17.15) gives $\tilde{\rho} \sim (3\tilde{t})^{1/3}$, leading to

$$\tilde{\rho} \sim F^{2/3} D^{-1/3} t^{1/3}. \tag{17.19}$$

As we shall see in the next section, the existence of these two regimes is supported by numerical simulations on the DDA model. In fact, there is a third regime, which the rate equations cannot account for: that of island coalescence, which eventually leads to a rapid decrease in the number of islands. This leads us to a discussion of what exactly is included into the rate equations, and what processes are neglected. At late times or large coverages the islands grow not only by capturing monomers diffusing on the surface, but many atoms land directly on the clusters. In (17.12)–(17.13) we neglect the fact that the islands grow, and they will cover the surface of crystal. The rate equations assume that the islands are point objects, which is a good approximation if the distance between them is much larger than their typical size. Thus they are valid only for the early times of the deposition process.

17.4 Results from simulations

Recently there has been a large number of relevant numerical results, many for models that differ only slightly from the DDA model [10, 35, 196, 198, 324, 440]. We show in Fig. 17.2 the variation of the monomer and island densities as functions of coverage, and find four distinct regimes:

- *Regime L:* At early times, the coverage and the typical island size are both small, so the predictions of the rate equations (17.12) and (17.13) should be valid. From (17.17), we see that there is a linear increase in the monomer density, while the island density increases as t^3. This regime is expected to hold if $t \ll (DF)^{-1/2}$. Indeed, as Fig. 17.2 illustrates, for early times we can see a much faster increase in ρ than ρ_1.

- *Regime I:* When the density of islands becomes comparable with the monomer density, the rate equations predict a slowing down in the increase of the island density. More precisely, we found in (17.19) that the *island* density ρ increases as $t^{1/3}$, while the *monomer* density ρ_1 decreases as $t^{-1/3}$. This indeed can be observed in Fig. 17.2.
- *Regime A:* As the size of the islands becomes comparable with the distance between them, the rate equations cannot tell us about the scaling properties of the system. The scaling behavior in this regime is quite complicated. The onset of this regime is signaled by a rapid decrease in the monomer density, and a plateau of the island density. In this regime, we observe a fattening of the islands by capturing the diffusing monomers, without further island creation. However, as the islands grow, they coalesce, and eventually percolate across the entire system.
- *Regime C:* Finally, the *number* of islands decreases drastically, since they will form a single huge cluster. The coverage approaches a full layer, and cluster formation on the top of this layer becomes relevant.

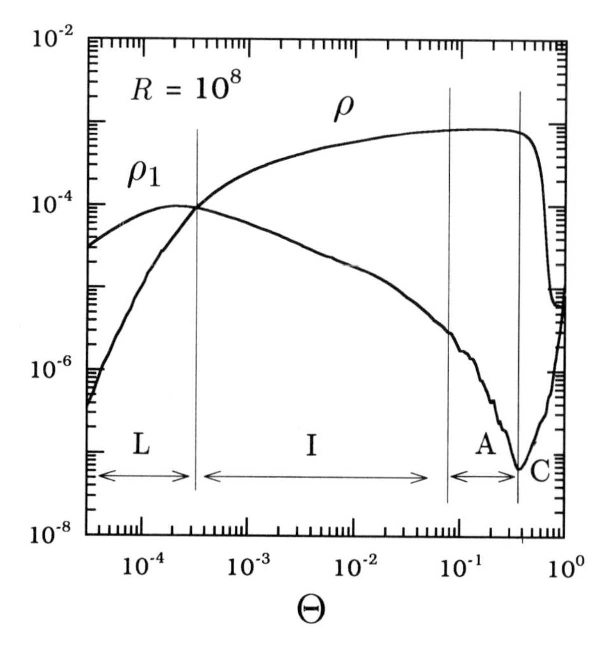

Figure 17.2 The island and monomer densities ρ and ρ_1 as functions of the coverage for $F/D = 10^8$. The regimes denoted by L and I correspond to the early time behavior. A and C denote the late time regimes. (After [10]).

The islands in regimes L and I resemble DLA (see §19.1). We assume that in these two regimes, deposition inside the cluster is negligible. Particles are deposited outside the clusters and diffuse randomly. If they reach the cluster, they stick. Thus the mechanism leading to the growth of the clusters is the same as for DLA. To obtain quantitative support for this observation, we can measure the fractal dimension of the clusters. Indeed, for early times $d_f \approx 1.7$, as is known to be the case for DLA [440, 196]. As clusters grow, particle deposition inside the cluster becomes relevant. As Fig. 17.1 illustrates, the clusters at late times become more and more compact, and measurements show that their fractal dimension increases toward the value $d_f = 2$.

17.5 Extensions of the DDA model

In this section, we discuss two extensions of the DDA model, one with activated diffusion and the other with cluster diffusion. As we show, the new ingredients enrich the scaling behavior and the observed morphologies.

17.5.1 DDA with activated diffusion

One of the deficiences of the DDA model is that it does not allow for detachment of atoms from the islands – the aggregation process is *irreversible*. This approximation is a good one at low temperatures. For real atoms, there is always a finite probability that a monomer can detach from the edge of an island, or that a dimer can break; increasing the temperature increases the probability of these break-away processes. Recently improved versions of the DDA model were considered that include the reversible nature of the microscopic processes [145, 378].

In the DDA model with activated diffusion, the unconditioned diffusion and sticking process is replaced by an activated one (see §12.2.3 and §15.2). The diffusion probability depends on the energy $E = E_0 + nE_N$, where E_N is the energy of the bond formed by the atom with its n nearest neighbors (so $n = 0, 1, 2, 3, 4$). The probability of a monomer moving is fixed by E_0, but if the monomer attaches to the edge of an existing island, forming, e.g., one bond, it must overcome the energy barrier of magnitude $E_0 + E_N$, so the probability of moving away decreases. The model reproduces the DDA model for low temperatures: in this regime once a bond is formed, it is unlikely that it will be broken, so all attachments are almost irreversible. However, at higher temperatures the atoms will have time to rearrange on the edge of the island, increasing the number of neighbors. This process results in more compact clusters.

This transition is illustrated in Fig. 17.3, where not the temperature but rather E_N is varied. The result is similar to changing the temperature. If E_N is large, the attachment to the island of a diffusing atom is almost irreversible, and the islands have a fractal structure. However, decreasing E_N increases the probability of an atom detaching from the island or diffusing along the island perimeter. Hence, the islands become non-fractal, acquiring a compact shape. While this model is quite different from the DDA model, especially in the small E_N region, it does show scaling properties that are similar to those of the DDA model. In particular the scaling law (17.4) is obeyed for a range of

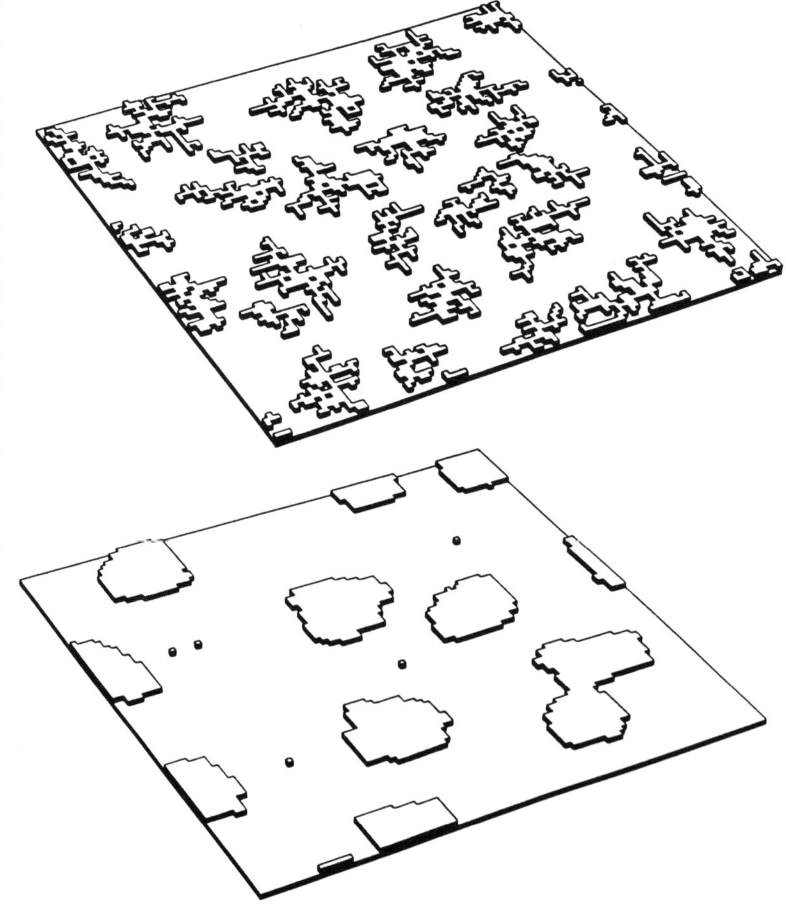

Figure 17.3 Typical island morphology at coverage $\Theta = 0.2$ for a 100×100 section of a 400×400 system obtained using Monte Carlo simulations with activated diffusion. For the top panel $E_N = 1.0$ eV, while for the bottom panel $E_N = 0.3$ eV. Note the transition from fractal to compact clusters as the bond energy decreases. (After [378]).

values of E_N, and the scaling of the monomer and island densities can be described by appropriately-modified rate equations.

While activated diffusion is probably the most natural way to generate the compact-to-fractal transition, other possibilities – such as including diffusion *only* along the edge of the islands – can also account for this transition [36].

17.5.2 DDA with cluster diffusion

The DDA model assumes that all clusters are immobile, and only monomers diffuse. However, there is experimental evidence that dimers, trimers, etc. can also diffuse. The probability for dimer diffusion is lower than for monomer diffusion. In general, the larger the cluster, the smaller its chance of diffusion. So if we wish to model cluster diffusion, we must include this dependence of mobility on cluster mass.

There are two qualitatively different cluster diffusion processes. First, there is *rigid* diffusion, for which all atoms of the cluster move at the same time, without altering the shape of the cluster. This is possible only for small clusters, such as dimers or trimers. Second, there is an effective diffusion of the entire cluster which occurs by migration of atoms along the edge of an island, which results in a mobility that is assumed to vary as [429]

$$D_n \sim 1/n^{\gamma}, \tag{17.20}$$

where n is the number of atoms comprising the island. Here we discuss a model that includes *rigid* diffusion†, but the diffusion constants obey (17.20) [196, 197].

We shall focus on the scaling in the DDA model with cluster diffusion at the percolation threshold,‡ which is reached when a large enough cluster develops to span the entire system. At the percolation threshold, there are *three* characteristic regimes, delimited by *two* crossover length scales ℓ_d and ℓ_2:

- *Regime I ($L < \ell_d$): 'Particle Diffusion Regime'.* In this regime, only one cluster is nucleated in the system (see Fig. 17.4). Since the

† Naturally, rigid diffusion of large clusters cannot take place in MBE. However, the method is an useful numerical approximation for understanding the effect of cluster diffusion on the scaling properties of the system. Moreover, rigid diffusion may be relevant to three-dimensional aggregation processes.

‡ A number of experimental techniques are useful for studying the properties of submonolayer films at the percolation transition, where one obtains a cluster that spans the system. In the DDA model this happens only in regime (c) of §17.4.

characteristic diffusion length of a single particle ℓ_d is larger than the system size L, every deposited particle attaches to the already existing cluster before the next particle is deposited.

- *Regime II* ($\ell_d < L < \ell_2$): *'Cluster Diffusion Regime'.* Now several clusters are nucleated, as can be seen in Fig. 17.4. The reason is that ℓ_d is now smaller than the system size. The spanning cluster is built mainly by the aggregation of the diffusing nucleating clusters.
- *Regime III* ($L > \ell_2$): *'Percolation Regime'.* At short times, many clusters are present in the system, and they are separated, as in regime II, by a distance ℓ_d. At the spanning time the system resembles a percolation network (see Fig. 17.4). The fractal dimension of the clusters is close to 1.9, corresponding to the value of percolation clusters.

A change of the flux affects the values of the two crossover lengths ℓ_d and ℓ_2. The crossover between regime I and II is given by (17.8) with $\psi_d = 1/4$, which is in agreement with the predictions (17.11) for diffusing clusters obtained if we choose $n^* = \infty$.

The second length scale, ℓ_2, also scales with the flux,

$$\ell_2 \sim \left(\frac{D}{F}\right)^{\psi_2}, \tag{17.21}$$

where $\psi_2 = 0.9 \pm 0.2$ for $\gamma = 1$. ℓ_2 can be interpreted as the length

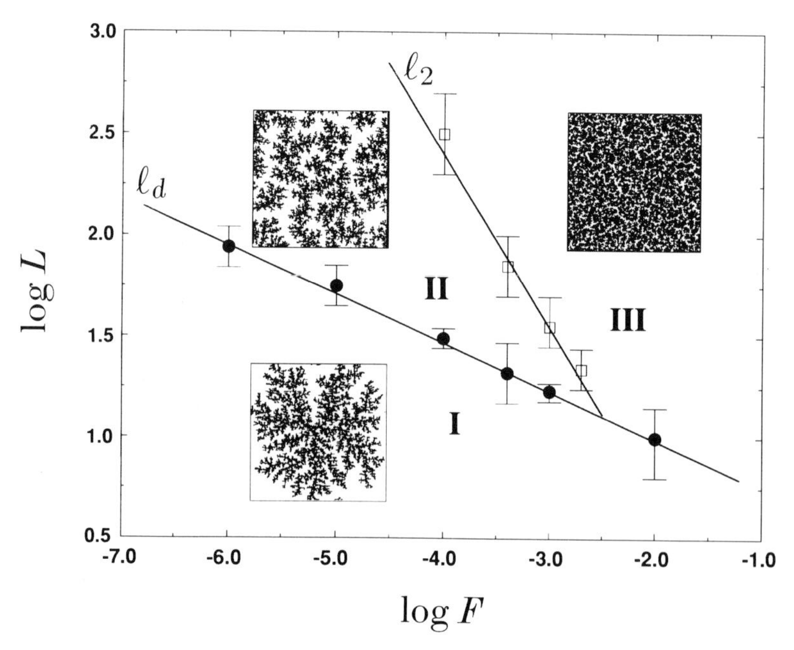

Figure 17.4 Morphology phase diagram for $\gamma = 1$ in the variables L and F. Shown is the dependence on the flux of the two length scales ℓ_d (•) and ℓ_2 (□). The lines separating the three regimes have been obtained by linear fits of the data for ℓ_d (slope 0.24 ± 0.02) and ℓ_2 (slope 0.9 ± 0.2). The insets show the typical morphologies at the percolation threshold for each of the three regimes. (After [197]).

scale determined by the competition between particle deposition and *cluster* diffusion. This analysis is confirmed by the fact that the second crossover ℓ_2 is not observed as $\gamma \to \infty$, so that only single particles are allowed to move.

We may now construct a 'morphology phase diagram' that serves to characterize the morphology of the system at the spanning time in terms of the two tuning parameters L and F (Fig. 17.4). The three regimes, I–III, are delineated by the two crossover lines $\ell_d(F)$ and $\ell_2(F)$, which intersect at a 'triple point' whose coordinates, (F_c, L_c), depend only on the diffusion exponent γ. Thus for a fixed system size L, two situations can arise, depending on the value of γ. (i) If $L << L_c(\gamma)$, then the system shows a direct transition from the single cluster regime to percolation as the normalized flux F increases. (ii) If $L >> L_c(\gamma)$, then regime II can also be observed for intermediate values of the normalized flux.

17.6 Experimental results

Island formation is a basic process in MBE and thin film growth. Thus unveiling the details of submonolayer deposition is a very important task, with numerous practical applications. With the availability of the STM, it is possible to observe directly the details of island formation. Lately, there have been a large number of observations that serve to provide qualitative support for the morphologies and mechanisms described in the previous sections.

Some versions of the arguments regarding the dependence of the diffusion length and island distribution on flux, temperature, and diffusion constant have been long used to determine the activation energy and the diffusion coefficient of monomers [319, 342]. However, these studies mostly considered a simple version of the scaling arguments discussed in the previous sections, arriving at the conclusion that the characteristic length scale in the system depends on the flux as $F^{-1/4}$. As we saw in §17.5, this is true only under very special circumstances – e.g., if island diffusion is relevant.

Compact or fractal islands have been observed in various systems, the actual island morphology depending on temperature, deposition rate, and binding energies. Studies include Fe deposition on Fe(001) [430], Ni on Ni(100) [236], Cu on Cu(100) [116, 508], Ag on Pt(111) [380], Ag on Ni(100), Si on Si(001) [324], Sb cluster deposition on amorphous carbon [200].

One of the first studies leading to the direct imaging of the fractal submonolayer island structures has been carried out by observing Au deposition on Ru(001) surfaces (and later extended to Cu on Ru(0001)) [192, 193]. The deposition rate varies between 0.2 and 2 ML/min, and the experiments are done at room temperature. Fig. 17.5 shows the morphology of the islands for various coverages obtained for a flux of 1.8 ML/min. For small coverages, the islands are compact, but they have a few branches.

The compactness of the islands indicates that the aggregation process is not completely irreversible, but there is some diffusion taking place along the edge of the islands, which smooths out the perimeter at short length scales. However, the appearance of branches indicates that edge diffusion is limited and is not able to overcome the instability generated by particle diffusion. At increasing coverages the fractal structure of the islands becomes apparent. The main virtue of the results presented in Fig. 17.5 is that they give qualitative support to the relevance of the DDA model for real systems, and thus confirm the picture presented in the previous sections: at low coverages we see a few islands that take on a DLA-like shape as the coverage increases. Further increase in the coverage results in the 'fattening' of these islands, which eventually coalesce at high coverage (see Fig. 17.1). It was possible to measure directly the fractal dimension of the structures from the digitized pictures. The result, $d_f = 1.72 \pm 0.07$, is in good

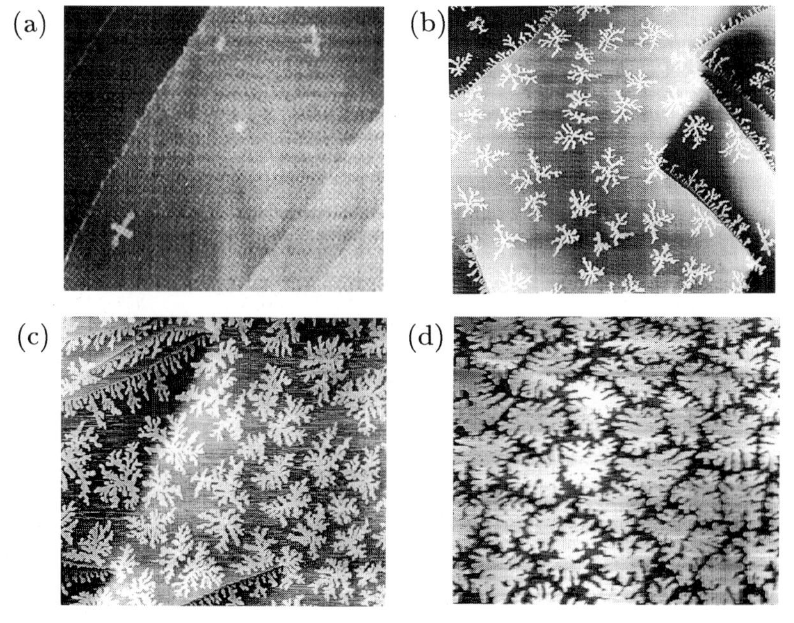

Figure 17.5 STM images of four films (Au on Ru(0001)) of varying coverages deposited at room temperature with a flux 1.8 ML/min. The coverages and the size of the shown sections are (a) Θ=0.03 ML (0.70 $\times$ 0.60 μm), (b) Θ=0.15 ML (1.31 $\times$ 1.20 μm), (c) Θ=0.37 ML (1.13 $\times$ 0.96 μm), (d) Θ=0.69 ML (1.32 $\times$ 1.13 μm). (After [193]).

agreement with the fractal dimension of DLA, as predicted by the DDA model [302].

As noted in §17.5, the shape of the islands depends on the temperature. At low temperatures, when nonequilibrium effects dominate (in the sense that edge diffusion and atom detachment is limited), the islands have a fractal structure. However, on increasing the temperature, the islands become more compact and assume their equilibrium shapes. This process is illustrated in Fig. 17.6, where we reproduce results obtained for Pt deposition on Pt(111) surfaces.

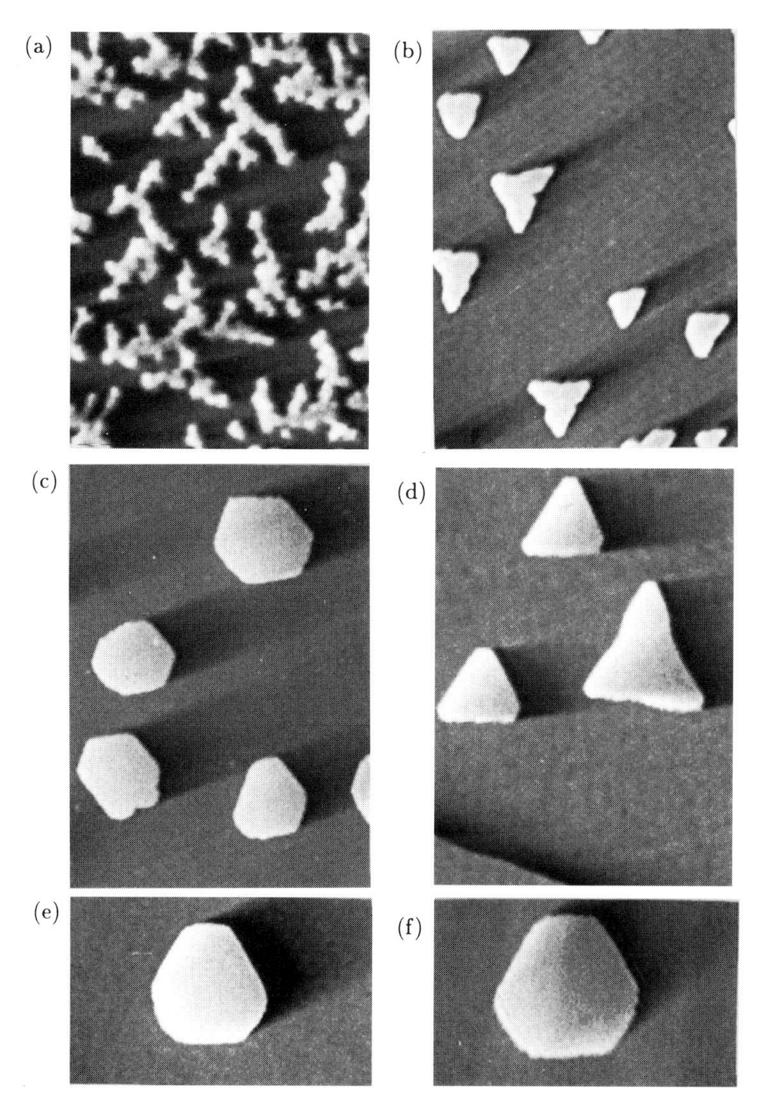

Figure 17.6 Island shapes on Pt(111) resulting at various surface temperatures T after deposition of an amount Θ with a a typical rate of 10^{-2} ML/s. The temperature, coverage and the size of the segments shown are (a) 200 K, Θ=0.2 ML, 280 × 400 Å; (b) 400 K, Θ=0.08 ML, 1300 × 1900 Å; (c) 455 K, Θ=0.14 ML, 770 × 1100 Å; (d) 640 K, Θ=0.15 ML, 2300 × 3300 Å; (e) 710 K, Θ=0.08 ML, 1540 × 1100 Å; (f) after deposition at 425 K, Θ=0.08 ML the sample was *additionally annealed* to 710 K for 1 min and then imaged (630 × 900 Å). (After [318]).

However, a number of quantitative measurements provide data for a more critical comparison. The first results were obtained during Si deposition on Si(001) surfaces. The measurement of the island density as a function of the temperature allows for the determination of the monomer diffusion constant [324, 326, 371].

Another set of experiments was carried out using Cu deposition on Cu(100) surfaces, and the interface was studied using HRLEED. One of the quantities investigated was the island separation, ℓ_d, as a function of the deposition rate F at a constant coverage Θ (see Fig. 17.7). At $T = 223$ K for both coverages, we see two different scaling regimes (Fig. 17.7(a)). For small flux, $\psi_d = 1/6$, in accord with the prediction (17.8). However, larger fluxes suggest $\psi_d = 1/4$. While the first result indicates that monomers are stable, corresponding to $n^* = 1$ in (17.11), the second one corresponds to island diffusion, which is unlikely for this system for the considered temperatures. The exact origin of ψ_d is still unknown. For $T = 263$ K, shown in Fig. 17.7(b), the

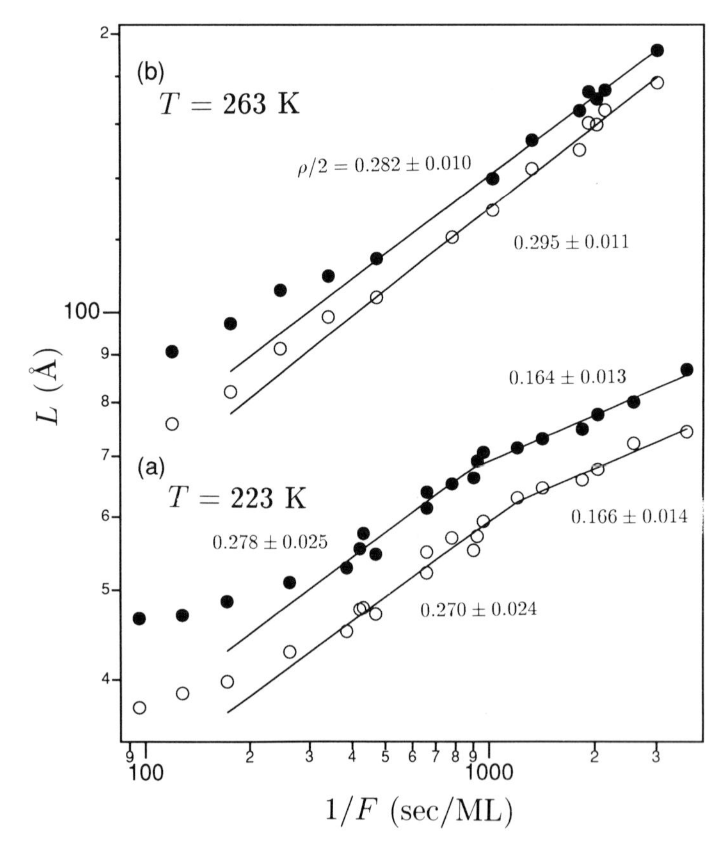

Figure 17.7 Island separation *vs.* inverse flux $1/F$ for Cu deposition on Cu(100). The slopes of the fitted lines are given in the figure. In (a), the substrate temperature is 223 K and the solid circles correspond to Θ=0.7 ML. In (b), the corresponding numbers are 263 K and 0.63 ML. Open circles correspond to Θ=0.3 ML. [508]).

scaling is consistent with $\psi_d = 3/10$, which is predicted by (17.11) with $n^* = 3$. This means that the smallest stable island at this temperature is the tetramer, and dimers could become unstable and dissociate.

The results presented thus far, while providing strong support for the theoretical considerations of the previous sections, are not comprehensive. More work is required to check the limits of the theoretical predictions, and the increasing experimental possibilities are expected to lead to even more detailed results in the near future.

17.7 Discussion

If we wish to understand the formation and morphology of rough interfaces, first we must have a correct picture concerning the early-time behavior. Thus the dynamics of island formation is an essential first step toward a complete understanding of deposition processes. In this chapter, we have presented the basic questions and the most important experimental and theoretical tools involved in these studies. However, we must emphasize that this is a new and rapidly-evolving field, so the future may bring significant progress. Even at this stage, we have an understanding of the basic processes involved in shaping island morphology and the most important quantities needed to be measured during these growth processes. Thus a basic picture that is supported by the available experimental investigations is beginning to arise.

Suggested further reading:

[10, 199, 455, 465]

Exercises:

17.1 Consider the one-dimensional version of the DDA model. Calculate ℓ_d. Write down the rate equations and try to obtain the various scaling regimes. Discuss the relevant differences between the one dimensional and two dimensional models.

17.2 Generalize the above results for arbitrary dimensions.

18 The roughening transition

The primary focus of this book is on interface roughening generated by various nonequilibrium deposition processes. However, crystal surfaces may be rough even under equilibrium conditions – with no atom deposition. Consider, e.g., a flat surface in equilibrium at a very low temperature. Thermal fluctuations do not have an observable effect on the shape of the crystal, and all atoms remain in their appropriate lattice positions. As temperature increases, the probability that an atom will break its bonds with its neighbors increases. Some atoms hop onto neighboring sites, thereby generating roughness on the atomic scale. At first glance, one might expect a gradual transition to a rough morphology, since the higher the temperature, the more the atoms wander on the surface – until the surface melts. Indeed this is a correct description of the short-ranged correlations between neighboring atoms. However, as we shall see, on much longer length scales there is a distinct (higher order) thermodynamic phase transition that takes place at a critical temperature T_R. For $T < T_R$ the crystal is *smooth*, corresponding to a flat facet, while for $T > T_R$ it is *rough*, implying a rounded crystal shape analogous to a liquid drop. This 'roughening transition' can be successfully described using ideas of statistical mechanics, the formalism being analogous to that developed in previous chapters.

18.1 Equilibrium fluctuations

The lowest energy state of the crystal corresponds to a flat surface. Fluctuations, which lead to broken bonds and displaced atoms, increase the total energy and the total surface area. The energy of a two-dimensional surface is proportional to the total area

$$\mathcal{H} = \nu \int d^2x \sqrt{1 + (\nabla h)^2}. \tag{18.1}$$

Here ν is the free energy in the absence of any fluctuations, and is called the 'stiffness,' or 'surface tension.' Assume that the surface is not too rough – i.e., that $|\nabla h| \ll 1$ everywhere. Then (18.1) can be expanded, giving

$$\mathcal{H} = \nu L^2 + \frac{1}{2}\nu \int d^2x \, (\nabla h)^2, \tag{18.2}$$

where L is the linear size of the system. Equation (18.2) describes the fluctuations of a *free* surface; it is the same as the Hamiltonian (C.1) describing the stationary properties of the EW equation. The saturated system (which corresponds to the equilibrium properties of the Hamiltonian) is rough at any temperature, the roughness increasing logarithmically with the system size for the two-dimensional interface. Thus (18.2) alone cannot account for the roughening transition mentioned in the previous section.

The most important difference between (18.2) and a real crystal is that (18.2) describes a *continuous* system, while the surface of a crystal is *discrete*, since it is made from discrete atoms. The height of a crystal does not change continuously, but in jumps equal to the lattice constant of the system, since the increase in the height can occur only by adding or removing particles from the surface. A simple way to account for this is to add a *lattice potential*, $V(h)$, to the energy (18.2) that discourages the continuous variation of the height and forces the interface to grow in discrete steps. To mimic the periodicity of the lattice, $V(h)$ must be periodic in h, the period being the lattice constant.

If we want to describe real crystals, we must choose a delta function as a potential that forbids any height but the discrete values fixed by the lattice. However, we obtain a mathematically more attractive model if we expand the periodic potential in Fourier components, retain its lowest order term, and set

$$V(h) = -V_0 \cos\left(\frac{2\pi h}{a_0}\right), \tag{18.3}$$

where a_0 is the lattice constant.

Thus the Hamiltonian describing the fluctuations of a discrete interface is

$$\mathcal{H} = \int d^2x \left[\frac{1}{2}\nu(\nabla h)^2 - V_0 \cos\left(\frac{2\pi h}{a_0}\right)\right], \tag{18.4}$$

where we neglected the constant νL^2 term in (18.2). A schematic representation of the difference between the interface described by the free Hamiltonian (obtained for $V_0 = 0$) and the full (18.4) is shown in Fig. 18.1. For $V_0 = 0$ the surface is rough. However, for $V_0 \neq 0$, every height that is not an integer multiple of a_0 has a nonzero additive energy. Thus the interface wants to 'lock in' to the discrete h values. Suppose that at point x the height is $h(x)$, and at $x+1$ is $h(x)+a_0$. Then the second term of the energy has a minimum, but between the two points there is a height difference a_0, leading to a nonzero gradient; this is again an energetically unfavorable configuration. Thus the first term in (18.4) favors keeping the surface smooth and the second acts to lock it into discrete lattice positions. The interface morphology for which (18.4) has a minimum corresponds to a flat interface whose height is an integer multiple of a_0.

18.1.1 Renormalization group treatment

The equilibrium properties of the system described by the Hamiltonian (18.4) can be studied using the RG method [265, 352]. The periodic potential acts as the nonlinear term in the KPZ equation, introducing a coupling between the different modes. However, at this point we are dealing with an equilibrium problem, so we need not use the dynamic RG formalism. Rather, Wilson's equilibrium RG is applicable, which begins by considering the partition function

$$Z = \int \mathcal{D}h \exp\left[-\frac{\mathcal{H}(h)}{k_B T}\right]. \tag{18.5}$$

The calculation is carried out using Fourier components

$$Z = \prod_{k=0}^{\Lambda} \int dh_k \exp\left(-\frac{1}{k_B T}\left[\sum_k \frac{1}{2}\nu k^2 |h_k|^2 + V(h_k)\right]\right). \tag{18.6}$$

Here Λ is the upper cutoff in k, corresponding to a small length scale.

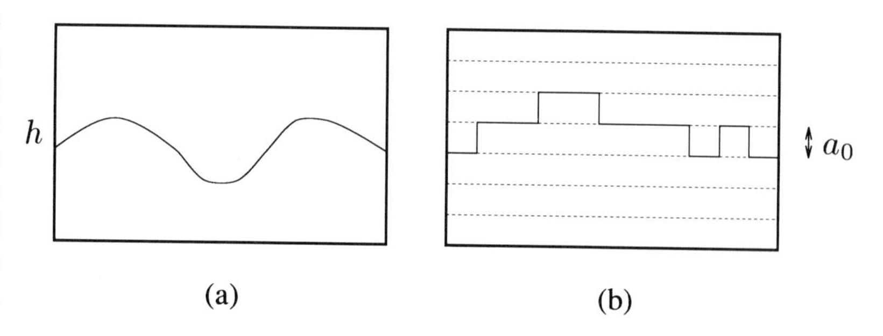

Figure 18.1 The difference between the interface described by the free Hamiltonian (18.2) and the crystal surface described by (18.4). (a) The height of the interface described by the free Hamiltonian (18.2) varies continuously. (b) The heights of the interface described by (18.4) are discrete heights, separated by the lattice constant a_0. The dotted lines correspond to the heights where the lattice potential has a minimum. The interface configurations of (a) and (b) will look much the same on longer length scales since the individual layers cannot be resolved.

The conceptual steps of the RG were described in Chapter 7:

- *Step 1:* Divide the fluctuations into two classes, the long wavelength modes, $k^< \in [0, \Lambda/b]$, and the short wavelength fluctuations, $k^> \in [\Lambda/b, \Lambda]$, and integrate over the short wavelength fluctuations.
- *Step 2:* Rescale the coordinates with $b = e^l$, such that the final cutoff will be the same as the original one.

The flow equations have the form

$$\frac{dy}{dl} = 2y\frac{x-1}{x}, \tag{18.7}$$

$$\frac{dx}{dl} = \frac{y^2}{2x}A\left(\frac{2}{x}\right). \tag{18.8}$$

Here

$$x = \frac{2\nu a_0^2}{\pi k_B T}, \qquad y = \frac{4\pi V_0}{k_B T \Lambda^2}, \tag{18.9}$$

and $A(z)$ is a rather complicated function of its argument; A is always positive, and its numerical value is $A(2) = 0.398$. The flow equations (18.7)–(18.8) have an impressive history, and they appear in many branches of physics. They originally were derived by Kosterlitz and Thouless for the XY-model [237].

In the following we analyze (18.7)–(18.8), in order to draw conclusions regarding the roughening transition of the system. First, we notice that (18.7) has a fixed point for $x = 1$, which defines the roughening temperature†

$$T_R = \frac{2\nu a_0^2}{\pi k_B}. \tag{18.10}$$

To see that T_R indeed separates two thermodynamically different phases, we calculate the flow near the critical point. In the vicinity of the critical point, $x = 1$, we assume that A varies slowly and that it is equal to $A(2)$. We have two qualitatively different flow trajectories, as shown in Fig. 18.2:

(a) If $T < T_R$, the rhs of (18.7) is positive, so y increases. As is shown in Fig. 18.2, under successive renormalization y diverges to infinity. Physically this means that the lattice potential (which is proportional to y according to (18.9)) is renormalized to a very large value, and at large length scales dominates over the surface tension. In this regime the interface is flat.

† We note that (18.8) is really an equation for the renormalization of ν. The ν appearing in (18.10) is the long wavelength or macroscopic $\nu(\infty)$ and not the bare ν that is given in (18.6).

(b) For $T > T_R$ the rhs of (18.7) is negative. Under successive renormalization y decreases to zero. The lattice potential V_0 *vanishes*, and thus the surface is described by the free Hamiltonian (18.2). With $y = V_0 = 0$, we have a logarithmically rough interface.

Thus we see that the system shows a true phase transition from a smooth phase ($T < T_R$) to a rough one ($T > T_R$). We call it the *roughening transition*, since it separates a smooth and a rough phase; in other fields it is sometimes called the Kosterlitz–Thouless phase transition.

The analytical form of the separatrix separating the two phases in phase space (see Fig. 18.2) can be calculated by dividing (18.7) by (18.8). This produces the differential equation for the trajectory $y(x)$

$$\frac{dy}{dx} = \frac{x-1}{y} \frac{4}{A(2/x)}. \tag{18.11}$$

This can be integrated in the vicinity of $x = 1$, where we can assume that $A(2/x)$ is constant and equal to $A(2)$. We thereby obtain the trajectory

$$y^2 = \frac{4\,(x-1)^2}{A} + C, \tag{18.12}$$

where C is an integration constant.

It is useful to examine an interface that has its 'bare' parameters (v, V_0 and Λ) fixed. In the parametrization (18.9), x/y is independent of temperature and depends only on the bare parameters. Thus an interface with fixed bare parameters corresponds to a straight line $y = C'x$, with $C' = 2\pi^2 V_0 / v\Lambda^2 a_0^2$, in the flow diagram of Fig. 18.2. The

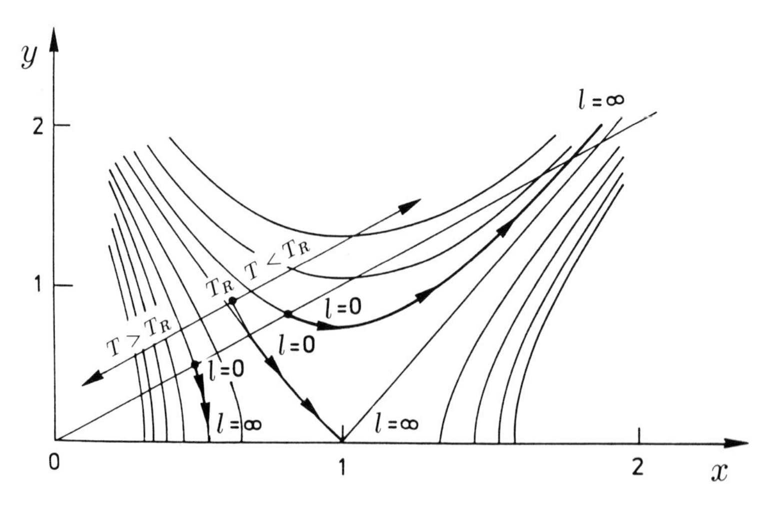

Figure 18.2 The RG flow diagram near the roughening transition. The separatrix, starting from $x = 1$, delimits the two phases: one for $T < T_R$ and the other for $T > T_R$. The oblique line starting from $x = y = 0$ corresponds to states sharing the same bare parameters, the different points of the line corresponding to different temperatures. Starting from points near this line, the flow may take us either to $y = 0$ or $y = \infty$, depending on which side of the separatrix we begin. (After [265]).

different points of the straight line correspond to states with differing temperatures. The crossing point of this line with the separatrix defines T_R.

We can follow various trajectories on the flow diagram. If we start with a temperature smaller than T_R, we see that y diverges; the interface is dominated by the lattice potential and is therefore flat. When we increase the temperature (which corresponds to moving to the left on the line $y = C'x$), our starting point moves to the $T < T_R$ side of the separatrix, where the flow takes us to $y = 0$ and to some nonzero x, corresponding to a rough surface. Thus, if we slowly increase the temperature, we should observe a phase transition at T_R.

18.1.2 Correlation length

The existence of a critical temperature implies the existence of correlations in the system. As the flow diagram of Fig. 18.2 indicates, if we start with arbitrary values of the bare parameters x and y, the flow will take them to their macroscopic values. When we start with a small y and $T < T_R$, y diverges – but only after an infinite number of successive renormalizations. The RG flow corresponds to a successive coarsening of the system, i.e., viewing the interface from large length scales.

Thus we come to the main question: what is the length scale at which y begins to dominate the behavior – i.e., the length scale at which the lattice potential determines the morphology of the interface? Reminding ourselves that the parameter l in flow equations (18.7)–(18.8) is related to the length scale of the system, we can estimate the value l_c at which the lattice parameter y becomes relevant. There is no firm criterion that can be chosen to be the threshold of the relevance of y, so we simply choose $y = 1$. Assuming that at l_c the lattice potential y becomes one, the renormalized cutoff, $\tilde{\Lambda} \equiv \Lambda/b$, defines a length scale which we interpret as the correlation length

$$\xi_{\parallel} \sim \frac{1}{\tilde{\Lambda}} \sim \frac{1}{\Lambda/b} \sim \frac{\exp(l_c)}{\Lambda}. \tag{18.13}$$

For length scales smaller than the correlation length ($l < \xi_{\parallel}$) the lattice potential is not built up yet ($y < 1$), and the surface is free – it is 'logarithmically rough,' controlled only by the surface tension. However, for $l \gg \xi_{\parallel}$ the lattice potential takes over, and the interface is smooth. Near the critical point [352]

$$\xi_{\parallel} \sim \exp\left[\frac{B}{\sqrt{T_R - T}}\right], \tag{18.14}$$

where B is a temperature-dependent constant. According to (18.14), $\xi_\parallel$ diverges as we approach the roughening transition.

Our next question concerns the surface roughness in the two different phases. We have seen that when $T > T_R$ the lattice potential vanishes. Thus it is only the system size that limits the increase in the fluctuations. The width of the interface in this regime increases slowly with L, as $w \sim \log L$.

We demonstrated above that the interface is also rough below T_R, and that it becomes flat only when the length scale is larger than $\xi_\parallel$. The role of the system size in limiting the interface width is controlled by the correlation length, $\xi_\parallel$. Hence $w \sim [\log(\xi_\parallel)]^{1/2}$ which, according to (18.14), diverges as $w \sim 1/(T - T_R)^{1/4}$.

While we think of the roughening transition as a smooth-to-rough transition, it may be difficult to see any change in the interface morphology at T_R (see Figs. 18.3 and 18.4). The reason for this is the diverging interface width at T_R. The smoothness of the interface for $T < T_R$ is a *large-scale* property, and the interface is smooth only for length scales larger than $\xi_\parallel$. As $T \to T_R$, $\xi_\parallel \to \infty$. Thus the transition may appear to be a gradual transition from a smooth to a rough interface, and only explicit calculation can identify the actual phase transition.

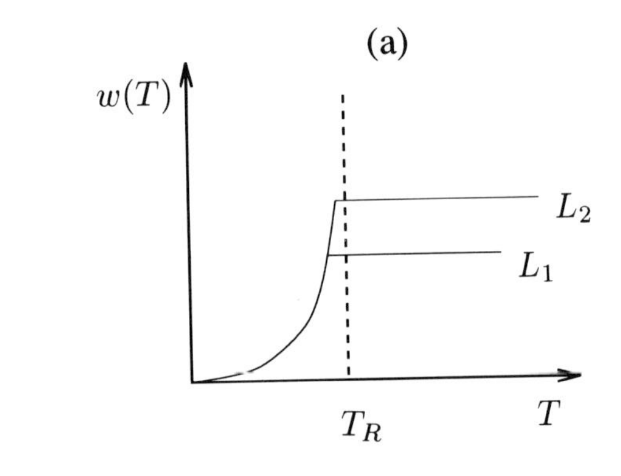

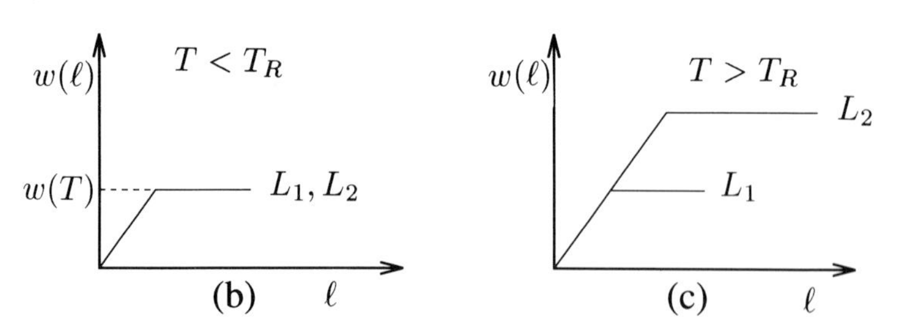

Figure 18.3 There is an important difference between the roughness for $T > T_R$ and for $T < T_R$. (a) The divergence of the width near T_R is limited by the system size, L. (b) Below T_R the roughness is independent of L, but depends on the temperature. (c) Above T_R, the saturated roughness increases as $\log L$.

18.2 Discrete models and experimental tests

The roughening transition described above has been confirmed numerically, and there is by now a large amount of experimental data indicating its relevance to real crystals. In this section we discuss some of the most important results.

18.2.1 Results from simulations

Most numerical results on the roughening transition were obtained using models of the solid-on-solid type, similar to the model discussed in Chapter 8. The interface is characterized by columns whose height h can take only integer values, the discreteness corresponding to the lattice potential in (18.4). The energy of the interface is given by

$$E = \frac{J}{2} \sum_j f(h_j - h_{j+\delta}), \tag{18.15}$$

where δ runs over the nearest neighbors of site j, and the sum is over all sites j of the interface. The function $f(x)$ can have two generic forms: $f(x) = |x|$ for what is termed the *absolute SOS model* and $f(x) = x^2$ for the *discrete Gaussian model.* The lowest energy of the

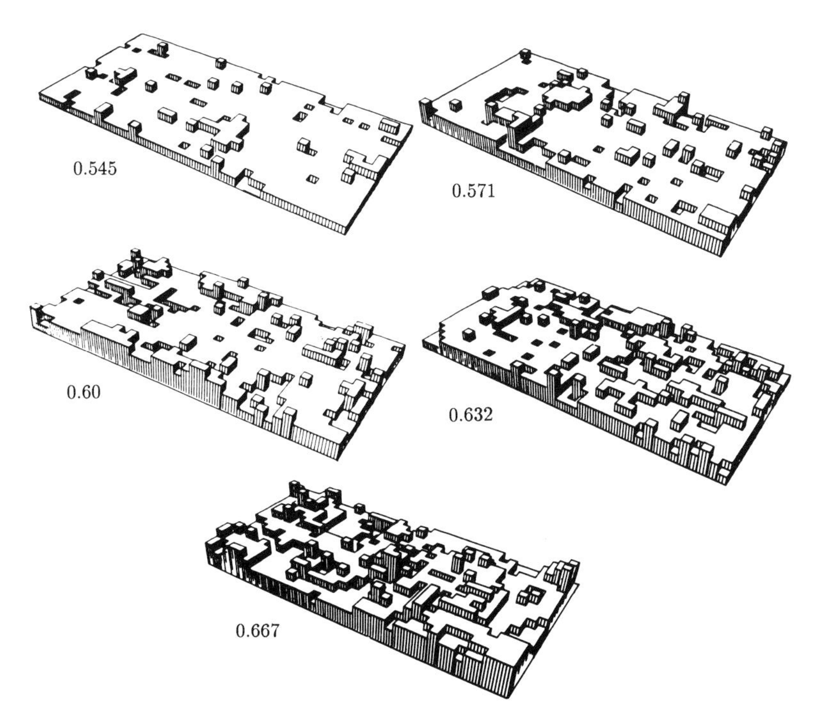

Figure 18.4 Morphology of the interface of the SOS model for various temperatures near the roughening transition T_R. The numbers near the figures represent the temperature in the units of the coupling constant, T/J. The roughening transition is at $T_R/J = 0.62$. (After [475]).

surface corresponds to a flat interface, and every height difference between the neighboring columns leads to energy penalties.

How can we identify the roughening transition in such a system? We saw in the previous section that the interface roughness diverges at T_R. To identify the correct roughening temperature, the height–height correlation function is commonly studied. However, a number of related quantities also show singularities at the roughening transition [475, 476]. One of them is the difference, Δc, in the concentration between the surface layer and the layer above it. For $T < T_R$, there are almost no particles in the second layer, so Δc is large. However, as the interface becomes rough approaching the roughening transition, Δc decreases and eventually vanishes at T_R (see Fig. 18.5). Also plotted in Fig. 18.5 is the inverse of the fluctuation in the *particle number*, which also appears to vanish at T_R. At high temperatures, thermal fluctuations act to increase the randomness of the system, generating a thermally rough interface.

Numerical simulations confirm the existence of the roughening transition for various SOS models, and the results are in good agreement with the theoretical predictions. Moreover, for one particular SOS model it is possible to obtain exact results. The body-centered SOS model has been shown to be equivalent to the 6-vertex model [38]. For the 6-vertex model a number of exact results are available, and it is known that the model exhibits a Kosterlitz–Thouless transition. Thus the mapping between the two models allows one to translate the exact results obtained for the 6-vertex model into the interface language, proving analytically the existence of the roughening transition.

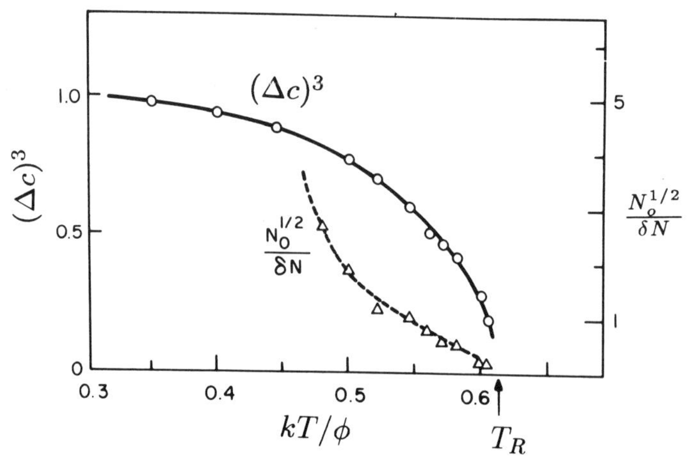

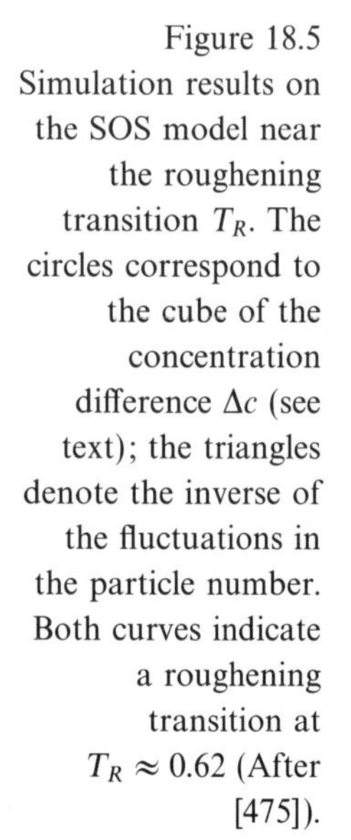
Figure 18.5 Simulation results on the SOS model near the roughening transition T_R. The circles correspond to the cube of the concentration difference Δc (see text); the triangles denote the inverse of the fluctuations in the particle number. Both curves indicate a roughening transition at $T_R \approx 0.62$ (After [475]).

Table 18.1 *Experimental results on the roughening transition for various metal surfaces. For comparison, in addition to T_R we give T_m, the melting temperature of the material. After [265].*

Surface	$T_R(K)$	$T_m(K)$	Reference
In(110)	290	420	[177]
Pb(110)	415	600	[495]
Ag(110)	910	1234	[52]
Cu(110)	> 900	1355	[503]
Ni(110)	> 1300	1720	[69]

18.2.2 Experimental results

For a large number of materials, T_R is close to or greater than the material's melting temperature, T_m. Hence it can be quite difficult to locate the roughening transition, or to distinguish the roughening transition from the melting transition. Despite this difficulty, there are a number of reliable experimental results giving evidence for the roughening transition in various materials. Table 18.1 summarizes some of the known results for fcc surfaces.

Two fundamentally different techniques are used to observe the roughening transition. In the first, one observes the shape of small crystals maintained in equilibrium with their vapor. Increase the temperature to T_R, and a given facet disappears. We notice that different facets may have differing roughening temperatures, indicating that it is possible for the same material to have a *series* of roughening temperatures, each one dependent on orientation of the interface. This is consistent with the theory developed in the previous section, since T_R is a function of material-dependent quantities such as ν.

In the second, more widely used, technique, one measures the height–height correlation functions using diffraction techniques or microscopy. While experimental results provide solid evidence for the existence of the roughening transition, understanding them requires a thorough knowledge of the diffraction techniques used in the measurements.

18.3 Nonequilibrium effects

Thus far we have considered the *equilibrium* aspects of the roughening transition. Apart from the theoretical question of understanding the effect of deposition in general, there are experiments that indicate the existence of a roughening transition *in the presence of deposition*. Hence in this section we discuss the main results of the nonequilibrium theory, with the goal of connecting to the equilibrium theory of the previous sections.

18.3.1 Experimental motivation

A particularly interesting system that exhibits a roughening transition is ^{4}He [143, 489]. The roughening temperature T_R is 1.28 K. To observe the roughening transition, the interface of the solid ^{4}He is placed in contact with the superfluid phase through a small hole in a box immersed in an experimental cell. The superfluid pressure is increased beyond its equilibrium value, creating a finite chemical potential difference between the fluid and the solid interface. As a result, the deposition of the He atoms onto the solid ^{4}He surface begins. One of the quantities measured is the growth velocity, v, of the interface for a given value of $\Delta\mu$, shown in Fig. 18.6. The data shows a transition in the velocity near the roughening temperature T_R.

To understand the origin of the change in the velocity, let us see what happens at the atomic level. The solid interface is in direct contact with its liquid phase. The fluid molecules frequently stick to the surface, diffuse on it, detach, and return to the liquid. For $T > T_R$, the interface is rough. The atoms of the fluid can easily find positions

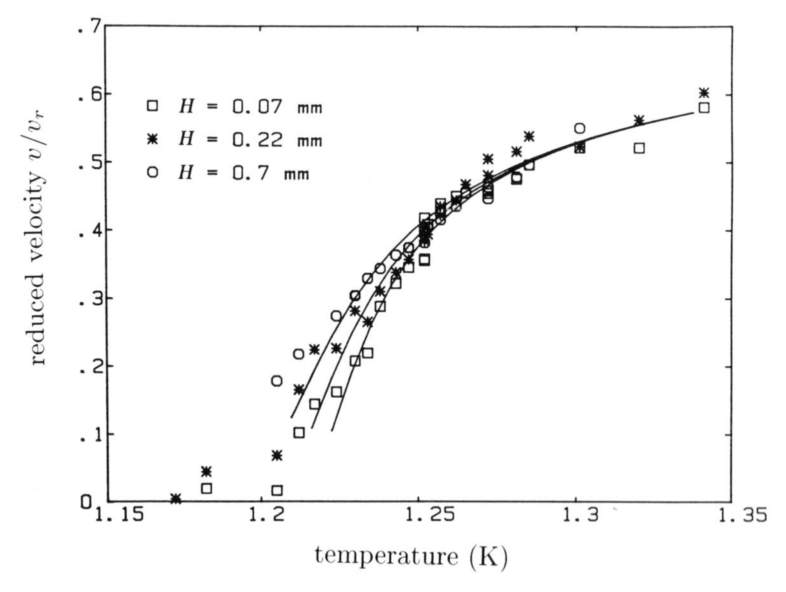

Figure 18.6 Reduced interface velocity v/v_r as a function of temperature for ^{4}He experiments. The different curves correspond to different values of the chemical potential difference, $\Delta\mu$ (the quantity H is proportional to $\Delta\mu$). The data indicate an increase in the velocity of the interface at roughening temperature. However, the transition becomes smoother and smoother as $\Delta\mu$ is increased. (After [143]).

on the surface where they acquire a large number of neighbors so that sticking is energetically favorable. For $T < T_R$, the surface is flat. The atoms of the liquid do not have good contact sites with the interface, and detach from the surface. As we shall see, growth can proceed at low temperatures only by the nucleation of islands of a sufficiently large size. Since nucleation is difficult to achieve, the growth velocity approaches zero.

18.3.2 Model and analytical results

Since the primary experimentally-observable quantity is the interface velocity, we must modify our formalism to include the dynamical properties of the system. This can be done using Eq. (C.2), which gives the equation of motion†

$$\gamma \frac{\partial h}{\partial t} = \nu \nabla^2 h - \frac{2\pi}{a_0} V_0 \sin\left(\frac{2\pi}{a_0} h\right) + \frac{\lambda}{2} (\nabla h)^2 + F + \eta(\mathbf{x}, t), \qquad (18.16)$$

where we included the KPZ nonlinearity $(\nabla h)^2$, a term whose relevance will be discussed below, and two others. The first two terms on the rhs are taken directly from Hamiltonian (18.4), the second being the lattice potential. As we saw in previous sections, this term is responsible for the roughening transition. The presence of the external force, F, indicates that there is particle deposition onto the surface. The randomness in the deposition process is incorporated into the noise, η, which is assumed to be uncorrelated.

We have already treated some of the limits of (18.16) in previous chapters. When $V_0 = 0$ it reduces to the KPZ equation, which we know leads to roughening. With $\lambda = 0$ we have a driven sine-Gordon equation. The fact that the KPZ nonlinearity may be relevant was suggested first by Hwa, Kardar, and Paczuski and has been discussed in detail subsequently [24, 190, 191, 320, 382].

Let us consider first the $\lambda = 0$ case. The experimental results differ from the roughening transition discussed in the previous chapter primarily in regard to interface mobility. The mobility can be defined as the ratio between the velocity and the driving force, expressed as a limit in which the driving force goes to zero. For the undriven interface ($F = 0$), there is a sharp transition at the roughening temperature from an interface pinned by the lattice potential to one that moves. Once the driving force is turned on, the transition becomes smooth, and the

† This equation with $\lambda = 0$ was first proposed and analyzed by Chui and Weeks [77].

interface has a finite velocity even for $T < T_R$ – coinciding with the result observed experimentally (Fig. 18.6).

For $T \ll T_R$ we are in the nucleation regime, discussed in 1951 by Burton, Cabrera, and Frank [68, 353]. The interface is expected to be flat. Single atoms are not stable on the interface, and fluctuations may generate islands of size r on the surface. There are two competing effects influencing the growth of the islands. It can be shown that the free energy is $E(r) = 2\pi r\epsilon - \pi r^2 \mathscr{F} a_0$, where ϵ is the surface energy of the island edge and $\mathscr{F} a_0$ is the force on a single step per unit length, being proportional to $\Delta\mu$. The critical island size (that minimizes the free energy) is $r_c = \epsilon/\mathscr{F} a_0$, and has an energy $E_c = \pi\epsilon^2/\mathscr{F} a_0$. An island smaller than r_c is unstable and shrinks. However if $r > r_c$, the island will grow, eventually completing a full monolayer. The interface at a given instant is covered by a large number of islands of differing sizes. The probability that an island of size r_c will be created is proportional to

$$\frac{1}{\tau} = \frac{1}{\tau_0} \exp\left(-\frac{E_c}{k_B T}\right). \tag{18.17}$$

After time t the number of islands of size r_c is $L^2 t / r_c^2 \tau$, and they are separated by a distance $\ell_1 \sim r_c(\tau/t)^{1/2}$. Each island grows radially by capturing atoms, the growth velocity being $v_i = \mu_e \mathscr{F} a_0$, where μ_e is the mobility of the island edge. Islands coalesce when their radius becomes equal to ℓ_1. The typical time for the coalescence process is $t \sim (r_c^2 \tau / v_i^2)^{1/3}$. Thus the velocity of the interface due to this nucleation process is

$$v = \frac{a_0}{t} \sim a_0 \left(\frac{\mu_e^2}{\epsilon^2 \tau_0}\right)^{1/3} (F a_0)^{4/3} \exp\left(-\frac{\pi\epsilon^2}{3\mathscr{F} a_0 k_B T}\right). \tag{18.18}$$

We can see that (18.18) confirms the experimental observations: (a) there is a nonzero growth velocity due to nucleation, and (b) the velocity increases with driving force in a highly nonlinear fashion.

If $T \gg T_R$, the interface is rough even in equilibrium. The existence of a large number of steps facilitates sticking of the fluid atoms, and the mobility of the interface approaches the mobility of the free interface.

In conclusion, the growth equation (18.16) with $\lambda = 0$ reproduces qualitatively and quantitatively the experimental results obtained for ^{4}He. Do we require the KPZ nonlinearity $(\nabla h)^2$? Apparently we do, because the RG flow calculated to second order shows that such a nonlinearity will be generated even if it is not originally included in the growth equation [382] – i.e., even if we start with $\lambda = 0$, a $\lambda(\nabla h)^2$

term is generated at large length scales, so a correct calculation must include it from the beginning. However, the analysis of the resulting flow equations shows that the nonlinear term does not greatly affect the mobility of the interface, the error in the mobility obtained by neglecting the nonlinear term being less than 10^{-3} [382].

The main effects of the driving force and the nonlinear term are similar: on large length scales, they cause the lattice potential to vanish, and so allow for a nonzero mobility of the interface even for $T < T_R$. Since the lattice potential vanishes, the interface morphology is determined by the free Hamiltonian. Thus in the presence of deposition or nonequilibrium effects the interface is rough for any temperature, the roughness being logarithmic if $\lambda = 0$, and possibly a power law if the KPZ nonlinearity is present.

18.3.3 Results from simulations

The most direct way to understand the influence of the driving force is to consider a model known to undergo a roughening transition, and to drive it out of equilibrium. Recently a number of simulations were reported for this problem [238, 239]. The model used is the body centered SOS model, known to be equivalent to the 6-vertex model.

Let us consider a typical model of the SOS family. In equilibrium, the probability that an atom evaporates or condenses on the surface is equal, resulting in zero growth velocity. In this case the chemical potential difference between the surface and the vapor is zero. Increasing this difference would result in an increased condensation probability, leading to a nonzero growth velocity for $T < T_R$. Computationally, this can be accomplished by choosing the condensation and desorption probabilities to be

$$P^c \equiv \frac{\exp(\Delta\mu/k_BT)}{1+\exp(\Delta E/k_BT)}, \qquad P^d \equiv \frac{1}{1+\exp(\Delta E/k_BT)}. \tag{18.19}$$

Here ΔE is the change in the total energy of the system resulting a the deposition or desorption event. If $\Delta\mu = 0$, the system is in equilibrium, while if $\Delta\mu > 0$, the interface grows by atom deposition.

Figure 18.7 shows the mobility of the interface as a function of temperature for various driving forces. The roughening transition is smoothed out if the system is not in equilibrium; even for $T < T_R$ the interface has a nonzero velocity that increases with the driving force. When the driving force is present, $w \propto \log L$ even for $T < T_R$, as discussed in the previous section. The change in the morphology of the interface is illustrated in Fig. 18.8, where we show the interface for $T = 0.7\ T_R$ and for several values of $\Delta\mu$. Note that even a small value of $\Delta\mu$ leads to a rough interface.

18.4 Discussion

In this chapter we focused on an important property of crystal surfaces: their equilibrium roughness. We showed that crystals can be rough, even under equilibrium conditions, if the temperature is larger than the roughening temperature T_R. The roughening temperature depends on the crystal's material properties, and can be calculated exactly once these properties are known.

Out of this discussion arises an obvious question: how do we relate the MBE models studied in the previous chapters to the phenomenology of the roughening transition? Where do these models work – above or below T_R? As Table 18.1 indicates, the roughening temperature of most materials is high – for many surfaces it is larger than the melting temperature and cannot be observed. Thus the kinetic models discussed in the previous chapters must work in the smooth phase – for $T < T_R$. According to the equilibrium theories, the surface should be smooth in this regime, only exhibiting a roughness that is small, temperature-dependent, and predictable. Why does roughening nevertheless appear?

As mentioned in §18.3, there is an important difference between the phenomenology of the roughening transition and atomistic nonequilibrium models. The latter are inherently *driven*, i.e., there is particle deposition on the surface. And, as demonstrated in §18.3.2, particle

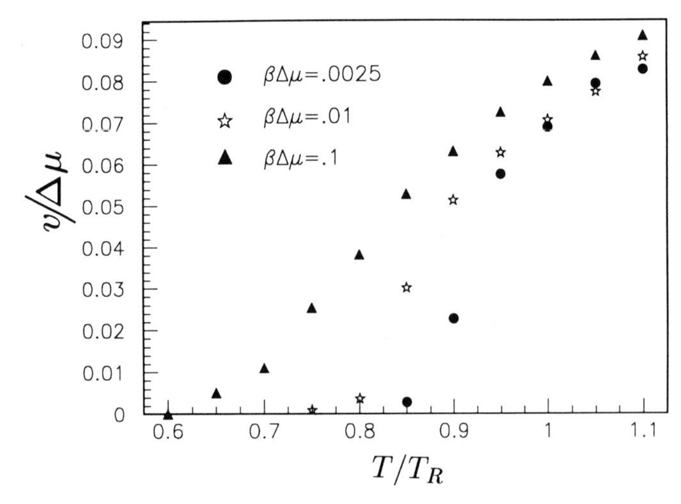

Figure 18.7 Interface mobility $v/\Delta\mu$, as a function of the reduced temperature in the vicinity of the roughening transition for the BCSOS model. The curves correspond to different values of the chemical potentials $\Delta\mu$, and $\beta \equiv 1/k_BT$. The transition becomes smoother as $\Delta\mu$ increases. (After [240]).

deposition washes out both the lattice potential and the roughening transition. When the deposition flux is non-zero, the interface is rough at any temperature. The velocity measurements may still show some behavior reminiscent of the roughening transition, but other quantities – such as the height–height correlation function – indicate that we are dealing with a rough interface. Thus deposition 'saves' the nonequilibrium models: it relaxes the requirement that all crystal surfaces should be smooth when the temperature is experimentally relevant.

There is yet another difference between MBE models and the models discussed in this section. The origin of the Laplacian term, $\nabla^2 h$, in the growth equation is desorption (cf. §13.3), while the discrete models that we discussed in this chapter all allow for both desorption and

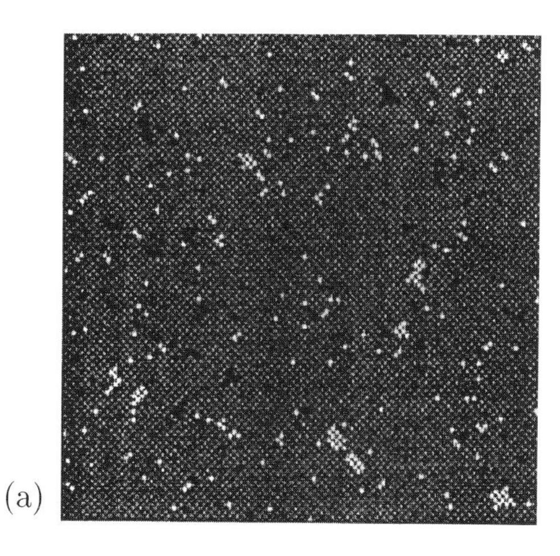

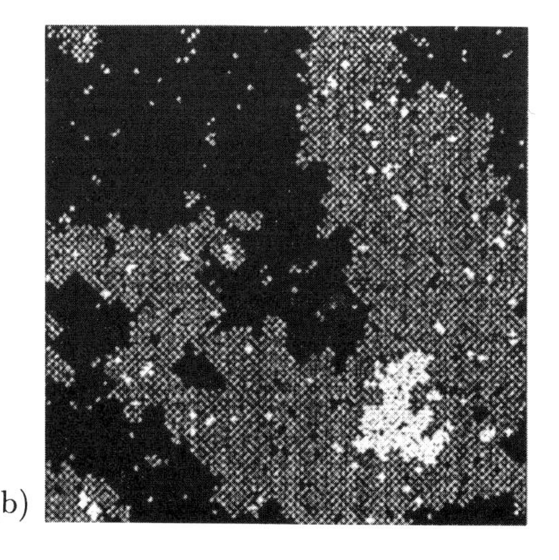

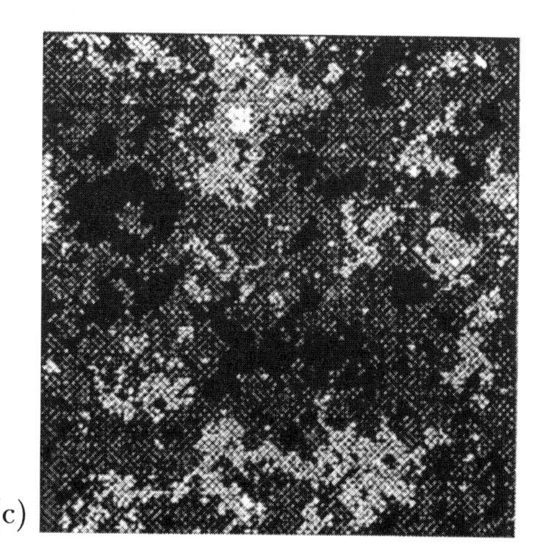

Figure 18.8 Grey scale depiction of the surface height at saturation for a system of size 128×128 and temperature $T = 0.7\ T_R$. In (a) the interface is in equilibrium, $\Delta\mu = 0$. In (b) and (c) the surfaces are out of equilibrium with $\Delta\mu = 0.1$ and 10, respectively. We note that driving the system out of equilibrium results in a rough interface. (After [239]).

deposition. However, under normal MBE conditions, desorption is negligible and the relevant microscopic processes are surface diffusion and deposition, as discussed in Chapter 12. No one has yet shown how the roughening transition is relevant to models in which desorption is disallowed, or how surface diffusion affects the roughening transition. These are open questions, and interesting results may emerge in the future.

Suggested further reading:

[265, 352, 475]

Exercises:

18.1 Derive (18.13) and (18.14) from (18.7)–(18.8).

18.2 Show that in the nonequilibrium SOS model discussed in §18.2.1 the growth rate and $\Delta\mu$ always have the same sign.

19 Nonlocal growth models

Most of this book deals with *local* growth processes, for which the growth rate depends on the local properties of the interface. For example, the interface velocity in the BD model depends only on the height of the interface and its nearest neighbors. However, there are a number of systems for which *nonlocal* effects contribute to the interface morphology and growth velocity. Such growth processes cannot be described using local growth equations, such as the KPZ equation; if we attempt to do so, we must include nonlocal effects. In this chapter we discuss phenomena that lead to nonlocal effects, and we also discuss models describing nonlocal growth processes.

19.1 Diffusion-limited aggregation

Probably the most famous cluster growth model is diffusion-limited aggregation (DLA) [417, 418, 480, 481]. The model is illustrated in Fig. 19.1. We fix a seed particle on a central lattice site and release another particle from a random position far from the seed. The released particle moves following a Brownian trajectory, until it reaches one of the four nearest neighbors of the seed, whereupon it sticks, forming a two-particle cluster. Next we release a new particle which can stick to any of the six perimeter sites of this two-particle cluster. This process is then iterated repeatedly. In Fig. 19.2, we show clusters resulting from the deposition of 5×10^5, 5×10^6, and 5×10^7 particles.

Early studies demonstrated that DLA clusters are fractal objects. There are many ways to measure the fractal dimension. The most straightforward method is box counting. For this we cover the cluster with balls of size ℓ and count the number of balls needed for complete coverage, which scales with ℓ as in (3.2). The fractal dimension of a two-dimensional DLA cluster has been calculated quite accurately, with the result $d_f = 1.71 \pm 0.01$ [174, 302, 304].

It is straightforward to generalize the DLA growth rule to higher dimensions, and to calculate the corresponding fractal dimensions; e.g., for three-dimensional DLA one finds $d_f = 2.5 \pm 0.1$. Despite the simplicity of the growth rule, we do not have a theory for DLA that can give the fractal dimension for even a single value of d. It is the nonlocal nature of the growth process that makes the problem intractably difficult to solve analytically.

The nonlocality of DLA is a result of the shadowing generated by the branches of the cluster. Consider the cluster shown on Fig. 19.1. If a new particle diffuses toward the cluster, it can stick on any part of the cluster. With higher probability it will be captured by the external parts of the arms, and with lower probability it will penetrate the cluster. Thus the growth rate in a given point depends on the entire geometry of the cluster, not only on the local morphology.

For DLA, the growth probability of a site perimeter is the probability that the random walker visits the site. This probability, u, obeys the steady-state diffusion equation

$$\nabla^2 u = 0, \tag{19.1}$$

where $u = 1$ at infinity, and $u = 0$ on the perimeter sites. The probability that a perimeter site will be visited is proportional to the flux on that site, i.e. $p \propto \nabla u$. The growth velocity, $\mathbf{v}_n$, obeys the continuity equation $\nabla \cdot \mathbf{v}_n = 0$. Since $|\mathbf{v}_n| \propto p \propto \nabla u$, (19.1) follows.

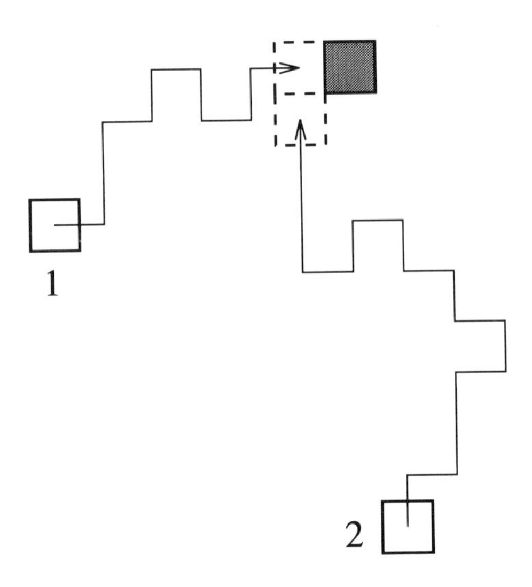

Figure 19.1 The growth rule for DLA. We place a seed on the plane (shaded particle) and release a particle (1) at a distant point from the seed. The particle follows a random walk, until it reaches a site next to the seed, whereupon it sticks. Next we release another particle (2) from a randomly chosen distant position, and let it diffuse until it reaches a site next to the cluster formed by the seed and particle 1. DLA is made by iterating this growth rule.

DLA clusters can be grown not by particle deposition, but calculating in every moment the Laplacian field *u*, and choosing a growth rate proportional to the local gradient of the field [345].

The aggregation process leading to the DLA cluster is widespread in nature. Branched structures reminiscent of DLA have been observed in viscous fingering [279, 346], electrodeposition [150, 211, 297], dielectric breakdown [345], dendritic and snowflake growth [223, 347, 348, 349, 415], chemical dissolution [91, 92, 335], thin film crystallization [271], geological phenomena such as disequilibrium silicate mineral textures [133], and even biological phenomena such as bacterial growth [39, 40, 138, 296, 460], neuronal growth [70], and fingering of HCl across the mucus lining of the stomach [43]. However, in every case there is

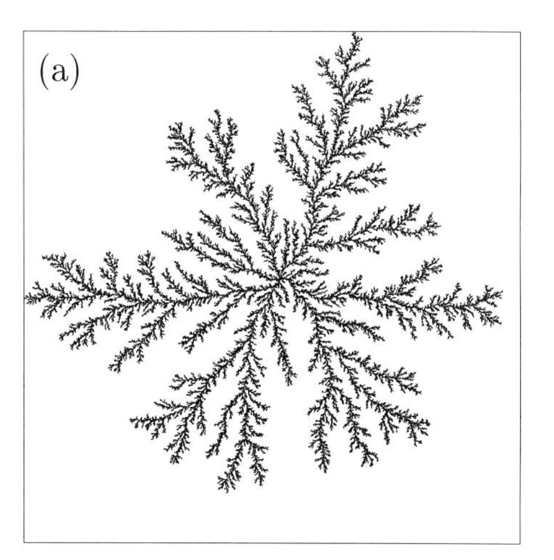

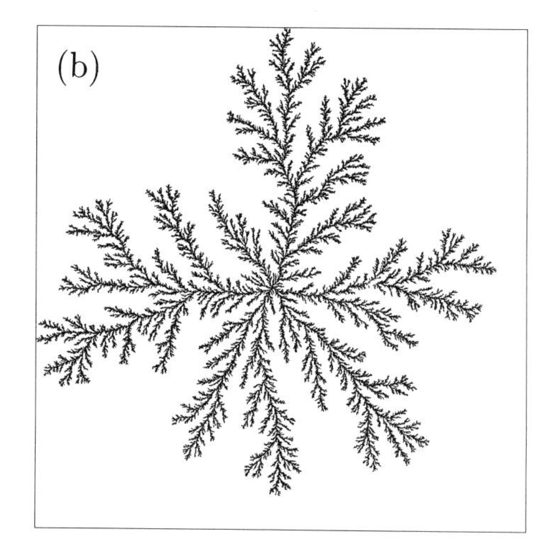

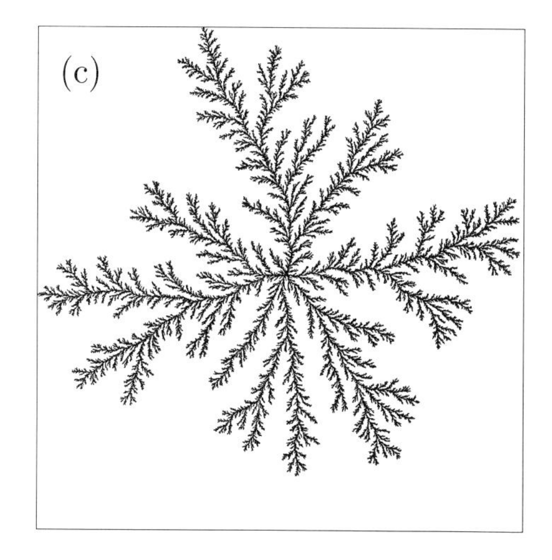

Figure 19.2 Comparison of three different off-lattice DLA clusters of differing masses: (a) 5×10^5 particles, (b) 5×10^6 particles, and (c) 5×10^7 particles. The size of the clusters has been rescaled so that each fits into a box of the same size, in order that the self-similarity of this archetypal random fractal is visually apparent. (After [356]).

another diffusing field. The role of u is played in viscous fingering by the pressure field, in electrodeposition by the electric field, in dendritic growth by the temperature, and in bacterial growth by the nutrient concentration.

The cluster shown on Fig. 19.2 was grown around a central seed. However, if we study deposition on a surface, it is more appropriate to use a strip geometry – instead of a seed we deposit on a line of seeds. The results of such a deposition process is shown in Fig. 19.3.

19.2 Sputter deposition

Sputtering is a common experimental method used to grow thin films. In sputtering an energized beam of particles is directed at a bulk specimen of the material. The beam particles collide with this target material and eject particles from its surface. The ejected particles travel ballistically until they deposit on another target situated in the direction of the beam. The particles arrive at this growing interface either from a fixed angle or from random angles, depending on the experimental setup.

Sputter deposition is used in the production of magnetic recording films and also in the production of the silicon films used in photovoltaic

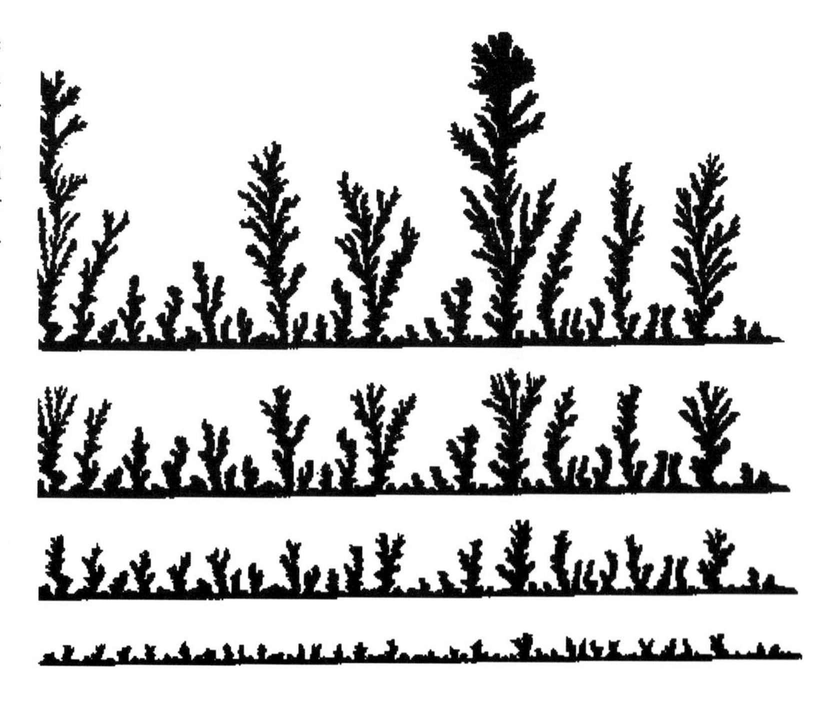

Figure 19.3 Zinc electrodeposition along a line for four successive times, increasing from bottom to top. (After [297]).

applications. Typically, the morphology of the films produced by sputtering is amorphous. In this section we discuss the various theories proposed to describe growth by sputtering. Since the morphology appears to depend on the deposition method – whether atoms arrive from random directions or have oblique incidence (see Fig. 19.4) – we discuss these two cases separately.

19.2.1 Random incidence

Sputter deposition is influenced by *nonlocal* effects due to geometrical shadowing. The local growth rate of the interface at a given point is proportional to its exposure to the deposition beam (see Fig. 19.4(a)).

The effect of shadowing is nicely illustrated by the *grass model* [22]. In this model each 'blade of grass' grows in proportion to the amount of light it receives at its tip (see Fig. 19.4(a)). On a cloudless day this amount is proportional to the solid angle of sky 'open' to the grass blade – i.e., not blocked by its neighbors. Figure 19.5(a) shows the result of the growth process, starting from an initial configuration of blades of grass with small but random differences in height. The shortest stalks are shadowed by their taller neighbors, and soon stop growing. The 'winners' compete among themselves, always sorting out the shortest. The final result is a self-similar distribution of the heights, which means that any small part of Fig. 19.5(a) rescaled will be indistinguishable from the original.

This simple grass model mimics the situation when atoms stick where they are first deposited. However, in many systems smoothing by surface diffusion competes with the instability generated by shadowing. A simple continuum model that includes surface diffusion and shadowing has been proposed by Karunasiri *et al.* [221]. They considered a single-valued interface (with no overhangs) that grows

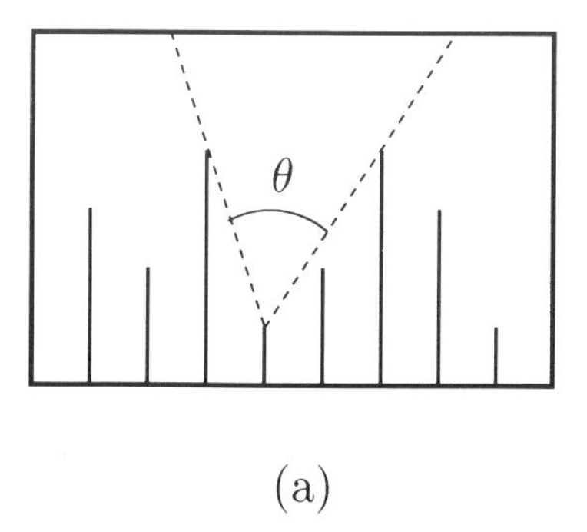

(a)

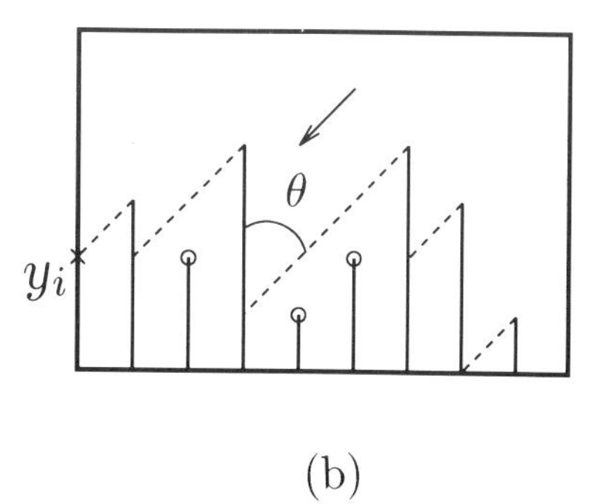

(b)

Figure 19.4 Schematic illustration of deposition with (a) random and (b) oblique incidence. In the grass model (a) particles arrive from random directions. Thus the growth rate of a column is proportional to the probability that its top is reached by a particle which follows a ballistic trajectory with randomly-selected direction. The growth rate is proportional to the exposed angle θ from which a particle can reach the tip. (b) In the needle model, particles arrive ballistically with all trajectories forming an angle θ with the vertical. Some columns (marked by circles) can never grow, being shadowed by the taller neighbors.

according to the growth equation

$$\frac{\partial h}{\partial t} = -K\nabla^4 h + F\theta + \eta(x,t). \tag{19.2}$$

The second term in the rhs is the deposition rate F and $\theta = \theta(x, \{h\})$, the exposure angle (see Fig. 19.4), that includes nonlocal effects in the growth process. Note that the exposure angle θ depends on the entire interface $\{h(x)\}$. This term shows how the local growth rate is proportional to the exposure: more atoms arrive on the 'hills,' which are more exposed to the beam, and fewer in the 'valleys' shadowed by the hills. Thus the second term is responsible for the instability observed already in the grass model. Numerical integration of (19.2) indicates that surface diffusion cannot balance the effect of shadowing. With a finite diffusion constant K, we obtain a morphology that is reminiscent of the one generated by the grass model without diffusion (see Fig. 19.5(a) and (b)).

Figure 19.5 suggests that, due to the shadowing effect, the interface has a columnar structure, with a characteristic thickness ξ of the formed columns [381, 499]. If the diffusion length of the deposited atoms is shorter than the characteristic length ξ, the morphology of the interface is dominated by wide columns (of lateral size ξ), with deep grooves between neighboring columns [23]. Shadowing from neighboring columns is important for the grooves, but *not* for the broad tops. Thus the broad tops evolve as if they are growing with constant velocity along the surface normal. This latter growth mechanism is the Huygens principle known from geometric optics [439]. In the model based on the Huygens principle, an initially rough

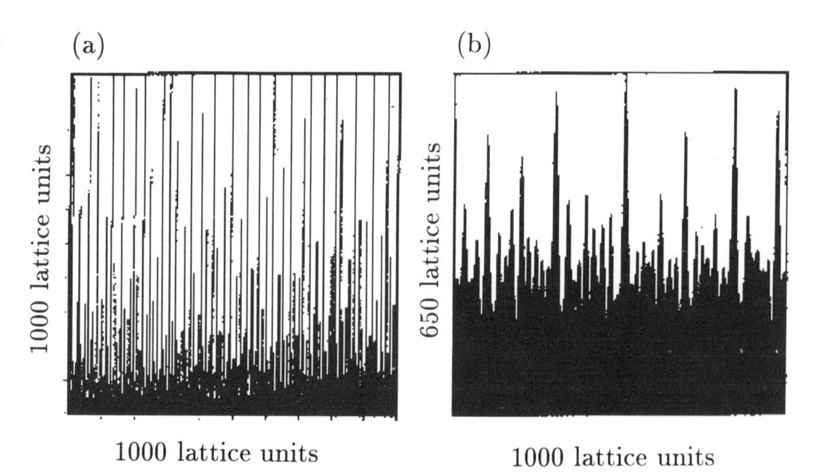

Figure 19.5 Grass model (a) without and (b) with surface diffusion. (After [221]).

interface grows with a constant velocity along the local surface normal (see Fig. 19.6). This leads to the formation of cusp singularities whose number decrease in time as the larger columns take over the smaller.

During this coarsening process the interface becomes smoother, and measuring changes in the interface roughness is no longer useful. The scaling of the surface is quantified instead by the mean size and the height distribution of the forming columns. Note that the Huygens principle for growth is very similar to the growth mechanism observed in the noiseless KPZ equation (see Chapter 25).

19.2.2 Oblique incidence

Deposition of atoms arriving ballistically with oblique incidence has been studied in detail both experimentally and theoretically. The main motivation is to explain the columnar structures observed in vapor-deposited thin films. In this section we discuss two models. The counterpart to the grass model is the needle model, which captures some essential features of the screening mechanism. The needle model is particularly attractive, since it can be mapped onto a particle diffusion problem, and we can obtain the scaling properties of the model exactly. A more realistic model is ballistic deposition with oblique incidence. In the last part of this section we summarize the scaling properties of the obtained columnar structure for such a model.

Needle model – The needle model for growth with oblique incidence is illustrated in Fig. 19.4(b) [246, 248, 311, 334]. In contrast to the grass model, for which at any moment all columns grow, albeit with different growth velocities, for growth with oblique incidence there are active and inactive columns. When one column shades another, the shaded column stops growing. The competition among the columns

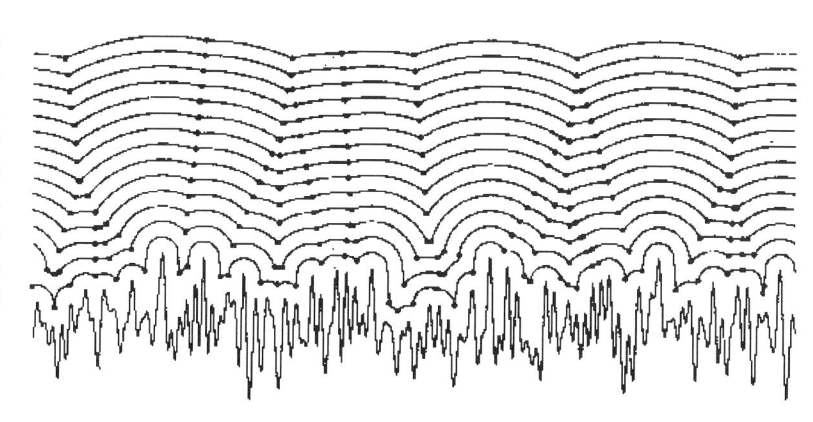

Figure 19.6 The evolution of a random initial surface according to the Huygens principle. The solid circles indicate the cusp singularities. (After [439].)

for the incoming particle flux leads to an algebraic distribution of the column heights

$$N(h) \sim h^{-\gamma}, \tag{19.3}$$

where $N(h)dh$ is the number of columns with height between h and $h+dh$ [307]. The growth velocity of the unshaded column is given by

$$\frac{dh}{dt} = \eta(x,t), \tag{19.4}$$

where $\eta(x,t)$ is an *uncorrelated* random force. While this growth equation looks simple, being reminiscent of (4.7) describing random deposition, a complication arises from the fact that only the *unshaded* columns follow it, while the shaded columns do not grow. At every time step one must decide which columns become shaded by the neighbors, which is a nonlocal operation, since shading may come from columns that are far apart.

A simple mapping onto a diffusion problem gives the exponent γ. Let us project the tip of the ith active column h_i onto the vertical axis $i = 0$, so $y_i = h_i - i\cot\theta$. We interpret y_i as the position of a particle that drifts in the positive y direction. The tips grow randomly according to (19.4), corresponding to a random drifting of the particles in the positive y direction. If two particles collide, one of the tips has been shadowed by another, in which case we remove one of the particles from the system. Periodic boundary conditions correspond to particles constrained to diffuse on a circular substrate.

Let us follow our particle model for a moment. Initially we have a random distribution of particles, corresponding to the heights of the active columns. We move the particles randomly in the positive direction. If two particles collide, we remove one of them. The number of particles will decrease in time, corresponding to the fact that the number of active columns decreases due to shadowing. The problem of a set of coalescing random walks has been studied in the literature [107, 413]. One main quantity is the concentration of particles, $c(h)$, as a function of time. Here time is defined to be the average height of the surviving columns, which is related to the survival probability for a column of height h, $P_{\text{surv}}(h)$, which can be calculated by assuming that every column of height larger than h is a potential shadower for columns with height h. Since $P_{\text{surv}}(h) = c(h)\cot\theta$, we find

$$P_{\text{surv}}(h) \sim \int_h^\infty N(h')dh' \sim h^{-(\gamma-1)}. \tag{19.5}$$

In the random walk model, it was found that for $t \gg 1$ the particle concentration decays as $c(h) \sim h^{-1/2}$ [107], corresponding to $N(h) \sim h^{-3/2}$. This gives $\gamma = 3/2$, independent of the angle of incidence [246].

Ballistic deposition with oblique incidence – Both the grass and the needle models are solid-on-solid models, neglecting overhangs or holes in the bulk of the growing material. When the growth takes place at low temperatures, however, overhangs may become relevant, and in this limit the most appropriate model to use is ballistic deposition with oblique incidence [303]. Figure 19.7 shows the columnar structure generated by the model. Much attention has focused on the internal structure of the formed columns. A column can be characterized by its height, $h(s)$ and its width, $w(s)$, which scale with the number of particles s forming the column, as

$$h(s) \sim s^{\nu_\parallel}, \qquad w(s) \sim s^{\nu_\perp}. \tag{19.6}$$

The number of columns with particle number s scales as

$$N(s) \sim s^{-\tau}. \tag{19.7}$$

The previous exponents are not independent, but are linked by the scaling relations

$$\nu_\parallel + \nu_\perp = 1, \qquad \tau + \nu_\parallel = 2. \tag{19.8}$$

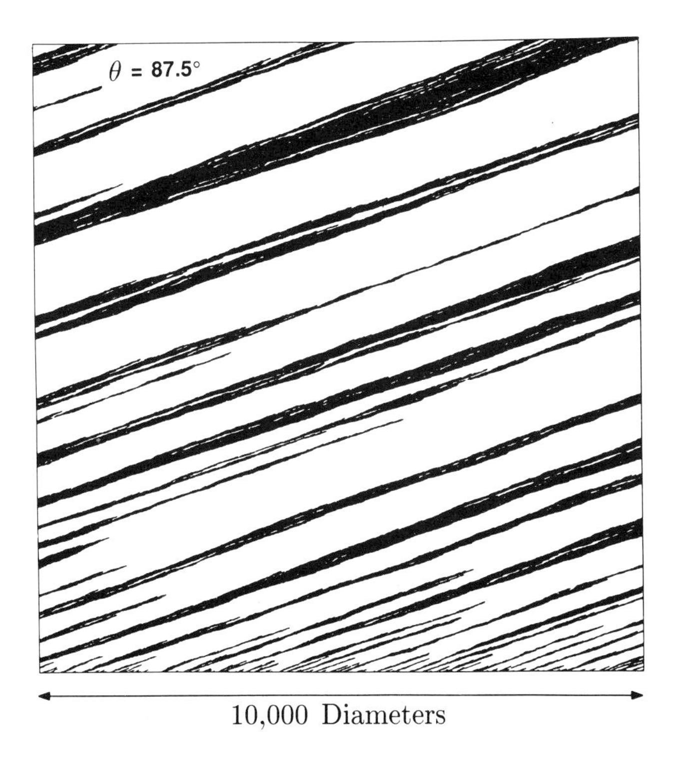

Figure 19.7 Columnar structure obtained by ballistic deposition with oblique incidence. (After [246]).

The first relation can be obtained from the fact that columns are compact, implying that $h(s)w(s) \sim s$. The second is obtained from finite-size scaling arguments [247, 307, 377]. We studied the ballistic deposition model in detail in Chapter 2 and §8.1. If we start the growth from a flat surface, the interface is characterized by parallel correlation length, which increases with time according to (2.13). Identifying $\xi_\parallel$ with the width of the columns $w(s)$, and time with the height of the clusters $h(s)$, we find that $z = \nu_\parallel/\nu_\perp$ [246]. The value of $z = 3/2$ in one dimension is given by the KPZ equation. Thus, using (19.8), we have for the exponents

$$\nu_\parallel = \frac{3}{5} \qquad \nu_\perp = \frac{2}{5} \qquad \tau = \frac{7}{5}, \tag{19.9}$$

which are quite close to the values obtained from simulations [307].

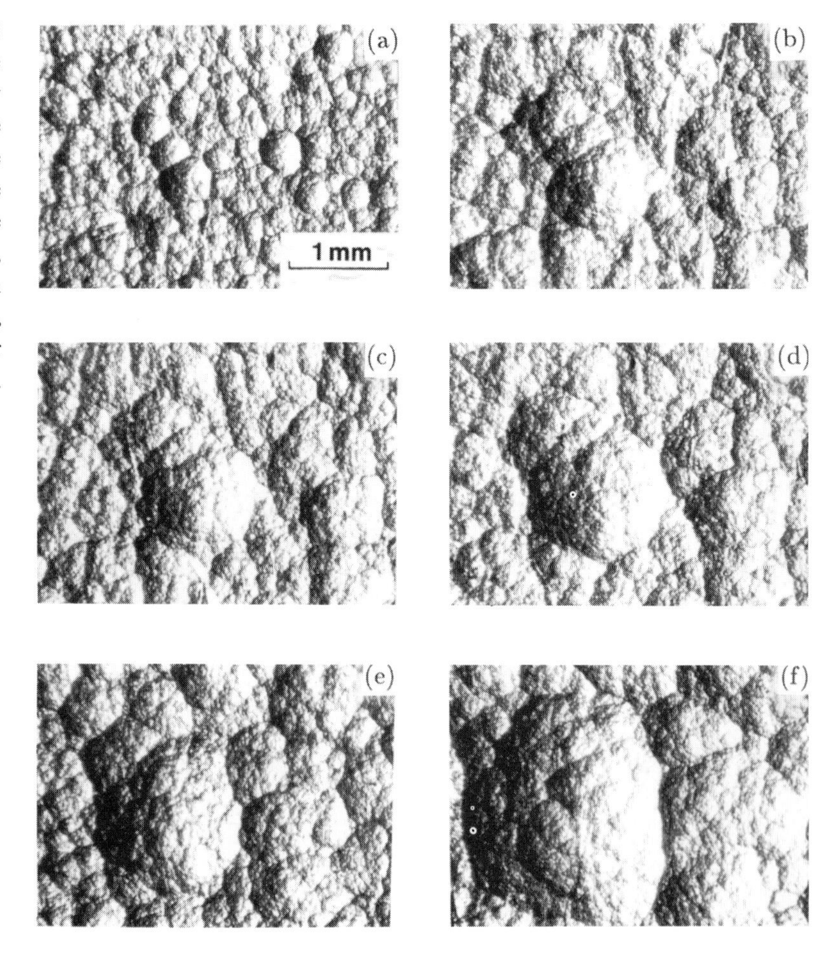

Figure 19.8 Evolution of the surface morphology of pyrolytic graphite as a function of the film thickness. The heights of the interfaces are 1, 6, 8, 10, 14, and 20 mm for (a)–(f), respectively. After [316].

19.3 Experimental results on sputter deposition

19.3.1 Random incidence

As discussed above, during sputter deposition one must consider the interplay between roughening generated by the random deposition of particles, smoothing by surface diffusion and nonlocal effects generated by shadowing. In particular, shadowing has a strong effect on the interface morphology, leading to the generation of a columnar structure.

Thin film morphologies obtained by sputtering can be partitioned into three main classes, depending on the ratio T/T_m between the substrate temperature T and the melting temperature T_m [332]. At low T, shadowing leads to low density columnar structures (Zone 1). At higher T surface diffusion leads to wider columnar structures (Zone 2), while at even higher T bulk diffusion generates large bulk like grains (Zone 3).

An illustration of the self-similar structure of the interface morphology generated by sputter deposition is given in Fig. 19.8. The pyrolytic graphite film is grown at 1900 C, corresponding to a ratio $T/T_m \approx 0.5$, so the growth belongs to Zone 1. Fig. 19.8 presents six micrographs showing the film at different stages of its morphological evolution. One can see that the large individual features ('bumps') grow in time, the larger structures taking over the smaller ones [316, 383].

Similar results have been obtained for growth of NbN films [321] sputtered at 300 C. At this temperature, the film is crystalline, but has a large defect density, suggesting that surface diffusion is reduced. To allow the observation of the time evolution of the surface morphology, AlN was deposited periodically as a marker and an insulating layer. AlN has the same crystalline structure and lattice constant as NbN, so its deposition does not affect the surface morphology. Fig. 19.9 shows the cross section of the film, which was created by deposition of 101 alternating layers of NbN and AlN. One can see the formation of large columns, which take over the smaller ones. The time evolution and the morphology is similar to the interface obtained using the Huygens principle, shown in Fig. 19.6.

Note that the structure of the interface resembles the time evolution of an initially rough interface governed by the noiseless KPZ equation. Indeed, measuring the exact profile of the columns reveals that they have a parabolic shape (see §25.1). The visibility of the consecutive interfaces allows one to study the time evolution of the interface width, and $\beta \approx 0.27$ is found.

Recently You *et al.* have studied gold sputtering onto a silicon substrate. STM and X-ray reflectivity measurements were used to characterize the interface [500]. The resulting scaling plots are shown in Fig. 19.10. The height–height correlation function was calculated using the interface profile from STM, and results in a roughness exponent $\alpha = 0.42 \pm 0.03$. The time evolution of the interface width leads to $\beta = 0.40 \pm 0.02$ at 300 K and $\beta = 0.42 \pm 0.02$ at 220 K. This latter value is much larger than the value $\beta \approx 0.27$ determined from the decorated technique. One reason for the discrepancy might be the considerable surface diffusion present at $\approx$ 300 K, compared to the reduced surface diffusion of the NbN films.

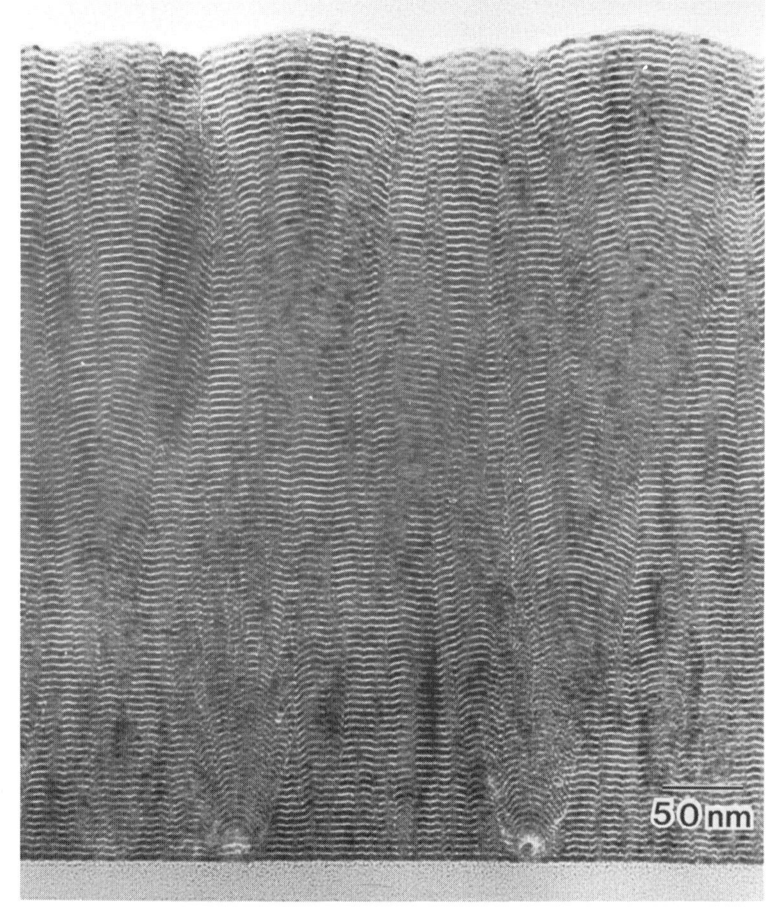

Figure 19.9 Cross-section transmission electron microscope micrograph of the AlN and NbN multilayer film. (After [321]).

19.3.2 Oblique incidence

A series of vapor deposition experiments of growth with oblique or near normal incidence was carried out by deposition of gold on glass substrates [148, 172, 388]. The angle of incidence of the arriving particles was set between 2 and 25 degrees from the surface normal, and the growth rate was 30 nm/s at the substrate temperature 298 K. Fig. 19.11 shows the STM images obtained from the films at different stages of growth. The figure reveals an interface formed by columns separated by voids, the height of the columns and their width increasing during deposition. The tops of the columns appear rather smooth and free of defects. The column size sets a characteristic length scale, which increases with time. Such a characteristic length scale shows up in the width as well, as shown in Fig. 19.12. The data are consistent with the existence of two distinct scaling regions,

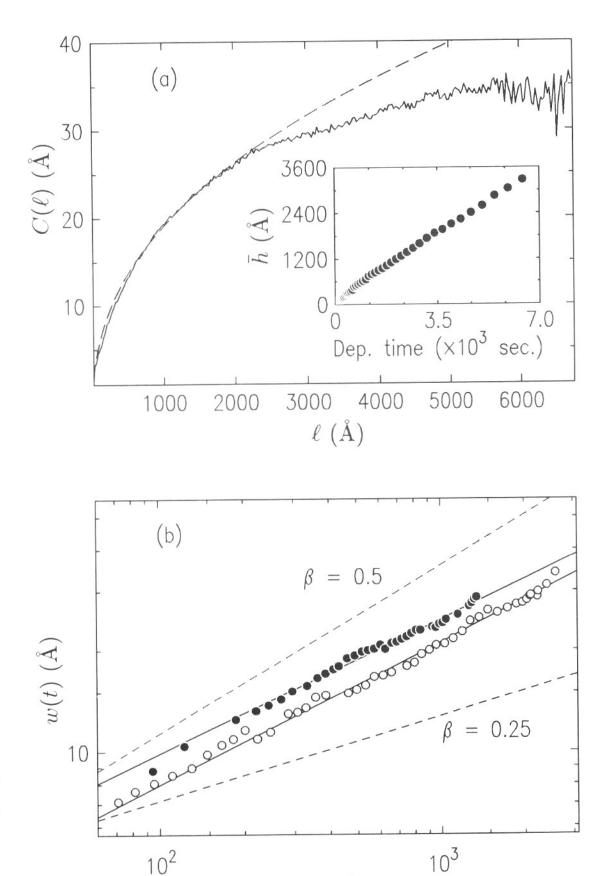

Figure 19.10 The scaling behavior of the interface roughness of a sputter-deposited gold film. (a) The scaling of the height–height correlation function. The dashed line is a power law fit to the data. The inset shows the increase in the average height with time. (b) The scaling of the surface width with time for films grown at 220 K (open circles) and 300 K (filled circles). The solid lines are the power law fits to the data. The two dashed lines are shown to guide the eye. (After [500]).

separated by a characteristic length scale $L_{\times} \approx 38$ nm, very close to the characteristic column size.

In order to explain the experimental results, we must review the microscopic effects expected to play a role in the formation of the column structure. The oblique incidence of the deposited atoms leads to shadowing, and the formation of the columnar structure. However,

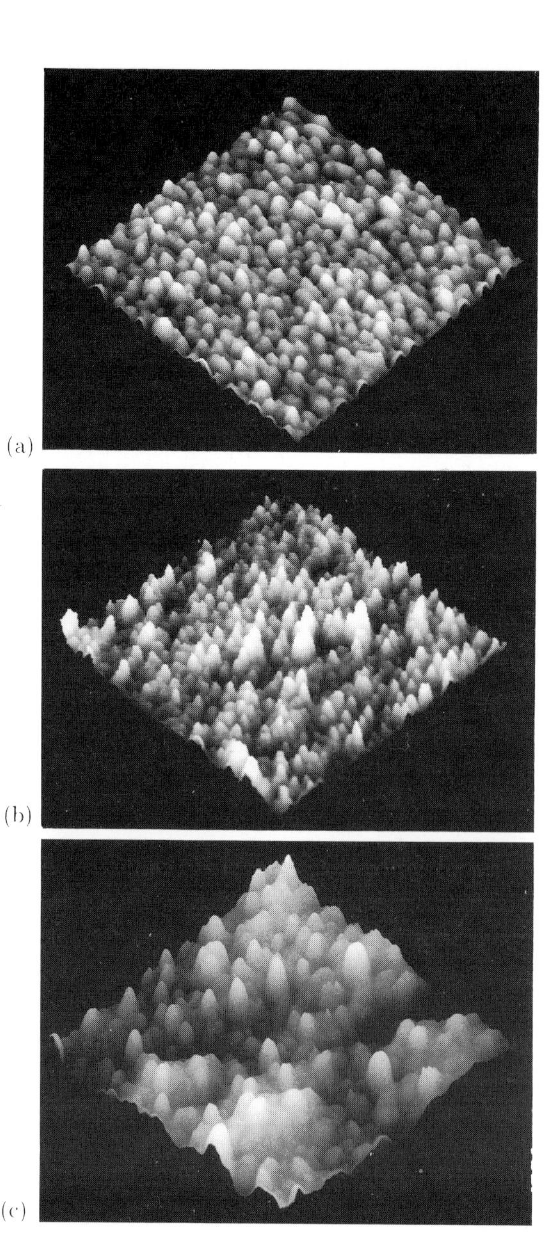

Figure 19.11 STM images of vapor-deposited gold films. The three films are in different stages of the growth process characterized by different values of the average height $\bar{h}$: (a) $\bar{h} = 30$ nm, (b) $\bar{h} = 160$ nm, and (c) $\bar{h} = 850$ nm. The sample sizes are 640×640, 600×600 and 510×510 nm^2, respectively. (After [172]).

the smoothness of the columns and the apparent absence of defects indicates that surface diffusion acts to smooth the surface at short length scales.

For short length scales, the correlation function can be described by a roughness exponent $\alpha \approx 0.69$–0.73, but this result must be corrected for 'subdividing artifacts' that increase the value to $\alpha \approx 0.89$. Presently, we can only speculate about the origin of this value. One can argue that the columns are smooth at short length scales. Thus the width measures the large local slopes, and the resulting value is a crossover to the value $\alpha = 1$. However, the values obtained are between $\alpha = 2/3$, predicted by the *nonlinear* growth equation (14.2) including surface diffusion, and $\alpha = 1$, predicted by the *linear* equation (13.9). Thus at short length scales the exponent may be evidence of diffusion-dominated roughening.

For $L \gg L_{\times}$, the correlation function leads to a roughness exponent $\alpha \sim 0.35 \pm 0.05$. From Table 27.2, we can see that the value closest to it is the prediction $\alpha \approx 0.38$ of the KPZ equation, suggesting that at large length scales the experiment may be described by the KPZ equation. Assuming that it is indeed the nonlinear surface diffusion that describes the short length scale behavior, this experiment seems to be an example of the crossover discussed in Chapter 14: the short length scale behavior is described by (14.2), while at large length scales the KPZ equation takes over. What is the origin of the KPZ nonlinearity? Probably void formation, since desorption is not expected to play a role at room temperatures.

The effect of surface diffusion at short length scales during columnar growth was nicely illustrated by the experiments of Tong *et al.*

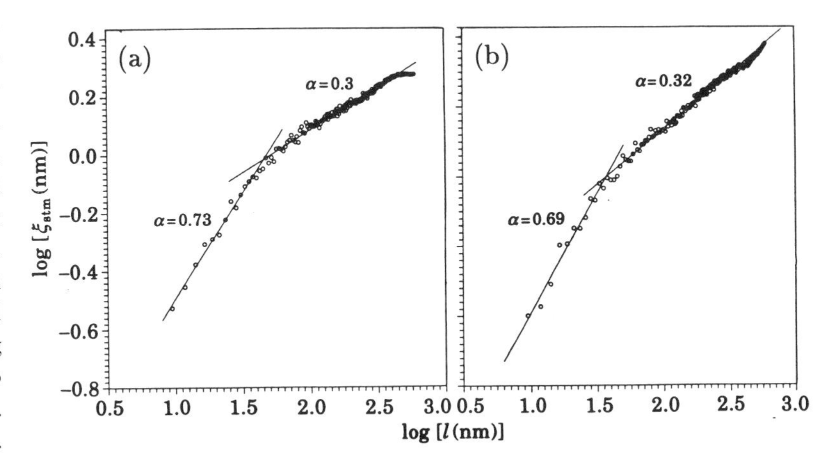

Figure 19.12 The scaling of the interface width for gold films, calculated from STM scans in two orthogonal directions, the (a) x and (b) y directions. The crossover length $L_{\times}$ can be estimated from the crossing point of the two straight-line fits. (After [388]).

[447, 449]. In their studies, CuCl was deposited on a $CaF_2(111)$ substrate. The STM picture reproduced in Fig. 19.13 reveals a columnar structure similar to the one seen in Fig. 19.11, indicating a coarsening of the columnar structure with deposition. Assuming that (19.2) describes correctly the competition between shadowing and surface diffusion, its Fourier transform predicts that for large moments ($q \to \infty$) the reciprocal space correlation function, $\langle |h(\mathbf{q}, t)|^2 \rangle$ is dominated by surface diffusion, being proportional to $1/q^4$, coming from the fourth order diffusion term. Indeed, as can be seen in Fig. 19.13, for large q the reciprocal space correlation function calculated from the scanned samples reveals a slope of -4 on a log-log plot, providing direct evidence that the short length scale behavior is dominated by surface diffusion (note that large q corresponds to small length scales). For large length scales (small q in the figure), the correlation function behaves differently, shadowing possible nonlinear terms influencing the scaling regime.

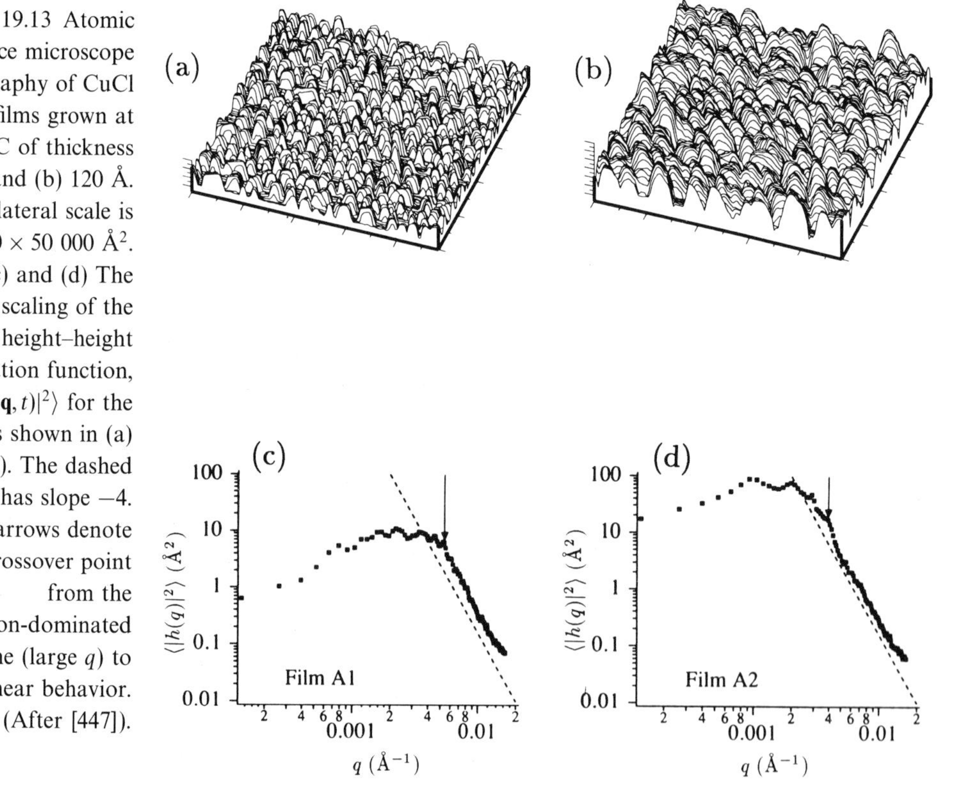

Figure 19.13 Atomic force microscope tomography of CuCl films grown at 110 °C of thickness (a) 60 and (b) 120 Å. The lateral scale is 50 000 × 50 000 Å^2. (c) and (d) The scaling of the height–height correlation function, $\langle |h(\mathbf{q}, t)|^2 \rangle$ for the films shown in (a) and (b). The dashed line has slope -4. The arrows denote the crossover point from the diffusion-dominated regime (large q) to nonlinear behavior. (After [447]).

19.4 Roughening by ion bombardment

19.4.1 Basic phenomena

In the previous sections, we focused on roughening generated by the random deposition of atoms. The inverse process, *sputter etching*, also results in rough interfaces. During sputter etching, the interface is bombarded by ions which hit the surface and transfer part of their energy to the surface atoms, 'kicking' them out. In order to release a surface atom, the incoming ion must have an energy larger than the bonding energy of the corresponding atoms. The larger the energy of the bombarding ion, the more atoms are released from the surface. The number of released atoms for every incoming atom is called the *sputtering yield*. The ion bombardment problem seems to be exactly the inverse of the deposition processes – indeed, they have a number of common features. The bombarding ions arrive randomly, acting as a random force. Shadowing may play a key role, since tips that are more exposed erode more easily than the shadowed valleys. Surface diffusion may act as a smoothing mechanism for higher temperatures. However, there is one main difference: in deposition processes, atoms arrive only on the surface, while the eroding ion can penetrate into the bulk of the material, releasing its energy for the surrounding atoms along its path. High energy ions can even locally melt the surface, releasing a large number of atoms.

In the transport theory of Sigmund [399, 400], the incoming ions spread their kinetic energy inside the solid by inducing cascades of collisions among the substrate atoms, or through other processes, such as electronic excitations. The scattering events take place within a certain layer of average depth a, whereas most atoms that are released are those located on the surface. While this latter fact suggests that sputtering may be treated as a purely surface phenomena, in fact the growth equation is determined by the interaction between the substrate and the bulk.

Figure 19.14 sketches a simple microscopic model for sputtering. Ions penetrate a distance a inside the solid, where they spread their energy radially with an assumed distribution. The actual shape of the distribution is determined by parameters such as angle of incidence, the structure of the sample and the features of the microscopic scattering processes. The velocity of erosion at a given surface point A is proportional to the total power reached there by the deposition of the ion at point O. A simple argument convinces us that the erosion velocity depends on the curvature of the interface [53]. Consider two interface morphologies, as shown in Fig. 19.14. For a uniform flux of incoming ions, one can see that point A receives more energy than point A', so the erosion velocity in A will be larger. This suggests that the erosion velocity is proportional to the negative Laplacian of the local interface height, so in the linear approximation the growth equation should be the EW equation with a negative ν. A more detailed analysis that considers nonlinear terms as well shows that the actual growth equation is the anisotropic KPZ equation [55, 90] (see Chapter 25), with the coefficients depending on parameters such as the penetration depth and the angle of incidence. Moreover, varying the angle of incidence, the signs of the coefficients λ_x and λ_y, ν_x and ν_y change, indicating that the scaling exponents describing the roughening process may depend on the angle of incidence.

A growth equation with negative ν is linearly unstable. However, surface diffusion induces an additional $-K\nabla^4 h$ term in the growth equation (see §13.1), that stabilizes the effect of the instability for short length scales. The competition between $\nu_x \nabla^2 h$ and $-K\nabla^4 h$ leads to a characteristic length scale, $L_R = \sqrt{K/|\nu_x|}$, that is seen in experiments as the wavelength of a ripple structure.

In conclusion, there are three interesting regimes. The short-range structure can be described in terms of the linear theory, and predicts a ripple structure. The long-wavelength structure, however, is described by the anisotropic KPZ equation. The scaling behavior is controlled by either the EW or the KPZ equation, depending on the angle of incidence of the incoming atoms and the properties of the eroded material. Next we review the pertinent experiments, starting with the short wavelength behavior, and going toward the asymptotic scaling.

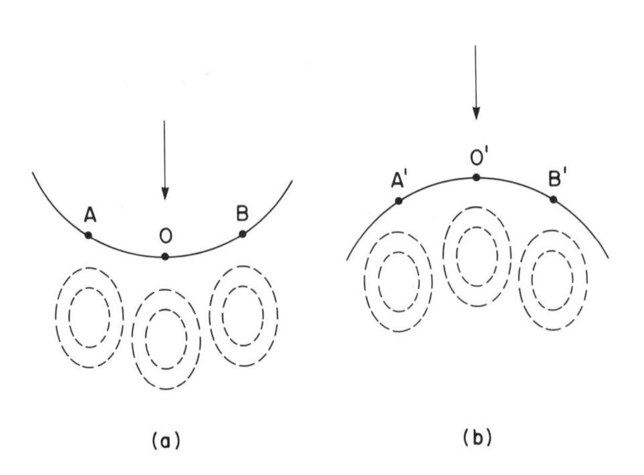

Figure 19.14 A normally incident ion beam striking (a) a valley and (b) a crest. The arrows indicate the beam direction. Contours of equal energy deposition (dotted lines) are shown for ions striking the surface at points O and O'. (After [53]).

19.4.2 Experiments on ion bombardment

Evidence for the ripple structure on the surfaces of SiO_2 and Ge has been provided in a series of studies by Chason *et al.* [72, 73, 74, 301]. We shall discuss here the results obtained on SiO_2 [301]. A low energy ion beam (Xe, H or He), with energies $\leq$ 1 KeV is directed towards a SiO_2 sample with an angle of incidence of 55° from normal. The typical incoming flux is 10^{13} $cm^{-2}s^{-1}$. The interfaces are analyzed using *in situ* energy dispersive x-ray reflectivity and *ex situ* atomic force microscopy (AFM).

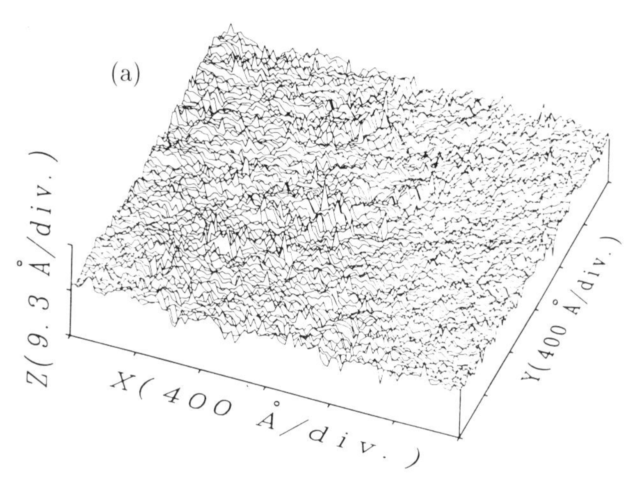

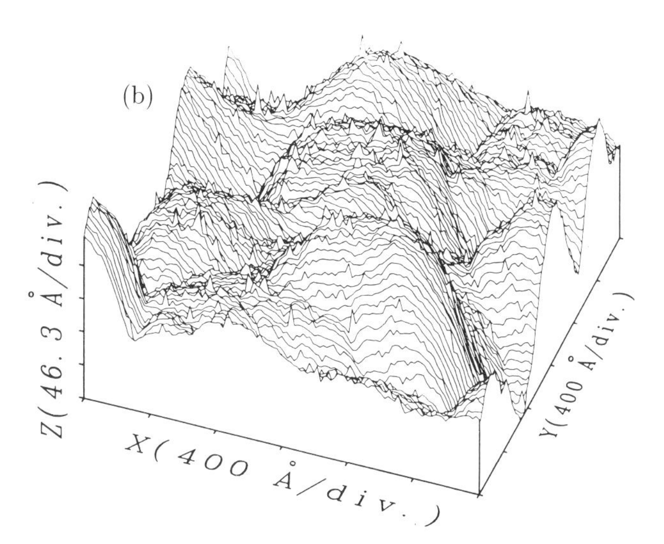

Figure 19.15 (a) STM topograph of a graphite surface after bombarding with a flux $F = 6.9 \times 10^{13}$ ions/cm^{2}s^{-1} and an ion fluence $Q = 10^{16}$ ions/cm^2 at room temperature; the sample size is 2400 Å ×2400 Å, and the vertical size is 18.6 Å. (b) Same but with $Q = 10^{18}$ ions/cm^2. The vertical size is 231.5 Å. (After [114]).

Bombarding the surface with 1 keV Xe ions, one finds that the interface roughness, determined from X-ray diffraction, increases linearly with the fluence (the fluence is the number of incoming atoms per surface area, and plays the role of time in these measurements). Thus $\beta = 1$, too large a value to be interpretable by continuum theories. Such a large value of β indicates the existence of an instability in the system. Indeed, as discussed above, the source of the instability is the negative surface tension. But the instability should be balanced by surface diffusion, leading to the appearance of the ripple structure. As Fig. 1.11(b) indicates, such a ripple structure can be seen if one inspects the AFM pictures of the interface. A similar ripple structure has been observed for Ge surfaces bombarded by Xe atoms [73].

These experiments provide evidence concerning the short-scale instability and ripple structure generated by the negative surface tension. However, the large-scale behavior may be described by the anisotropic KPZ equation. Evidence for algebraic scaling of the local width was provided by the measurements of Eklund *et al.* and Krim *et al.* [114, 242].

In the experiments of Eklund *et al.*, pyrolytic graphite was bombarded by 5 keV Ar ions, which arrived with an angle of incidence of 60°. The experiments were carried out for two flux values, 6.9×10^{13} and 3.5×10^{14} ions/cm^2, and the total fluences obtained were 10^{16}, 10^{17} and 10^{18}. The etched graphite was examined using STM. Figures 19.15 and 1.11(a) show the STM topograph of the graphite surface for three different fluences. We observe that large scale features de-

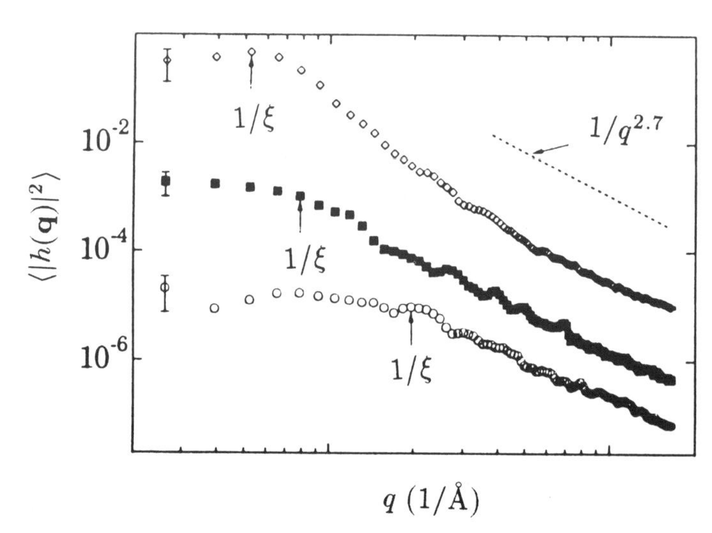

Figure 19.16 The height–height correlation function $\langle |h(\mathbf{q}, t)^2| \rangle$ for the interfaces shown in Fig. 19.15 and 1.11, with $Q = 10^{16}$ ions/cm^2 (○) $Q = 10^{17}$ ions/cm^2 (▪), $Q = 10^{16}$ ions/cm^2 (◇). The crossover vector $q_\times = 1/\xi$ is marked on the figure, and the $1/q^{2.7}$ dependence is found for large q. (After [114]).

velop with continuous bombardment, the interface becoming highly correlated and rough. The scaling properties can be probed using the height–height correlation function, whose Fourier transform is shown in Fig. 19.16. The data indicate a scaling region for large q, and saturation for small q values (corresponding to large length scales). The results indicate a dynamic exponent z in the range 1.6–1.8, and a roughness exponent in the range 0.2–0.4. The exponents are consistent with the predictions of the anisotropic KPZ equation for the strong coupling regime, whose exponents are $z \approx 1.6$ and $\alpha \approx 0.38$.

A somewhat larger roughness exponent has been measured for samples of iron bombarded with 5 keV Ar arriving with angle of incidence of 25°. The interface morphology was observed using STM, and the height–height correlation function results in a roughness exponent $\alpha = 0.53 \pm 0.02$ [242]. The mechanism leading to such a roughness exponent is not yet understood in terms of the continuum theories, since for two dimensions the growth equations predict 0.38, 2/3 and 1 (see Table 27.2), all values far from the observed value.†

19.5 Discussion

While many growth models are local, as some of the phenomena discussed in this chapter illustrate, there are a large number of growth processes that are governed by nonlocal effects. Nonlocal growth models are more difficult to work with than local growth models. A typical example is DLA: despite the simplicity of the model, there is no theory that can predict its measured fractal dimension.

The origin of nonlocality in the examples discussed is shadowing: the growth rate at a given point of the interface is determined by remote sites that capture the influx of particles. Usually this leads to an instability, and the continuum growth equations are no longer applicable. In some situations we are able to write down such a growth equation, but they usually serve only to predict the features of the instability, and do not lead to stable growth morphologies, as the local growth models do.

Suggested further reading:

[311, 395, 456]

† In some limits the anisotropic KPZ equation predicts logarithmic scaling. We are not aware of any experimental work observing this scaling regime.

Exercises:

19.1 Consider the following variant of the needle model: the initial column heights are selected randomly from some probability distribution $P(h_i(0))$. The columns are allowed to grow with constant velocities, given by

$$\frac{dh}{dt} = [h_i(0)]^{\alpha}, \tag{19.10}$$

for a short time interval δt, and shaded columns are removed from the list of active columns. The column growth velocities are then recalculated from (19.10). (a) Assuming that initially the columns are uniformly distributed between [0,1], show that the exponent γ defined in (19.3) is given by $\gamma = 1+\alpha/2$. (b) Calculate γ if $P(x) = (1+v)(1-x)^v$. (c) Similarly, calculate γ if the heights are power-law distributed, following (23.1).

19.2 The most frequently used method of determining the fractal dimension of a DLA cluster is measuring the density-density correlation function $n(\mathbf{r}) \equiv \langle \rho(\mathbf{r}+\mathbf{r}') - \rho(\mathbf{r}') \rangle_{\mathbf{r}'}$, where $\rho(\mathbf{r})$ is one (zero) if the site at position $\mathbf{r}$ is occupied (empty). How do we expect $n(\mathbf{r})$ to scale with $|\mathbf{r}|$?

20 Diffusion bias

We have seen that surfaces grown by MBE are rough at large length scales. Moreover, the dynamics of the roughening process follows simple power laws that are predictable if one uses the correct growth equation. In our previous discussion, we neglected a particular property of the diffusion process, the existence of the Schwoebel barrier, biasing the atom diffusion (see §12.2.4). In this chapter we show that this diffusion bias generates an instability, which eventually dominates the growth process. The growth dynamics do not follow the scaling laws discussed in the previous chapters and the resulting interface is not self-affine.

20.1 Diffusion bias and instabilities

We saw in §12.2.4 that the existence of an additional potential barrier at the edge of a step generates a bias in the diffusion process, making it improbable that an atom will jump off the edge of the step. Next we investigate how one can incorporate this effect into the continuum equations.

A nonzero local slope corresponds to a series of consecutive steps in the surface (see Fig. 20.1). Suppose an atom lands on the interface and begins to diffuse. If it reaches an *ascending* step, it sticks by bonding with the atoms of the step. If it diffuses toward the edge of a *descending* step, there is only a small probability the particle will jump down the step, since the edge barrier will reflect the particle back. Thus there is an asymmetry in the diffusion process: the particles will tend to move uphill on a stepped surface (to the right on Fig. 20.1), generating an average current†

† The uphill current does not mean that the particles jump up the step. They are, in fact, 'captured' by the ascending step; however, they exhibit a greater probability of moving towards the ascending step than towards the descending.

$$\mathbf{j} \propto \nabla h, \qquad (20.1)$$

which, according to (13.2), results in the growth equation

$$\frac{\partial h}{\partial t} = -|\nu|\nabla^2 h + \eta(\mathbf{x}, t), \qquad (20.2)$$

for which the Laplacian has a *negative* coefficient. However, a growth equation with a negative Laplacian is unstable, so the existence of an edge barrier leads to an instability in the growth equation.†

The result of this instability can be seen by inspecting the diffusion process taking place on a surface. Suppose a fluctuation creates a 'bump' on an otherwise flat surface. Since newly-arriving atoms will generate an uphill current by moving upward rather than downward, the bump will grow larger and the local slope will increase. This process does not stop; the steeper the slope, the stronger the current [in accord with (20.1)], resulting in an exponential growth of the bump.

20.2 Nonlinear theory

We noted above the presence of an instability, but did not address the fate of the surface over a long time period. In this section we will discuss the long-time behavior [205]. There is an important parameter in the problem: the diffusion length of the deposited atoms, ℓ_d, which is the typical distance traveled by an atom before it is incorporated into an island or step. Atoms do not diffuse indefinitely on a surface; they form islands. As discussed in detail in Chapter 17, the diffusion length of an atom depends on the deposition flux F as $\ell_d \sim (D/F)^{\psi_d}$, where ψ_d is an exponent whose value is determined by the actual mechanism of island formation.

Suppose a surface has a small slope $m \equiv |\nabla h|$ (see Fig. 20.2). A small slope corresponds to a large terrace width ℓ, since $\ell \sim 1/m$. If

† The fact that the growth equation is unstable can be seen if we Fourier transform (20.2) with respect to space, resulting in $\partial_t h(\mathbf{k}, t) = |\nu|k^2 h(\mathbf{k}, t) + \eta(\mathbf{k}, t)$. This equation when integrated leads to an exponential increase in $h(\mathbf{k}, t)$ for every mode k.

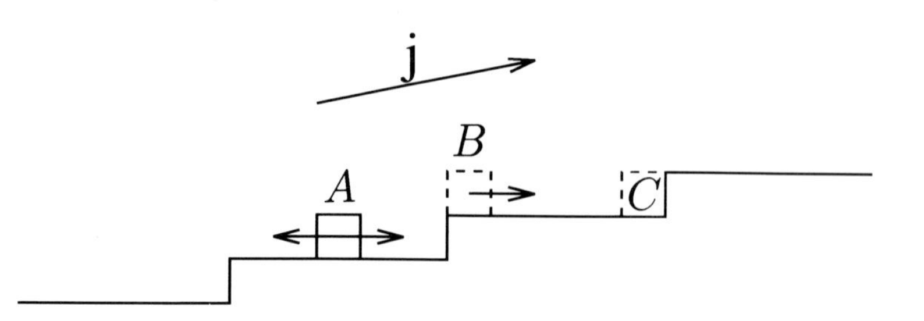

Figure 20.1 Illustration of the uphill current generated by the edge potential on a stepped surface. Although atom A has an equal chance of moving to the left or to the right, if it moves to the left it will reach the edge of the step and be reflected back by the edge potential (see Fig. 12.5) (e.g., atom B is more likely to move to the right than to the left). If A moves to the right, it will meet the ascending step and stick (e.g., atom C). As a result, since atoms move with a higher probability to the right than to the left, there is overall *uphill* current, $\mathbf{j}$.

the terrace width is larger than ℓ_d, islands form on the steps. Only those atoms within a strip of size ℓ_d from the edge will reach the edge of the upper step. Other atoms will become attached to the edge of the islands and will not contribute to the uphill current. The current is proportional to the number of atoms landing in the strip ℓ_d near the terrace, giving

$$j(m) \sim F\ell_d^2 m. \tag{20.3}$$

If, on the other hand, m is so large that $\ell < \ell_d$, all atoms will reach the terrace. If the terrace barrier is infinite so that no particle is able to jump off the step, then the uphill current is given by the number of atoms landing on the surface,

$$j(m) \sim F\ell \sim F/m. \tag{20.4}$$

Although we do not know the exact form of the current $j(m)$, we can approximate it using a simple formula that interpolates between (20.3) and (20.4) [205] (see Fig. 20.3)

$$j(m) \sim F\ell_d^2 \left(\frac{m}{1 + (\ell_d m)^2} \right). \tag{20.5}$$

The growth equation describing the time evolution of the interface can be obtained from (20.5), giving [182, 205]

$$\frac{\partial h}{\partial t} = -\nabla \left(\frac{\nabla h}{1 + (\nabla h)^2} \right) - K\nabla^4 h + \eta(\mathbf{x}, t) \qquad \rightarrow (20.6)$$

where, in order to simplify the notation, we assume the constants to be equal to one, and we have incorporated a surface diffusion term (without which the equation is ill-behaved).

The growth equation (20.6) is nonlinear, and is different from the continuum equations discussed in the previous chapters. The growth

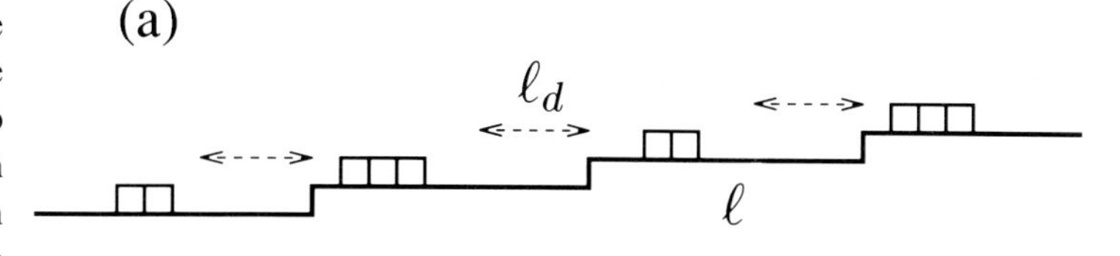

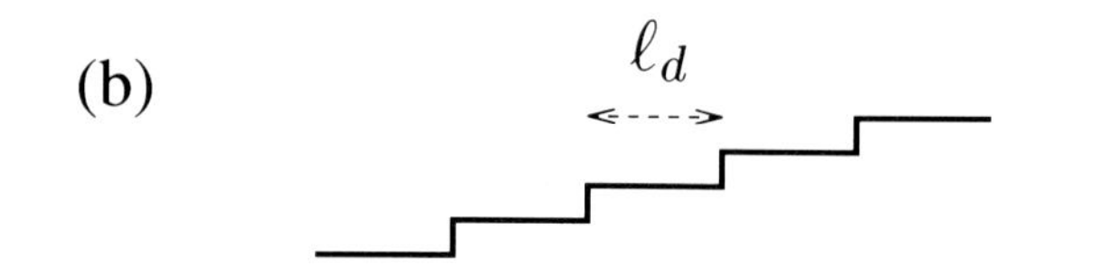

Figure 20.2 Cross-section of a stepped interface illustrating the factors influencing the uphill current. (a) If the slope m is so small that the diffusion length ℓ_d of the deposited atoms is smaller than the terrace size ℓ, islands will form on the terraces. Only those atoms within a strip of width ℓ_d (shown by arrows) will reach the ascending step and thus contribute to the uphill current. (b) If the terrace size is smaller than ℓ_d, all of the atoms will reach the edge.

equations discussed thus far were derived in the small slope approximation. Indeed, if it were true that $|\nabla h| \ll 1$, then we could expand (20.6) and obtain the KPZ equation with a negative Laplacian. This equation describes the early time evolution of the interface, for which $|\nabla h| \ll 1$. However, due to the instability, the interface will develop larger slopes, so the small slope expansion must break down. We expect that (20.6) should describe correctly both the early- and late-time behavior of the interface, but due to its unusual form, scaling arguments and RG cannot be used.

Let us consider a flat surface on which particles are deposited. The deposition process will lead to the formation of islands that, in turn, generate nonzero slopes on the surface. According to (20.5), an uphill current should develop, leading to the further increase of these slopes. As a result, the surface develops a number of 'bumps' or 'mounds.' The mounds are not stable, and their lateral and vertical size increases as the local slopes increase. Larger mounds will cover smaller ones during this process, leading eventually to a single large mound on the surface. Since the current is always positive, this coarsening process does not stop. Eventually the increase in the slope of the mounds will decrease; this occurs when the slope exceeds m^*, the value of slope where the current has a maximum (see Fig. 20.3).

We conclude that, since the current is positive for all m, the mounds do not have a characteristic slope and the local slopes increase indefinitely. A characteristic slope is selected only if the current becomes zero at a critical slope m_c (see the dashed line on Fig. 20.3). In this case, once the slope of the mound reaches m_c, the uphill current van-

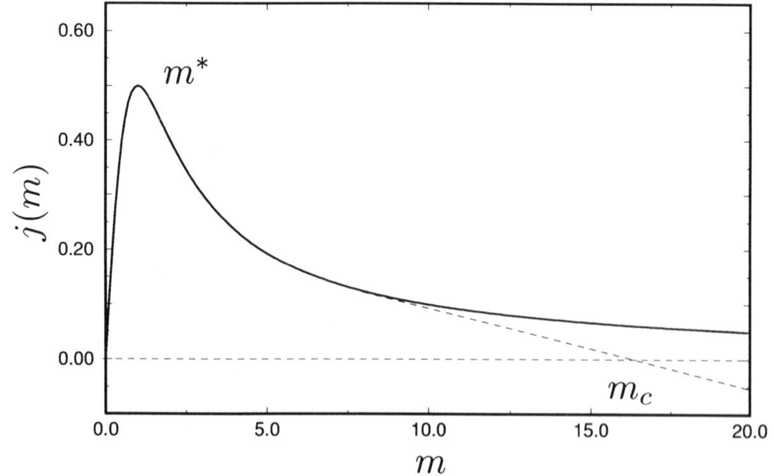

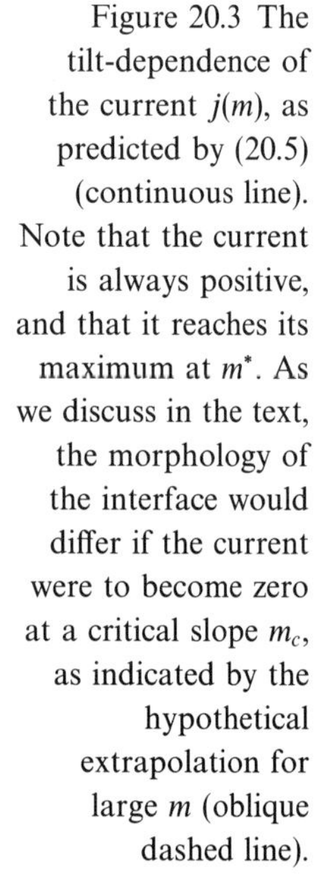
Figure 20.3 The tilt-dependence of the current $j(m)$, as predicted by (20.5) (continuous line). Note that the current is always positive, and that it reaches its maximum at m^*. As we discuss in the text, the morphology of the interface would differ if the current were to become zero at a critical slope m_c, as indicated by the hypothetical extrapolation for large m (oblique dashed line).

ishes and the slope stops increasing. In the presence of such a current the final surface is formed by pyramids with a characteristic slope m_c, similar to facets on the surface. Recently it has been suggested that lattice effects may lead to the selection of a characteristic slope of the surface [398].

Probably the most important difference between the roughening generated by the Schwoebel barrier and the kinetic roughening previously discussed is that the former does not follow the scaling law (2.8), which describes the dynamics of most growth processes. The reason for this lies in the unstable nature of the growth, which generates an exponential increase in the interface roughness for early times. Thus the exponent β is not well-defined in this case.

The saturation phenomenon has a different origin as well. The interface is formed by adjacent mounds that have a relatively smooth shape, and so is not self-affine. Over time these mounds increase, and the saturated state is reached when a single mound covers the entire interface. However, the width will reach saturation only if there is a characteristic slope for the mounds, while in the interface characterized by the current (20.5), the slope of this single mound will continue increasing indefinitely.

20.3 Discrete models

On the basis of simple physical considerations, we postulated the general form of the current given by (20.5). In this section we show that numerical simulations incorporating the Schwoebel barrier confirm the general form of this current. Suppose we have a deposition model with activated diffusion of the type discussed in §15.2. We can simply and

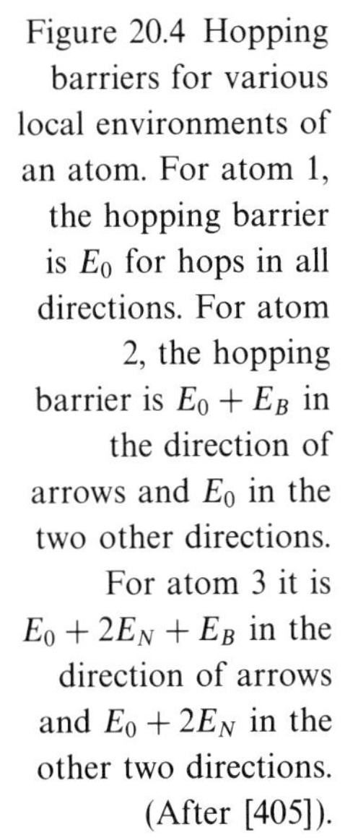
Figure 20.4 Hopping barriers for various local environments of an atom. For atom 1, the hopping barrier is E_0 for hops in all directions. For atom 2, the hopping barrier is $E_0 + E_B$ in the direction of arrows and E_0 in the two other directions. For atom 3 it is $E_0 + 2E_N + E_B$ in the direction of arrows and $E_0 + 2E_N$ in the other two directions. (After [405]).

efficiently incorporate the Schwoebel barrier by considering an energy barrier for hopping that depends not only on the nearest neighbors, but on the next-nearest-neighbors as well [205, 264, 404, 405, 406].

The hopping barrier in (15.2) is comprised of a substrate term E_0, a contribution E_N from each nearest neighbor in the same plane, and a step barrier E_B. Thus $E_D = E_0 + nE_N + E_B$, where $n = 0, \ldots, 4$ is the number of nearest neighbors (see Fig. 20.4). The vicinity of a step is detected by counting the number b_1 of next-nearest neighbors in the plane beneath the hopping atom before a hop and the number b_2 after a hop. The barrier E_B is nonzero if $b_1 > b_2$. A control parameter in the model that reflects the relevance of the Schwoebel barrier is $S_c \equiv 1 - \exp(-nE_B/k_BT)$. If $E_B = 0$, we have $S_c = 0$ and S_c increases towards 1 as the diffusion barrier, E_B/k_BT, increases.

Starting from a flat interface, Monte Carlo simulations using the hopping barrier described above indicate the formation of large mounds that increase and coarsen over time (see Fig. 20.5). We can use this discrete model to test the correctness of the current (20.5). By tilting the interface we can calculate the dependence of the current on the local slopes (see Appendix A.4). The results of such a calcula-

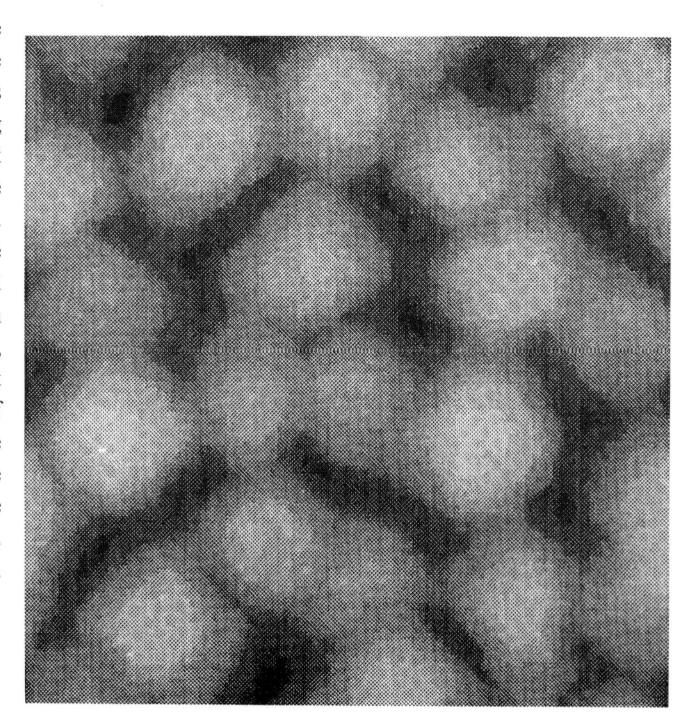

Figure 20.5 Interface generated by Monte Carlo simulations using a hopping energy that incorporates the Schwoebel barrier. The figure shows the interface generated after the deposition of 50 monolayers, starting from a flat (singular) interface of size 200 × 200. The parameters of the simulation are $S_c = 0.4$ and $\ell_d = 16$. (After [205]).

tion for the model are shown on Fig. 20.6, and we see that the current has the form predicted by (20.5).

20.4 Experimental support

Since most materials have a nonzero Schwoebel barrier, one should be able to observe the predicted instability and mound formation experimentally. Indeed, recently a number of experimental investigators have reported observing these large scale structures on growing interfaces. For example, in the late stage of growth of GaAs on GaAs(001), atomic force microscopy (AFM) studies have revealed the existence of large elongated mounds (see Fig. 20.7) [205]. The elongated shape of the mounds is caused by the anisotropy in the sticking probabilities and the diffusion process. The observed mounds are 8 nm high and have a typical size of 0.5 μm × 1.5 μm in horizontal directions, revealing that the typical angle of inclination is around 1 or 2 degrees for samples grown at 550 C.

A different study, on metal surfaces, arrived at similar conclusions regarding the late-time morphology of the interface [115]. During the deposition of Cu on Cu(100), the interface was observed to evolve into a state characterized by pyramid-like structures. Helium scattering studies indicate that the sides of these pyramids correspond to a Cu(115) surface at 200 K and a Cu(113) at 160 K. For short times, the interface width was observed to increase as a power of time –

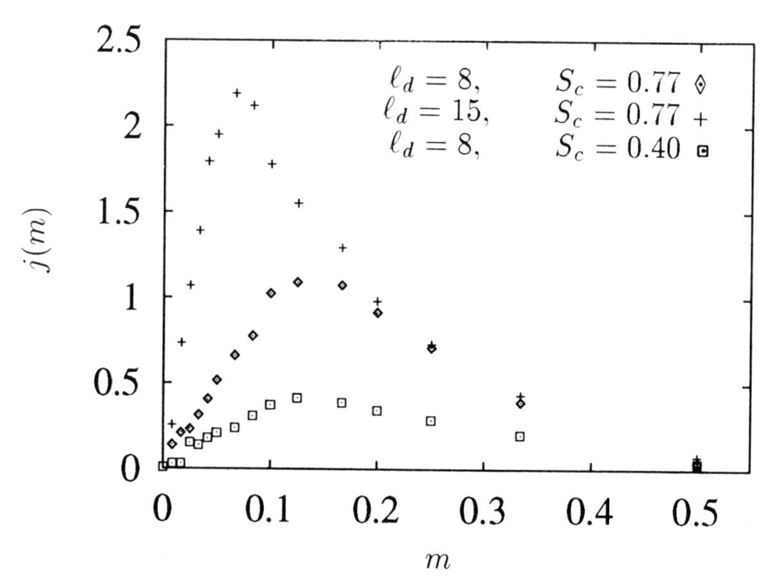

Figure 20.6 The tilt dependence of the current $j(m)$, as measured for the model with activated diffusion incorporating the Schwoebel barrier. The different curves correspond to different simulation parameters, as indicated in the figure. Note that the shape of the curves is very similar to the prediction (20.5), shown in Fig. 20.3. (After [205]).

more precisely, as a function of the number of monolayers deposited, which corresponds to time in the system – and β was found to be extremely sensitive to temperature. For $T = 200$ K, $\beta = 0.56$, while for $T = 160$ K, $\beta = 0.26$. For $T = 200$ K, $\beta > 1/2$, suggesting that there is an instability in the system (consistent with our finding in the previous section that early-time growth must be exponential).

For $T = 200$ K, a clear indication of the faceted structure has been found. However, at 160 K the diffraction measurements indicate that the pyramid-like structures are less well developed. The exponent β is much smaller as well, perhaps an indication that, at this temperature, the development of the diffusion-bias-generated instability is much slower, and kinetic theories may apply in the description of early-time behavior.

While the previous experiments were interpreted in terms of the diffusion bias, a number of earlier measurements had already observed the development of large mounds on the surface. For example, AFM measurements indicated the formation of these mounds for GaAs(001) deposition [407], and highly anisotropic mounds have been observed during the growth of InP(100) films as well [86].

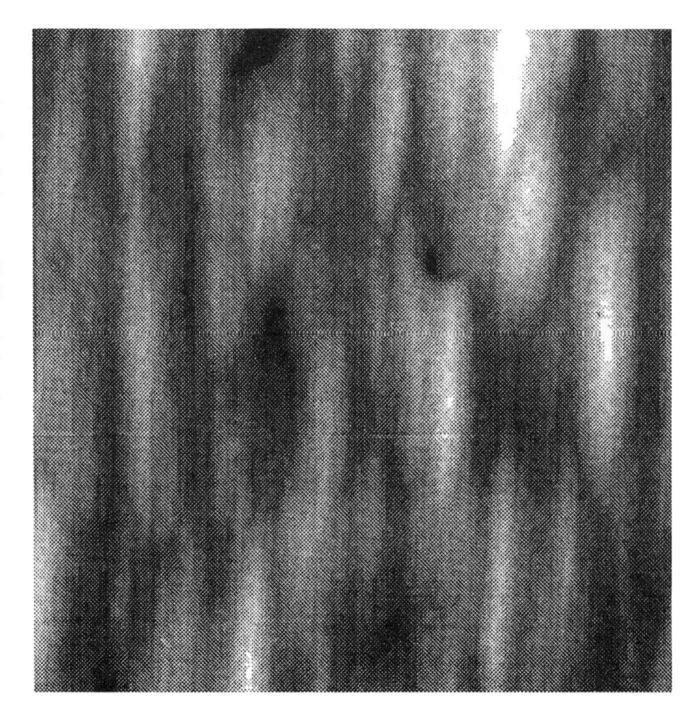

Figure 20.7 AFM image of a GaAs surface after deposition of 500 bilayers. The elongated shape of the mounds is due to the anisotropic diffusion constant of the deposited adatoms (After [205]).

20.5 Discussion

Most materials possess a nonzero Schwoebel barrier. However, the nonzero Schwoebel barrier may vanish accidentally if some unusual circumstance occurs. Thus the final morphology of much interface growth by MBE is dominated by large-scale structures, which we have called mounds or 'islands', that do not scale in a way we would expect for the nonequilibrium models discussed in previous chapters. In particular, the early-time behavior does not follow a power law, and the resulting interface is not self-affine.

These facts could lead us to the rather disappointing conclusion that studying various growth mechanisms and continuum growth equations is fruitless, since none of them applies to real MBE surfaces. If there is a nonzero Schwoebel barrier, its effect will determine the final morphology of the interface. However, reaching the final state may be a very lengthy process. The size of the Schwoebel barrier varies from material to material. The transient time for the mounds to develop may be very long if the barrier is small. During this transient time, the mechanisms leading to the continuum equations dominate, and one should be able to observe signs of the kinetic roughening process.

The existence of the Schwoebel barrier affects our interpretation of experimental results. We must take into account its existence, and determine carefully whether the roughening we observe is generated by the diffusion bias or by kinetic roughening.

There are a number of open questions. One concerns the existence of the critical slope m_c. The simplest current, given by (20.5), does not predict any critical slope. However, some experiments indicate the development of stable critical slopes. Thus further work is needed to understand the origin of the selection mechanism leading to a stable slope, and the dependence of the characteristic angle on the properties of the material. Another intriguing challenge is how to relate mound formation to equilibrium faceting, a question studied in great detail for the case of vicinal surfaces [478].

Suggested further reading:

[205, 463]

Exercises:

20.1 Find the free energy which generates the growth equation (20.6). Show that the time-independent solutions of the evolution equation can be thought of as an equation of motion moving in a potential valley. What is the form of the potential?

PART 5 Noise

21 Diffusive versus deposition noise

In Chapters 12–15, we discussed in detail the properties of growth processes dominated by deposition and surface diffusion. We saw that one origin of randomness is the stochastic nature of the deposition flux, which generates a *nonconservative* noise in the growth equations. A second component of randomness on a crystal surface comes from the activated character of the diffusion process. As we show in this chapter, this type of randomness generates a *conservative* component to the noise, leading to different exponents and universality classes.

While diffusive noise is certainly expected to be present in MBE, there is no experimental evidence for the universality class generated by it. For this reason, we separate it from the discussion of the simple MBE models. This chapter can be considered to be a theoretical undertaking of interest in its own right, investigating the effect of the conservation laws on the universality class.

21.1 Conservative noise

An important contribution to randomness on a crystal surface arises from the activated character of the diffusion process. Equation (13.5) results from a deterministic current that contributes to interface smoothing – i.e., only particle motion that aids smoothing is included. But, as discussed in Chapter 12, particle diffusion is an activated process in which all possible moves – each with its own probability – are allowed. Because the nature of the diffusion process is probabilistic, an inherent randomness is always present. Our equation therefore requires an additional term, $\eta_d(\mathbf{x}, t)$, to account for the diffusive noise.

The diffusive noise η_d is qualitatively different from the *deposition* noise η. The difference lies in the conservative character of η_d: it never changes the number of atoms existing on the surface. If η_d changes the height of the interface at a given point, it does so by moving particles into or away from that position, but any increase (decrease) in the height is accompanied by a decrease (increase) in another nearby position on the interface. Thus the integral of the diffusive noise over the entire interface must remain zero at every moment

$$\int d^d\mathbf{x}\ \eta_d(\mathbf{x}, t) = 0. \tag{21.1}$$

This property differs from the zero-average requirement of deposition noise: Eq. (13.6) shows that the deposition noise has a zero average if we integrate over many realizations of the noise, but may differ from zero if the number of realizations is small.

It can be verified that any noise term with the properties

$$\langle \eta_d(\mathbf{x}, t) \rangle = 0 \tag{21.2}$$

and

$$\langle \eta_d(\mathbf{x}, t)\eta_d(\mathbf{x}', t') \rangle = -2D_d\nabla^2\delta(\mathbf{x} - \mathbf{x}')\delta(t - t') \tag{21.3}$$

automatically satisfies the conservation requirement for conservative noise discussed above.

21.2 Linear theory

Adding diffusive noise to the growth equation (13.15), we obtain a more general linear growth equation for MBE growth

$$\frac{\partial h}{\partial t} = -K\nabla^4 h + \nu\nabla^2 h + F + \eta(\mathbf{x}, t) + \eta_d(\mathbf{x}, t), \qquad \rightarrow (21.4)$$

where F is the deposition flux. Before discussing the scaling predicted by (21.4) in detail, we must have a closer look at the competition between the nonconservative deposition noise η and conservative diffusive noise η_d. Rescaling the two noise terms using $\mathbf{x} \rightarrow b\mathbf{x}$ and $t \rightarrow b^z t$, we obtain $\eta \rightarrow D^{1/2}\eta b^{-(d+z)/2}$ and $\eta_d \rightarrow D_d^{1/2}\eta_d b^{-(d+z+2)/2}$, and find a characteristic length scale

$$L_2 = \left(\frac{D_d}{D}\right)^{1/2} \tag{21.5}$$

which separates two scaling regimes. Comparing them, we note that for $L \ll L_2$ the diffusive noise dominates, while in the hydrodynamic limit $L \gg L_2$, the deposition noise dominates.

21.3 Scaling regimes

For $L \gg L_2$, the diffusive noise η_d is negligible compared to the deposition noise η, and we may set $D_d = 0$. In this limit, we recover the growth equation (13.15). The possible scaling regimes with the corresponding exponents were discussed in §13.2. New scaling behavior is expected only for $L \ll L_2$, when η_d dominates. To simplify, we take $D = 0$. However, we have two different scaling regimes, depending on whether L is larger or smaller than L_1, introduced in Eq. (13.16).

- *Regime I: $L \ll L_1$*
  In regime I, $-K\nabla^4 h$ determines the scaling behavior, the relevant growth equation being
  $$\frac{\partial h}{\partial t} = -K\nabla^4 h + \eta_d(\mathbf{x}, t). \tag{21.6}$$
  Scaling arguments give
  $$\alpha = \frac{2-d}{2}, \qquad \beta = \frac{2-d}{8}, \qquad z = 4. \tag{21.7}$$
  For $d = 1$, $\alpha = 1/2$ and $\beta = 1/8$, while for the physically-relevant case $d = 2$ we have $\alpha = \beta = 0$. Thus for $d = 2$, the roughness is at most logarithmic in the regime in which surface diffusion and diffusive noise determine the dynamics of the interface.
- *Regime II: $L \gg L_1$*
  In regime II, we consider $K = 0$, and we have the Edwards–Wilkinson equation (5.6) with conservative noise
  $$\frac{\partial h}{\partial t} = \nu\nabla^2 h + \eta_d(\mathbf{x}, t). \tag{21.8}$$
  After rescaling the growth equations, we obtain for the scaling exponents
  $$\alpha = -\frac{d}{2}, \qquad \beta = -\frac{d}{4}, \qquad z = 2. \tag{21.9}$$
  Thus the EW equation with conservative noise predicts a negative roughness exponent, implying a smooth interface.
- *Estimating L_2*
  A simple calculation demonstrates how L_2 can vary [445]. Diffusive noise arises from surface diffusion. Hence the diffusion constant is proportional to the diffusion probability of the atoms, which can be estimated using the Arrhenius law $D_d \sim \omega_D \exp(-E_0/kT)$, where ω_D is the Debye frequency. Using $\omega_D \sim 10^{13}$ s^{-1}, $E_0 = 1$ eV as the activation energy of diffusion, and a deposition flux $F \sim 1$ Å/s, we obtain $L_2 \ll 1$ Å at room temperature, but $L_2 \sim 1000$ Å at

500°C. The room temperature estimate is smaller that the typical interatom spacing, the measurable scaling being determined only by the deposition noise. The estimate for 500°C, however, is equal to a typical experimental probe, so in this case the observed scaling behavior will be determined by the diffusive noise.

21.4 Nonlinear theory

Not only the linear, but also the nonlinear growth equations are modified by the presence of conservative noise. If both noise terms – conservative and nonconservative – are present, for length scales smaller than the crossover length L_2, the growth process is dominated by η_d. In this scaling regime the relevant diffusive growth equation becomes

$$\frac{\partial h}{\partial t} = -K\nabla^4 h + \lambda_1 \nabla^2 (\nabla h)^2 + \eta_d(\mathbf{x}, t), \qquad \rightarrow (21.10)$$

where the correlations of noise η_d are given by (21.2) and (21.3). This equation has been introduced and investigated by Sun *et al.* [432]. Scaling arguments indicate that the nonlinear term is relevant, and the dynamic RG procedure results in the flow equations

$$\frac{dK}{dl} = K\left[z - 4 + K_d \left(\frac{\lambda_1^2 D_d}{K^3}\right)\left(\frac{4-d}{4d}\right)\right], \qquad (21.11)$$

$$\frac{dD_d}{dl} = D_d[z - 2\alpha - d - 2], \qquad (21.12)$$

$$\frac{d\lambda_1}{dl} = \lambda_1[z + \alpha - 4]. \qquad (21.13)$$

As before, $K_d \equiv S_d/(2\pi)^d$, and S_d is the surface area of the d-dimensional unit sphere. Under rescaling, the flow of the coupling constant $g_d^2 = \lambda_1^2 D_d/K^3$ is governed by

$$\frac{dg_d}{dl} = \frac{2-d}{2} g_d + K_d \frac{3(d-4)}{8d} g_d^3, \qquad (21.14)$$

indicating that the critical dimension of the model is $d_c = 2$.

For $d \geq 2$, the nonlinear term is driven to zero. Hence the nonlinear term is irrelevant for $d \geq 2$, including the physically relevant dimension $d = 2$. For $d = 1$ the nonlinear term is relevant, and we have two scaling relations arising from the non-renormalization of the noise and the nonlinear term λ_1:

$$z - 2\alpha - d - 2 = 0, \qquad (21.15)$$

and

$$z + \alpha - 4 = 0, \tag{21.16}$$

which give for the scaling exponents

$$\alpha = \frac{2-d}{3}, \qquad \beta = \frac{2-d}{10+d}, \qquad z = \frac{10+d}{3}. \tag{21.17}$$

Since $d_c = 2$, the only physical dimension in which (21.7) holds is $d = 1$, for which we obtain $\alpha = 1/3$, $\beta = 1/11$ and $z = 11/3$. For $d \geq 2$, we find the exponents (21.7) predicted by the linear theory (21.6).

21.5 Discussion

The introduction of conservative noise into the growth equation leads to new universality classes. The related scaling exponents can be calculated analytically, using scaling arguments and RG methods. Conservative noise is implicitly included in the growth models with activated diffusion (see §15.2). Moreover, a number of models have been studied that do not have deposition, but the randomness has a conservative character [376, 432].

Suggested further reading:

[376, 432]

Exercises:

21.1 In order to apply the RG method to a system with conserved noise, we often have to use the extended form for the noise correlation function

$$\langle \eta(\mathbf{x}, t)\eta(\mathbf{x}', t')\rangle = (-2D_d\nabla^2 + D_2\nabla^4)\delta(\mathbf{x}-\mathbf{x}')\delta(t-t') \tag{21.18}$$

instead of (21.3). Explain why.

21.2 Analyze the flow equations (21.11)–(21.13). Find their fixed points, analyze their stability, and calculate the corresponding exponents.

22 Correlated noise

The notion of universality suggests that the exponents determined in the previous chapters are unique, since they belong to the only possible growth equation with a set of given symmetries. Any model or experiment will show these exponents if the hydrodynamic limit has been reached. But there is one term in all these equations that we have largely ignored thus far, one whose form is *not* fixed by symmetry principles: the noise η. We have assumed that this noise is *uncorrelated* – i.e., that it has a Gaussian distribution. But is this the only type of noise possible in a physical system?

What if the magnitude of the noise *in a given point* is not independent of the magnitude of the noise in a different point – i.e., what if there are *spatial* correlations in the noise? What if the magnitude of the noise at a given time is not independent of the noise at a different time – i.e., what if there are *temporal* correlations in the noise? In many experimental situations, we know little about the nature of any noise that may be present, so to consider it Gaussian is perhaps an unrealistically simple assumption.

One possibility is that the correlation length of the noise is *finite*, i.e., that the different events 'know' about each other only if they are within a *finite* spatial separation ξ or temporal separation τ. In this case, we can rescale our system with a finite parameter (ξ or τ), so that after re-scaling the noise events will be completely uncorrelated.

In this chapter, we consider the possibility that the noise has *long-range correlations*, i.e., events that are arbitrarily distant may still influence each other. These events, even after re-scaling the noise, display long-range correlations. The strength of their influence decays as a power law of the distance between the two events, compared to short-range-correlated noise, for which this decay is exponential.

Noise with long-range correlations is remarkably ubiquitous in nature, although its origin is not understood.

22.1 Introducing correlated noise

If the noise has long-range *spatial* correlations, the delta function in the spatial coordinates of Eq. (4.9) must be replaced by a term that decays as a power of the distance

$$\langle \eta(\mathbf{x},t)\eta(\mathbf{x}',t')\rangle \sim |\mathbf{x}-\mathbf{x}'|^{2\varphi-d}\delta(t-t'), \qquad \rightarrow (22.1)$$

where φ is an exponent characterizing the decay of spatial correlations. If, on the other hand, the noise has long-range *temporal* correlations, then

$$\langle \eta(\mathbf{x},t)\eta(\mathbf{x}',t')\rangle \sim \delta^d(\mathbf{x}-\mathbf{x}')|t-t'|^{2\phi-1}, \qquad \rightarrow (22.2)$$

where ϕ characterizes the decay of temporal correlations. In general, for a process correlated both in space and time we have

$$\langle \eta(\mathbf{x},t)\eta(\mathbf{x}',t')\rangle \sim |\mathbf{x}-\mathbf{x}'|^{2\varphi-d}|t-t'|^{2\phi-1}. \qquad (22.3)$$

For analytical calculations it is useful to consider the Fourier transformed form of (22.3)

$$\langle \eta(\mathbf{k},\omega)\eta(\mathbf{k}',\omega')\rangle = 2D\, k^{-2\varphi}\omega^{-2\phi}\delta^d(\mathbf{k}+\mathbf{k}')\delta(\omega+\omega'). \qquad (22.4)$$

In this form, the connection between the decay of spatial and temporal correlations and exponents φ and ϕ is more apparent. As we can see, the correlation in frequency decays as $1/\omega^{2\phi}$. If $\phi = 0$, the noise is uncorrelated and 'white,' while $\phi = 1$ for Brownian motion.

22.2 Linear theory with correlated noise

How does correlated ('colored') noise influence the growth exponents? We can answer this question for the linear theory. The exact values of the scaling exponents can be obtained either by using scaling arguments or by solving the growth equation exactly (see Chapter 5).

Let us consider the growth of an interface according to the EW equation (5.6), with spatially- and temporally-correlated noise (22.3). Rescaling the equation, using $\mathbf{x} \to b\mathbf{x}$, $h \to b^\alpha h$ and $t \to b^z t$, and multiplying both sides by $b^{z-\alpha}$, we obtain

$$\frac{\partial h}{\partial t} = \nu b^{z-2}\nabla^2 h + b^{\varphi-d/2+\phi z+z/2-\alpha}\eta. \qquad (22.5)$$

In order to have scale invariance we must set the exponent of b equal to zero. We obtain the scaling exponents

$$\alpha = \psi + 2\phi + \frac{2-d}{2}, \qquad z = 2. \qquad \rightarrow (22.6)$$

According to (22.6) the presence of correlated noise changes the scaling exponents. For uncorrelated noise ($\psi = 0$ and $\phi = 0$), we recover the exponents of the linear theory (5.16). The correlated noise modifies the scaling exponents of the linear theory, increasing the roughness exponent.

The scaling exponents (22.6) can also be obtained by directly solving the growth equation in Fourier space. Repeating the calculations of §5.4, we obtain for the correlation function of the EW model with correlated noise,

$$\langle h(\mathbf{x}, t) h(\mathbf{x}', t') \rangle = \frac{D}{2\nu} |\mathbf{x} - \mathbf{x}'|^{2-d+2\psi+4\phi} f\left(\frac{\nu |t - t'|}{|\mathbf{x} - \mathbf{x}'|^2} \right) \quad [\phi < 1/4]. \tag{22.7}$$

Numerical support for the variation of the exponents if the noise is correlated is provided by the simulations of Meakin and Jullien [308, 309]. They studied the roughening of an interface generated by random deposition with restructuring, a model that belongs to the EW universality class (see §5.1). To generate spatial correlations, particle n was deposited in position $x_n = x_{n-1} \pm \delta x$, where x_{n-1} is the coordinate of particle $n-1$, and δx is selected from a power law distribution

$$P(\delta x > x_0) = x_0^{-\psi/2} \quad [x_0 > 1]. \tag{22.8}$$

Figure 22.1 shows a portion of the interface generated by the model with correlated noise. As a consequence of the correlated character of

Figure 22.1 A typical interface of random deposition with restructuring, grown with spatially correlated noise with $\psi = 0.5$. The figure shows a portion of the interface of size $L = 16\ 384$ (After [308]).

the noise, large correlated slopes characterize the surface (cf. Fig. 22.1 with Fig. 5.2).

22.3 KPZ equation with spatially-correlated noise

Because the KPZ equation is nonlinear, an exact solution is not possible. The situation is even more complicated when correlated noise is present. The scaling exponents can be obtained using dynamic RG calculations. The calculations proceed along the lines presented for the KPZ equation with uncorrelated noise [314]. Hence we do not present the details here, but rather concentrate on the results. One important result is that Galilean invariance is not destroyed by spatially-correlated noise, so the scaling relation (6.13) remains valid.

Let us first discuss the case of spatially-correlated noise. Analytical results are available for one dimension only. There are three different regimes as a function of the value of ψ,

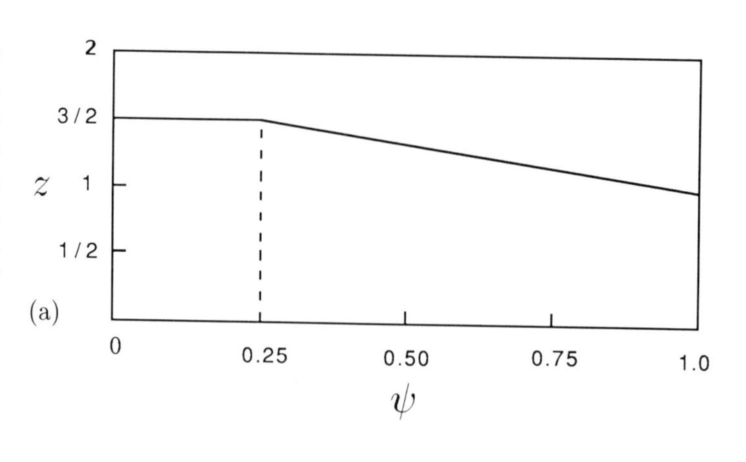

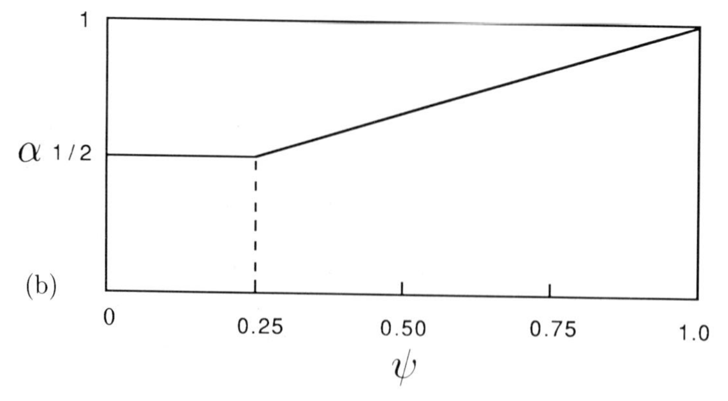

Figure 22.2 Dynamic exponent z and roughness exponent α as a function of the correlation exponent ψ for a one-dimensional interface (After [314]).

- (i) For $\psi < \psi_o$, where

$$\psi_o = \frac{d(d-2)}{8(d-3/2)} = \frac{1}{4} \qquad [d=1], \tag{22.9}$$

long-range correlations are irrelevant because the behavior is controlled by a fixed point determined only by the uncorrelated noise. Thus we have the well-known exponents of the KPZ equation, $\alpha = 1/2$ and $z = 3/2$.

- (ii) For $\psi > \psi_o$ the spatially-correlated noise changes the exponents to

$$z(\psi) = 1 + \frac{d+1-2\psi}{3}, \qquad \alpha(\psi) = 1 + \frac{2\psi - d - 1}{3}. \quad \to (22.10)$$

The values of z and α for $d = 1$ as functions of ψ are sketched in Fig. 22.2.

The presence of spatial correlations in noise can increase the roughness exponent (and decrease the dynamic exponent z). If we identify a universality class by its exponents, then we have a family of continuously-changing universality classes, the exponents being fixed by the value of ψ that describes the decay in the noise correlations.

- (iii) Finally, for $\psi > \psi_c$, where

$$\psi_c = \frac{d+1}{2} = 1, \tag{22.11}$$

the roughness exponent α given by (22.10) becomes larger than 1. In this regime $\nabla h \to b^{\alpha-1}\nabla h$ grows under rescaling, so higher order nonlinearities become relevant. Hence the KPZ equation must be modified to include higher powers of (∇h), such as $(\nabla h)^4$, $(\nabla h)^6$, etc.

Considerable effort has been invested in checking predictions (22.10) using numerical simulations. The numerical investigations concentrated on discrete models (e.g, BD [11, 308, 309, 364]), SOS models [11, 291], and direct solving of the KPZ equation with correlated noise [364]). Although it is generally agreed that the scaling exponents vary continuously with ψ, as predicted by the theory, different investigations disagree on the extent to which the results follow relations (22.10). Peng *et al.* found systematic deviations from the predicted values, while others found better agreement with (22.10) – apart from some deviations at the critical point [309, 364].

As is the case with uncorrelated noise, the scaling exponents for $d \geq 2$ cannot be obtained using dynamic RG analysis. However, based on the fact that correlated noise modifies the scaling exponents in one dimension, we expect this effect to hold in higher dimensions as well. Numerical results obtained for BD with correlated noise indicate that correlated noise modifies the scaling exponents in two dimensions [309].

Before concluding this section, we must mention that the RG results are obtained using one loop calculations, and it is possible that further corrections to the exponents might exist if one calculates higher loops. Motivated by the approximative nature of the RG method, a number of investigations attempted to find alternative methods to determine the scaling exponents in the presence of correlated noise. In the language of directed polymers in random media (which can be mapped onto the KPZ equation), Halpin-Healy investigated an equivalent problem using functional RG methods, finding complete agreement with the RG predictions [158].

Zhang, using a replica method, found the somewhat different results [505]

$$\beta = \frac{1+2\psi}{3+2\psi} \qquad [0 < \psi \leq 1/2] \tag{22.12}$$

and

$$\beta = \frac{1+2\psi}{5-2\psi} \qquad [1/2 < \psi \leq 1]. \tag{22.13}$$

A critical study regarding the correctness of the approximations involved in these approaches has not yet been carried out, so further analytical and numerical work may help provide insights.

22.4 KPZ equation with temporally-correlated noise

For spatially-correlated noise, the exponents can be obtained by the RG method for one dimension. However, for temporally-correlated noise the situation is more complicated. The most serious difficulty arises from the fact that for growth with temporally-correlated noise, Galilean invariance does not hold. As a result, the coefficient λ of the non-linear term in the KPZ equation will be renormalized and the exponent identity (6.13) will no longer hold. For growth with both temporal and spatial correlations the scaling relation

$$z(1+2\phi) - 2\alpha - d + 2\psi = 0 \tag{22.14}$$

is predicted to be valid, which follows from the non-renormalization of the temporal component of (22.2). This relation is valid for the case of only temporal correlations ($\psi = 0$), thus providing a scaling relation between α and z. But in the absence of a second scaling relation, we cannot obtain the scaling exponents. Medina *et al.* studied the RG flow equations numerically, obtaining approximative results for the exponents [314]. They found that the correlated noise is relevant only for $0.167 < \phi < 0.5$. In this region the scaling exponents are

$$\alpha(\phi) = 1.69\ \phi + 0.22, \qquad \beta(\phi) = \frac{(1+2\phi)\alpha(\phi)}{2\alpha(\phi)+1}. \qquad \rightarrow (22.15)$$

To check these exponents, Lam *et al.* studied BD with temporally-correlated noise [262]. The results for α and β (Fig. 22.3) are in good agreement with (22.15), small discrepancies being observed only in the vicinity of the transition point $\phi_c \approx 0.16$. But at ϕ_c the results deviate from the scaling relation (22.14) as well. This deviation suggests that the source of discrepancy is not in the numerical solution of the RG equation, but rather in some uncontrollable crossover effects in the simulation [262].

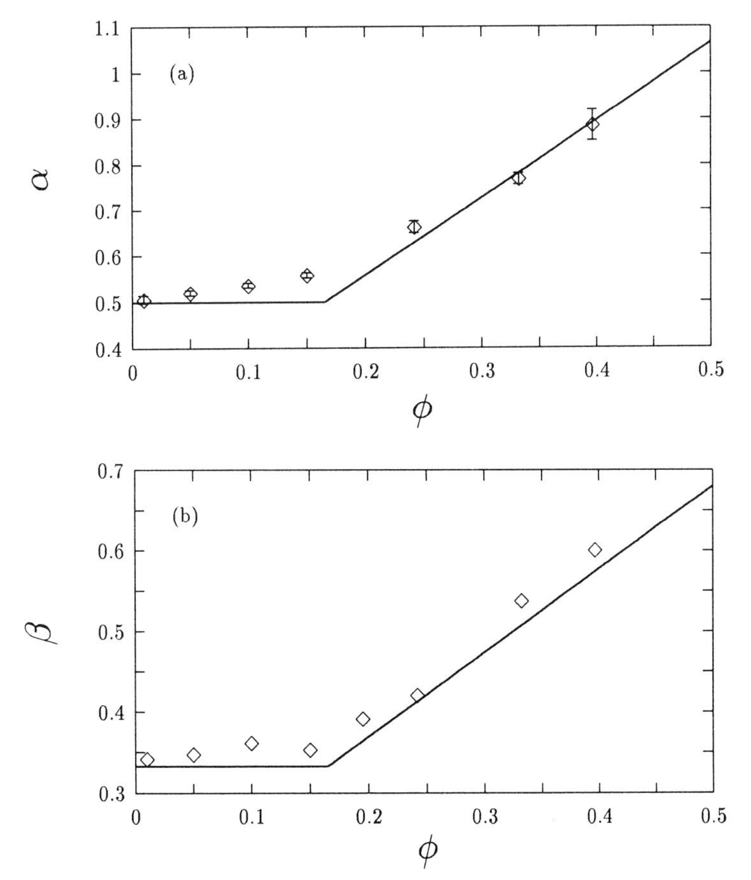

Figure 22.3 The scaling exponents for growth with temporally-correlated noise. The solid lines correspond to the RG prediction (22.15). The symbols correspond to numerical results obtained from ballistic deposition with temporally-correlated noise (After [262]).

22.5 Discussion

In this chapter, we have shown that the scaling exponents of the growth models or the continuum equations will change if we relax the uncorrelated noise hypothesis.

We have focused only on the effect of correlated noise on the EW and KPZ equations. However, correlated noise changes the exponents for other growth equations as well, e.g., for (14.2) [263].

Spatially- or temporally-correlated noise modifies the universality class, allowing the continuous tuning of the scaling exponents within certain limits. It follows that if in some experimental situations the measured scaling exponents are larger than the values predicted by the KPZ equation, one source of the different exponent values could be the correlated nature of the noise. This can be the result of the interaction of the experimental system with other underlying processes that generate correlated noise.

Suggested further reading:

[314, 364]

Exercises:

22.1 Calculate the scaling exponents for (14.2), with spatially correlated noise given by (22.1).

22.2 Calculate the scaling exponents for (21.10), with spatially correlated noise given by (22.1)

23 Rare events

In the previous chapter, we discussed how correlated noise might change the roughness and dynamic exponents in growth processes. If, on the other hand, the *amplitude* of the noise follows a power-law distribution, the scaling exponents may change *even in the presence of uncorrelated noise* as noted by Zhang [506, 507]. For such a noise, the probability of obtaining an event of size η is

$$P(\eta) \equiv \begin{cases} \mu\eta^{-(\mu+1)} & [\eta > 1] \\ 0 & [\eta < 1]. \end{cases} \qquad \rightarrow (23.1)$$

Here μ is an exponent parametrizing the decay in the amplitude of the noise. The noise events otherwise are uncorrelated in space and time, i.e., the appearance of large or small noises is independent of the noise in the neighborhood.

Evidence for such a power-law distributed noise has been obtained in experiments on fluid flow in porous media (see §11.1). Horváth, Family and Vicsek measured the probability distribution of the noise events by measuring the histogram

$$\eta \equiv \frac{\tilde{h}(x,t_2) - \tilde{h}(x,t_1)}{\Delta h}, \qquad (23.2)$$

where $\tilde{h}(x,t) \equiv h(x,t) - \langle h(x,t)\rangle_x$, and $\Delta h \equiv \langle h(x,t_1)\rangle_x - \langle h(x,t_2)\rangle_x$ is the average shift between the interfaces measured at time t_1 and t_2 [180]. The measured noise distribution is shown in Fig. 23.1. The existence of the power law distribution in the effective noise events is in agreement with the finding that quenched noise determines the anomalous roughening process. Quenched noise leads to an apparent power-law distribution if the noise is measured using Eq. (23.2) [350].

With Gaussian noise, the probability of having a very large noise event ('rare event') decreases exponentially with the size of the event (see Fig. 23.1). For power-law noise this decrease follows a power law, which is much slower than the exponential. As a result, there is a non-negligible probability of obtaining a noise with very large amplitude. The difference can be seen in Fig. 23.2, where we compare a sequence of noise events obeying a Gaussian distribution with a sequence of events obeying a power-law distribution. We can see the abundance of very large events for the power-law noise, compared with a rather uniform distribution of Gaussian noise. In addition, the nonuniformity of the power-law noise increases as μ decreases, while for large μ the noise approaches Gaussian noise in its effects (as μ increases, the probability of finding a large event becomes very small). In this chapter, for simplicity, we denote the scaling exponents of the EW and KPZ equations [(5.16) and (6.4)] as EW and KPZ exponents respectively, and denote them with the subscripts EW and KPZ.

23.1 Linear theory

As noted above, the power-law distribution (23.1) affects only the *amplitude* of the noise, and not its correlation function (4.8) and (4.9). As a result, it is not possible to treat this problem analytically using RG methods. On the other hand, an approximate 'mean field' approach for the linear theory does provide estimates of the scaling exponents [254].

Consider a portion of the interface with linear size ℓ. In the time

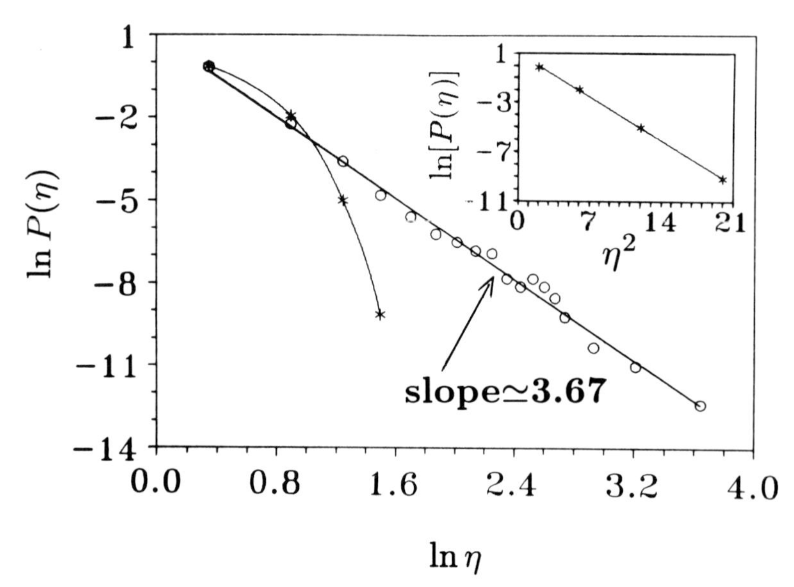

Figure 23.1 The scaling of the distribution $P(\eta)$ with η. The circles correspond to experimental results, while the stars are the result of computer simulations. Inset: The logarithm of $P(\eta)$ plotted against η^2 for the restricted SOS model, demonstrating that the noise distribution in the computer simulations of this model is Gaussian. (After [180]).

interval τ, there are

$$N \equiv \ell^d \tau \tag{23.3}$$

random events within this domain. Our goal is to estimate the expectation value of the largest of these events. From (23.1), we have

$$\text{Prob}\{\eta_{\max}(N) < x\} = \left[\text{Prob}\{\eta < x\}\right]^N = (1 - x^{-\mu})^N \tag{23.4}$$

from which, for large N, we obtain

$$\langle \eta_{\max}(N) \rangle \sim N^{1/\mu}. \tag{23.5}$$

In the linear growth equation (5.6), the smoothing process is governed by the surface tension $\nu\nabla^2 h$. As discussed in Chapter 5, the relaxation mechanism is conservative, so a large event generated by the noise spreads on the surface with decreasing amplitude such that the total 'mass' of the interface remains unchanged. As a result the amplitude of the large events $\eta_{\max}$ decreases to $\xi_\perp \sim \eta_{\max}/\ell^d$, where $\xi_\perp$ is the correlation length perpendicular to the interface, and is proportional to the local width (see §2.4). The characteristic time for the spontaneous

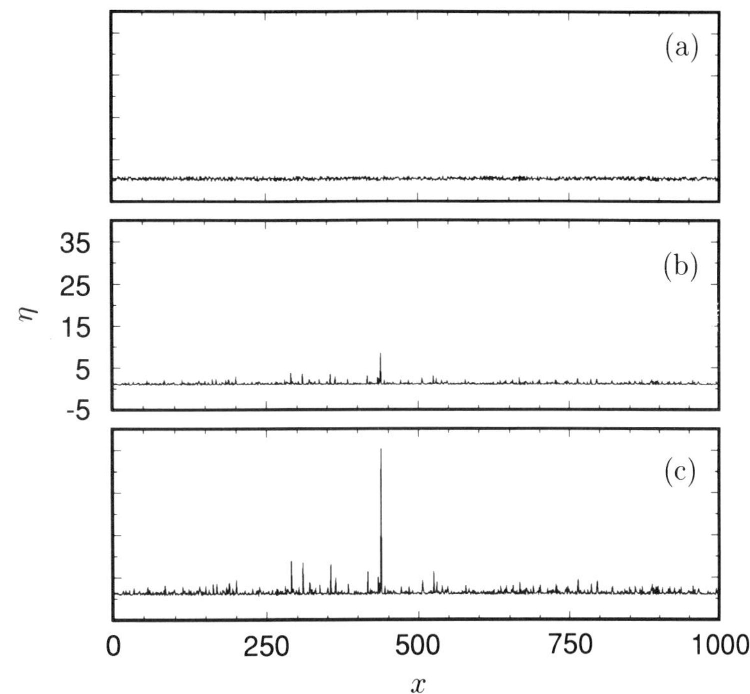

Figure 23.2 A sequence of noise events whose amplitude follows (a) a Gaussian distribution, and (b) a power-law [(23.1)] distribution, with $\mu = 5$. (c) Same as (b), but with $\mu = 3$. The same scale is used in all figures so that one can see the differences in the noise amplitudes: while in (a) there are no large events, (c) is characterized by large bursts. For (b), μ is large, and the distribution resembles the Gaussian case (a).

creation of a large event can be estimated using (23.3) and (23.5),

$$\tau_s \sim \frac{N}{\ell^d} \sim \frac{\langle \eta_{\max} \rangle^\mu}{\ell^d} \sim \xi_\perp^\mu \ell^{d(\mu-1)}. \tag{23.6}$$

The time τ_s, necessary to build up a roughness of size $\xi_\perp$ must be balanced by the smoothing effect of the linear term. The characteristic time scale associated with the relaxation mechanism can be estimated using dimensional analysis applied to the equation of motion $\partial_t h \sim \nabla^2 h$, and we find

$$\tau_a \sim \ell^2. \tag{23.7}$$

In the saturated regime the linear term balances the roughening effect of the noise, so we have $\tau_s \sim \tau_a$. Using (23.6) and (23.7), we find

$$\xi_\perp \sim \ell^{[2-(\mu-1)d]/\mu}. \tag{23.8}$$

Since $w \sim L^\alpha$ implies $\xi_\perp \sim \ell^\alpha$, the roughness exponent is

$$\alpha = \frac{2+d}{\mu} - d. \tag{23.9}$$

Let us for the moment concentrate on the $d = 1$ case. The exponent for the EW equation is $\alpha_G = 1/2$ (see (5.16)). On the other hand, (23.9) predicts that for large μ the roughness exponent decreases to zero. For large μ, we expect not much difference between Gaussian and power-law noise, since the unique property of the power-law noise – the abundance of large events – vanishes. Thus for large μ the roughness exponent should cross over to the Gaussian exponent (5.16). To obtain a critical value $\mu = \mu_c$ at which the crossover between (23.9) and the Gaussian exponents should appear, we can set the roughness exponent (23.9) to be equal to the Gaussian exponent (5.16). We thereby find that for arbitrary dimension d, a crossover to $\alpha_G(d) = (2-d)/2$ takes place at $\mu_c = 2$, independent of d. We thereby obtain the mean field exponent for the linear growth process driven by a power-law distributed noise,

$$\alpha = \begin{cases} \frac{2+d}{\mu} - d & [\mu \leq 2] \\ \frac{2-d}{2} & [\mu \geq 2]. \end{cases} \quad \rightarrow \tag{23.10}$$

Numerical support for (23.10) is provided by the simulation of the discretized EW equation (5.6), providing exponents which smoothly cross over from the small-μ results to the Edwards–Wilkinson exponents as μ is increased [261]. Discrepancies between the numerical results and the exponents (23.10) observed in the vicinity of μ_c were attributed to logarithmic corrections arising from the fact that the power-law distribution does not converge to the stable laws [259, 261].

23.2 Nonlinear theory

The scaling arguments we developed for the linear theory can be repeated for the KPZ equation as well, with small modifications due to the nonconservative character of the growth process [254, 505]. The characteristic time for the spontaneous creation of a large event can be estimated using (23.3) and (23.5),

$$\tau_s \sim \frac{N}{\ell^d} \sim \frac{\langle \eta_{\max} \rangle^\mu}{\ell^d} \sim \frac{\xi_\perp^\mu}{\ell^d}. \tag{23.11}$$

Since the growth process is not conservative, the rare events are not renormalized by the spreading of the total mass, so we choose $\xi_\perp \sim \eta_{\max}$.

The time τ_s necessary to build up a roughness $\xi_\perp$ must be balanced by the smoothing effect of the *nonlinear* term, since the nonlinear term determines the scaling behavior. From the dimensional analysis of nonlinear growth equation $\partial_t h \sim (\nabla h)^2$, we thereby obtain for the characteristic time scale

$$\tau_a \sim \ell^2/\xi_\perp. \tag{23.12}$$

In the saturated regime the nonlinear term balances the roughening effect, so

$$\xi_\perp \sim \ell^{(d+2)/(\mu+1)}, \tag{23.13}$$

from which the roughness exponent is

$$\alpha = \begin{cases} \frac{d+2}{\mu+1} & [\mu < \mu_c] \\ \alpha_{\mathrm{KPZ}} & [\mu > \mu_c]. \end{cases} \quad \rightarrow (23.14)$$

The critical noise exponent, μ_c, can be obtained by equating the μ-dependent roughness exponent (23.14) with the corresponding value obtained with Gaussian noise. For $d = 1$, with $\alpha_{\mathrm{KPZ}} = 1/2$, we obtain $\mu_c = 5$. For higher dimensions, since no exact Gaussian exponents are available, we must use the numerically determined exponents. Based on a mapping to Lévy flights, Havlin *et al.* presented an alternative derivation of (23.14) [164], which agrees with simulation results down to $\mu = 1$.

An important unresolved question concerns the value of the exponent β. In deriving (23.14), we assumed that the nonlinear term dominates the dynamics, equilibrating the noise. Assuming scale invariance, as in Chapter 6, we obtain $\alpha + z = 2$, the scaling relation (6.13) linking the two independent exponents. Whether these scaling arguments give the correct value of the exponents must be confirmed by numerical simulations. In the following, we limit ourselves to $d = 1$.

Zhang has measured the scaling exponents as a function of μ.

He found anomalously large exponents that decrease from $\alpha \approx 1$ to $\alpha \approx 0.56$ as μ increases [506]. He also found that $\alpha + z \approx 2$ for different values of μ, supporting the validity of scaling relation (6.13) in the presence of power-law-distributed noise. Following the work of Zhang, a number of models have been investigated with the goal of determining the exact μ dependence of the scaling exponents [6, 51, 62, 260]. In particular, a systematic study that confirmed the validity of the mean field prediction (23.14) was carried out for the ballistic deposition model by Buldyrev *et al.* [62]. The power-law noise was introduced by dropping – instead of particles – 'rods,' whose size follows a power-law distribution. The rare events have a spectacular effect on the morphology of the aggregate, as shown in Fig. 23.3.

The μ-dependence of the scaling exponents α and β is shown in Fig. 23.4(a). The simulation results agree with the mean-field curve. At $\mu_c = 5$, the exponents saturate, confirming the existence of a critical μ, above which power-law noise is no longer relevant. The exponents are consistent with the scaling relation $\alpha + \alpha/\beta = 2$ (Fig. 23.4(b)). Lam and Sander have argued that the mean field prediction (23.14) is in fact exact [259], and have proposed the existence of strong logarithmic

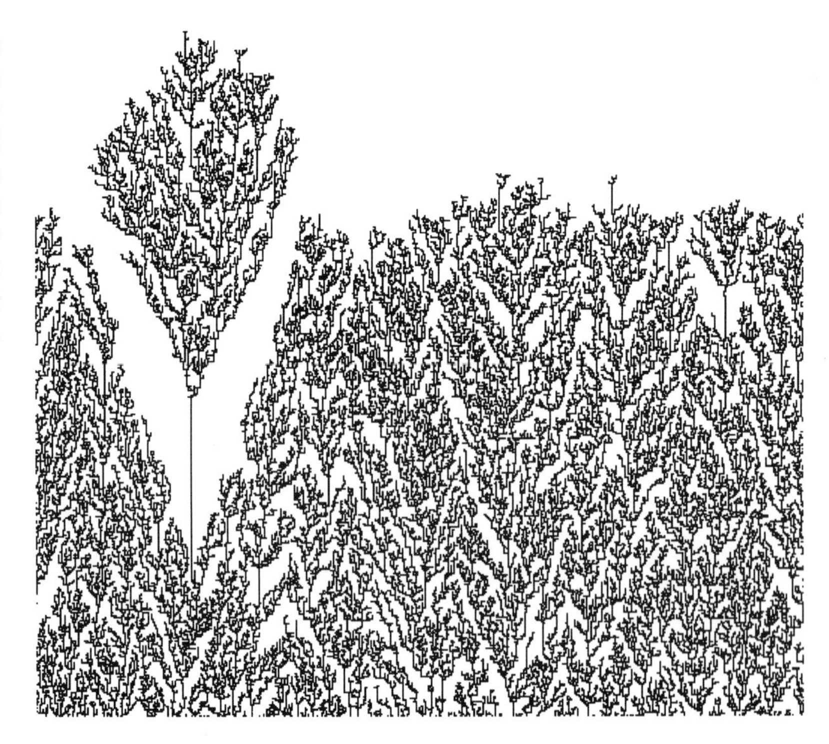

Figure 23.3 Ballistic deposition with power-law noise for $\mu = 3$. The unusual 'tree' morphology is the result of a large rare event: a very long rod falling on the surface. (After [62]).

corrections to the scaling law that might be the source of discrepancy in previous numerical efforts [6].

Extensive numerical simulations on a Connection Machine parallel supercomputer were carried out to check the effect of the power-law noise on the roughening of two-dimensional interfaces [51]. Within the error bars the simulations revealed good agreement with the mean field result $\beta = 2/(\mu - 1)$. Note that using $\beta = 1/4$ (see §7.5), we obtain $\mu_c = 9$ for $d = 2$.

23.3 Multi-affinity

Another interesting aspect of roughening dominated by power-law noise is that higher moments of the height–height correlation function do not scale with the same exponents, i.e., these interfaces exhibit *multi-affinity* (see Chapter 24). Figure 23.5 shows the scaling of the

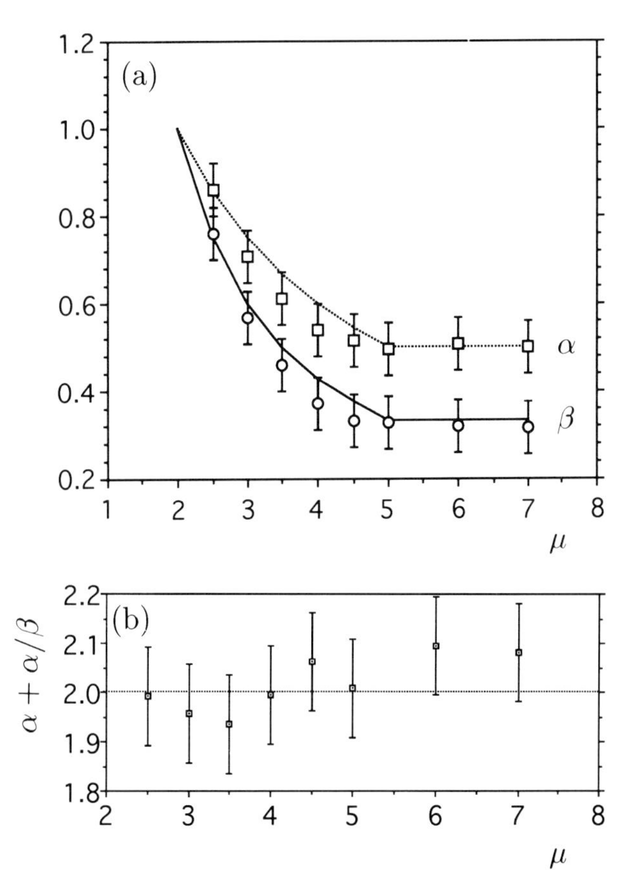

Figure 23.4 (a) Comparison of the numerically obtained exponents α ($\square$) and β ($\circ$) with the prediction of the mean field argument. The two lines shown correspond to the prediction of (23.14). Note that the simulations are consistent with the existence of a critical value $\mu_c = 5$. (b) Check of the scaling relation $\alpha + \alpha/\beta = 2$ using the numerically determined exponents. (After [62]).

q-th order correlation functions $c_q(\ell)$. We see there is a characteristic crossover length scale up to which the different moments scale with a q dependent exponent α_q. After the crossover length scale $\ell_\times$, the curves become parallel, scaling with the mean-field scaling exponent (23.14).

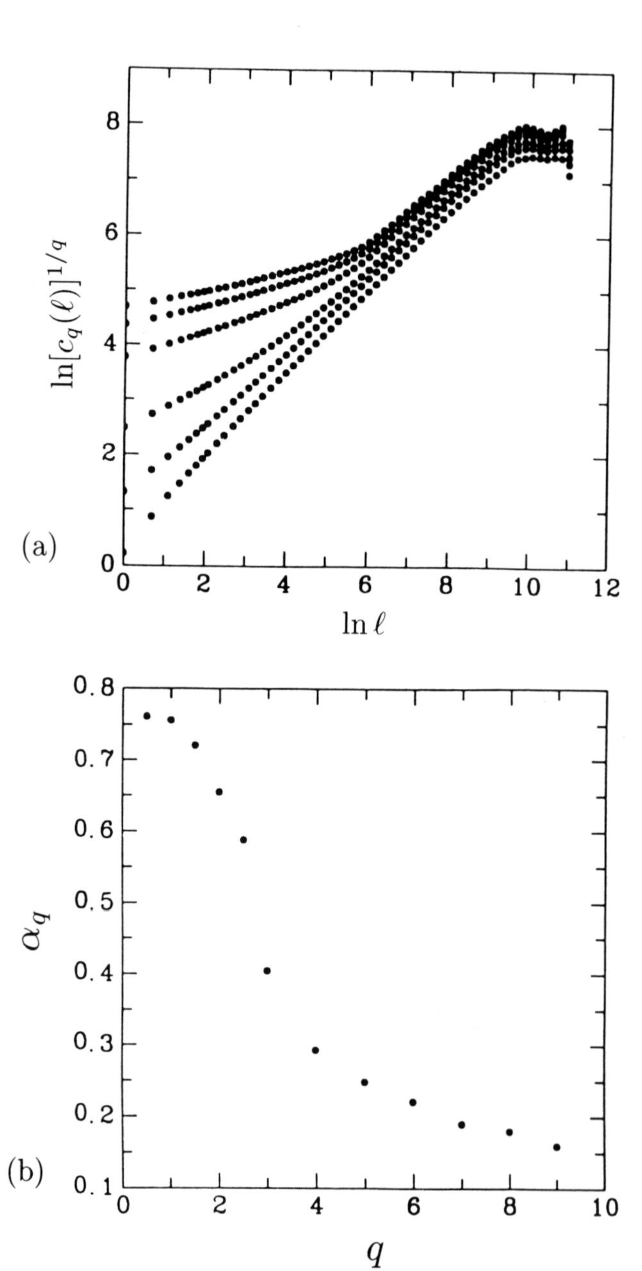

Figure 23.5 (a) The q-th order correlation functions $c_q(\ell)$ after the surface width has saturated. The system size is $L = 2^{16}$, and time is $t = 602\,890$ sweeps. (b) The variation with q of the q-th order roughness exponent α_q with q as predicted by the scaling of the correlation function. (After [30]).

The crossover length $\ell_\times$ can be directly linked to rare events [25]. If a large rare event η appears, the perturbation generated by it decreases linearly in time by spreading along the interface. Thus a large event affects the scaling on a length scale that is proportional to the size of rare event η. But since there is a finite expectation value for a very large rare event (see (23.5)), there is a finite distance $\ell_\times$ over which it affects the scaling of the correlation function. For $\ell \gg \ell_\times$ the effect of the rare events diminishes, marking the transition to self-affine scaling for the higher moments of the correlation function. Thus the crossover length is fixed by the expected value of large fluctuations, $\langle \eta_{\max} \rangle$, and depends on the value of μ.

The values of α_q calculated for $\ell < \ell_\times$ are shown in Fig. 23.5. The existence of this nontrivial scaling is attributable to the rare events in the noise, and the growth dynamics seems not to contribute to its development. Similar multiscaling has been observed in the temporal fluctuations of the saturated interface width of models driven by power-law distributed noise [25].

23.4 Discussion

We have seen that scaling arguments and extensive numerical simulations support Zhang's original proposal that power-law distributed noise changes both the roughness and the dynamic exponent of a growing interface. Before concluding, however, it is perhaps worthwhile to address the following question: *Why* does power-law distributed noise change the scaling exponents? To appreciate the importance of this question, we note that in *equilibrium* statistical physics the introduction of a power-law distributed noise does not affect the scaling exponents. It seems, that *nonequilibrium* systems, such as interfaces, are particularly sensitive to the nature of the noise. Power-law distributed noise introduces rare events – 'large bursts' – which change the morphology of the system over quite large length scales. The dynamics of these models is too slow to immediately suppress these rare events, so they dominate the scaling behavior of the system. On this fact is based the quite successful mean field argument leading to the exponents (23.14). The EW equation, an intrinsically *equilibrium* model, is also sensitive to the nature of the noise distribution.

Suggested further reading:

[62, 260]

Exercises:

23.1 Derive (23.5).

24 Multi-affine surfaces

Thus far, we have seen that many surfaces are self-affine, and their scaling can be characterized using a single number – the roughness exponent α. This is because the statistical properties characterizing the surface are invariant under an anisotropic scale transformation (3.4). For characterizing the scaling properties of some surfaces, however, the roughness exponent by itself is not sufficient.

In this chapter, we discuss a class of surfaces whose scaling properties are describable only in terms of an infinite set of exponents. Since this class is an extension of self-affine surfaces, we will call them *multi-affine surfaces* [33, 34]. We shall see that the multi-affine behavior is reflected in the existence of an entire hierarchy of 'local' roughness exponents, i.e., the roughness exponent for multi-affine interfaces changes from site to site.

A number of models studied recently in the literature lead to multi-affine surfaces. For example, if the noise in the system has a power-law distribution, the interface is multi-affine for length scales shorter than a characteristic length scale $\ell_\times$, set by the noise (see §23.3). Similarly, the temporal fluctuations of the interface width in the saturated regime leads to a multi-affine function [25]. Quenched noise, present in many experimental systems, leads naturally to power-law distributed noise, so one expects that multi-affine behavior might be present in such systems as well. Indeed, Sneppen and Jensen [411] studied the temporal properties of the SOD model discussed in §10.2, and found this type of multi-affine behavior. Moreover, Krug found that simple models, including surface diffusion that was introduced to mimic MBE, lead to multi-affine interfaces [245]. The mathematical formalism used to characterize multi-affine functions has its root in the multifractal formalism, introduced to describe the scaling properties of some complex fractal objects [137, 160, 171, 199, 269, 286, 305, 313, 419].

Multi-affine interfaces are frequently described as *turbulent*, because the local velocities in turbulent systems exhibit a similar multiscaling behavior [17, 136, 315, 343]. Indeed, simple deterministic models such as the one described in Fig. 24.1 can be used to approximate the experimentally obtained spectrum of exponents [209, 458].

24.1 Hierarchy of scaling exponents

As discussed in Chapter 3, for self-affine functions we can choose a particular point on the surface and rescale its neighborhood according to (3.4), using for α the roughness exponent of the interface. If we use the correct α, the surface obtained should be indistinguishable from the original. In principle, this concept can be used as a method for determining α, since by tuning the rescaling factor until we obtain a similar surface, we can record the value of α required to obtain good scaling. To reduce the subjectivity of our visual judgement, we perform the procedure a number of times, each time using a different part of the surface, and average over the calculated values. It is assumed that each value we obtain will not differ significantly from any other.

Suppose we use this method to determine α for a particular interface, but find different parts of the surface require significantly different roughness exponents to obtain interfaces that are similar to the whole. In this case the surface does not have a well-defined roughness exponent α, and there is a good chance that the surface is multi-affine.

Determining whether a given interface is multi-affine is difficult if only the rescaling method is used. A more objective method is to measure the qth order height–height correlation function [34] defined by†

$$c_q(\ell) \equiv \langle |h(x) - h(x')|^q \rangle_{x}. \qquad |x - x'| \equiv \ell \qquad \rightarrow (24.1)$$

The exponent hierarchy α_q is defined through the relation

$$c_q(\ell) \sim \ell^{q\alpha_q}. \qquad \rightarrow (24.2)$$

We call an interface *multi-affine* if α_q varies with q. For self-affine surfaces α_q is independent of q, and is equal to the roughness exponent α.

† Note that $c_2(\ell) \equiv [C(\ell)]^2$, where $C(\ell)$ is defined in Eq. (A.4).

24.2 A deterministic multi-affine model

Next we discuss two examples: a deterministic *multi-affine* function for which α_q can be calculated exactly, and continuous Brownian motion, which is *self-affine*. Figure 24.1 shows the construction of a deterministic multi-affine function, a generalization of the deterministic self-affine function discussed in §3.3. We divide the horizontal part into four equal intervals, and replace it with four segments such that the height of the first segment is b_1. The second segment is left unchanged, and we denote by b_2 the height of the third segment. This model reduces to the self-affine example of Fig. 3.4 if $b_1 = b_2 = 1/2$.

Now we calculate $c_q(\ell)$, defined by (24.1), for this discrete model.

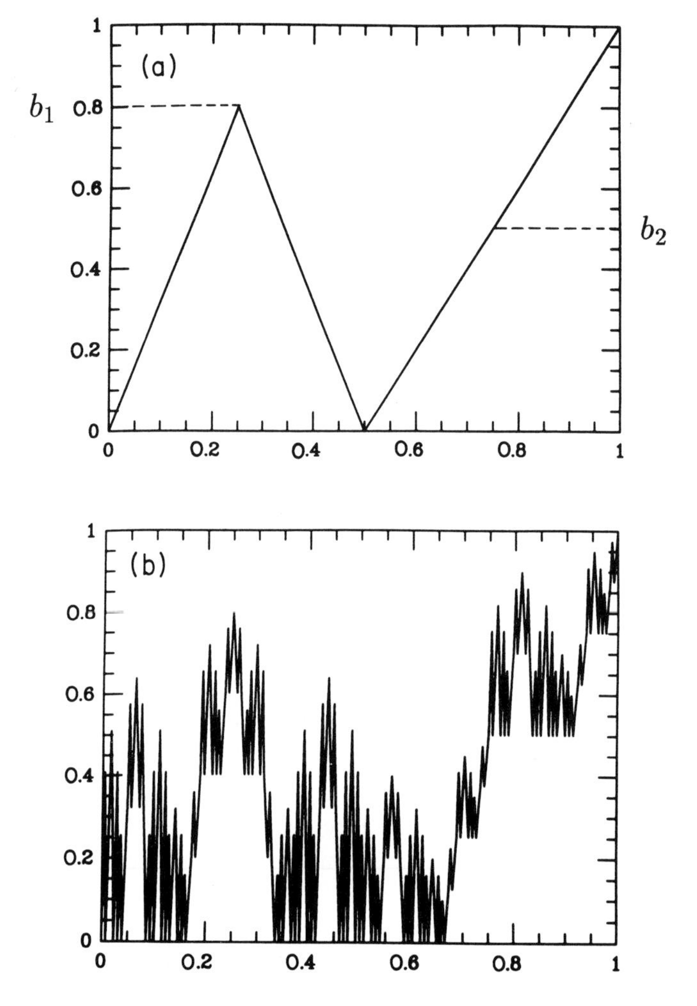

Figure 24.1 Construction of a deterministic multi-affine surface. In each step of the recursion, the intervals obtained in the previous step are replaced with the properly rescaled version of the initial configuration shown in (a) (see also Fig. 3.4). In (b) we show the function after the fourth iteration. The function becomes multi-affine in the $k \to \infty$ limit. (After [34]).

After the kth iteration, the number of intervals along the horizontal x axis is 4^k and the length of each is equal to 4^{-k}. In particular, after the first iteration we have four segments, whose heights $\Delta h \equiv |h(x) - h(x + \Delta x)|$ from the left are b_1, b_1, b_2 and b_2, where $1 - b_2 = b_2$. In the second iteration ($k = 2$) we have $\Delta x = 1/16$. Partitioning the first segment, whose height was b_1 in the preceding step, we obtain *four* segment heights: $b_1 b_1$, $b_1 b_1$, $b_1 b_2$, and $b_1 b_2$, which can be written $b_1(b_1, b_1, b_2, b_2)$. The sum of the heights of all sixteen segments after the second iteration can be obtained from $(2b_1 + 2b_2)^2$. In general, after the kth step, the number and height of the intervals of horizontal size 4^{-k} can be obtained from $(2b_1 + 2b_2)^k$, implying that $N(\Delta h)$, the number of intervals for which $\Delta h = b_1^n b_2^{k-n}$, is given by

$$N(b_1^n b_2^{k-n}) = 2^k C_k^n, \tag{24.3}$$

where $n = 0, ..., k$. Thus we have for the correlation function

$$c_q(\Delta x) \equiv \sum_{\Delta h} N(\Delta h)(\Delta h)^q$$

$$= \frac{1}{4^k} \sum_{n=0}^{k} 2^k C_k^n \left[b_1^n b_2^{(k-n)} \right]^q = \sum_{n=0}^{k} 2^{-k} C_k^n b_1^{nq} b_2^{(k-n)q}, \tag{24.4}$$

with $\Delta x = 4^{-k}$. Since (24.4) can be written as

$$c_q(\Delta x) = \left[\frac{(b_1^q + b_2^q)}{2} \right]^k, \tag{24.5}$$

from (24.2) we have for the scaling exponents

$$\alpha_q = \frac{\ln[(b_1^q + b_2^q)/2]}{q \ln(\frac{1}{4})}. \tag{24.6}$$

Thus α_q is not independent of q in this model, and the surface it generates [shown in Fig. 24.1(b)] is multi-affine. However, we can verify that $\alpha_q = 1/2$, independent of q if $b_1 = 1/2$ (remember that $b_2 = 1/2$ by definition). Thus the model of Fig. 3.4 is self-affine, and not multi-affine.

In general, we can obtain evidence about the multi-affinity and the scaling exponents of a given surface by measuring $c_q(\ell)$ directly. Figure 24.2 shows the exponents measured on a surface generated using the multi-affine model for $b_1 = 0.8$, and compares them with the exact result (24.6). Note that we cannot calculate numerically the negative powers of $c_q(\ell)$ if the surface has arbitrarily small height differences, since a negative power of a very small number leads to a divergent correlation function. But in most cases measurements of the positive moments will strongly indicate whether the interface is multi-affine.

24.3 Brownian motion

The simplest *self-affine* function is continuous Brownian motion, discussed in §3.3. It is instructive to demonstrate exactly how it is *not* multi-affine.

Using (3.9) we have for the height–height correlation function

$$c_q(t) \equiv \int_0^\infty P(x,t)\, x^q dx = \frac{(2t)^{q/2}}{\sqrt{\pi}} \int_0^\infty u^q e^{-u^2} du = I\ t^{q/2}. \qquad (24.7)$$

Here we have performed a change of variables $u \equiv x/\sqrt{2t}$; the integral denoted I is given by $I = \Gamma[(q+1)/2]$ for $q > -1$, but diverges for $q < -1$. However, for $q > -1$ we have $\alpha_q = 1/2$, independent of q. In conclusion, continuous Brownian motion is self-affine, not multi-affine.

24.4 Local dimensions

We stated above that multi-affine functions have scaling exponents that differ from point to point. This contention is supported by the exact model as well. In the self-affine case, we were able to magnify the box in Fig. 3.4(c) anisotropically, and obtain a structure that is identical to the whole. We could have put the box anywhere on the structure to achieve this. This is no longer true for the function shown in Fig. 24.1. The slopes for the first two segments differ from those of the third and the fourth segments, so it is not possible to rescale the system locally with the same factor everywhere and obtain an identical

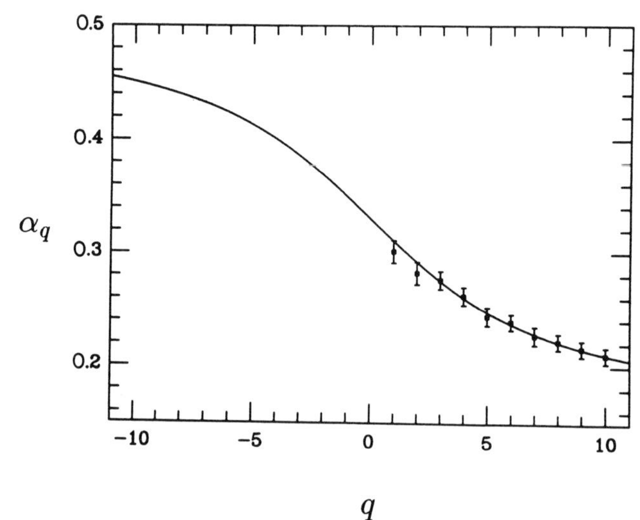

Figure 24.2 The α_q spectrum for the deterministic multi-affine model calculated from (24.6) with $b_1 = 0.8$. The theoretical result is compared with the exponents determined for the model numerically using a surface generated after nine iterations. (After [34]).

function. If we wanted to rescale different parts of this multi-affine function, we would need to use different scaling factors.

Hence with every point x_i we can, in general, associate a *local* scaling exponent γ_i to characterize the local singularity of the interface

$$|h(x_i) - h(x_i + \ell)| \sim \ell^{\gamma_i}. \tag{24.8}$$

This relation, in fact, describes the scaling of the height–height correlation function locally, so γ_i acts as a 'local roughness exponent.'

The number of points on the surface, $N(\gamma)d\gamma$, that have the singularity exponents in the range $(\gamma,\ \gamma + d\gamma)$ is found to scale with the length of the partition we use to divide the interface,

$$N_\gamma(\ell) \sim \ell^{-h(\gamma)}. \tag{24.9}$$

Here $h(\gamma)$ is the fractal dimension of the subset of points having the same exponent γ. Note that if the surface is self-affine, γ_i is the same for every point, and thus $h(\gamma) = 1$. In the limit $x \to 0$, $N \sim x^{-1}$ and the correlation function $c_q(\ell)$ can be written

$$c_q(\ell) \sim \frac{1}{N}\int \ell^{q\gamma'} N_{\gamma'}(\ell)\rho(\gamma')d\gamma' \sim \frac{1}{N}\int \ell^{1+q\gamma'-h(\gamma')}\rho(\gamma')d\gamma'. \tag{24.10}$$

Here, instead of integrating over ℓ, as we did in (24.4), we integrate over γ. The function $\rho(\gamma)$ incorporates all quantities that are independent of ℓ.

For a system to be continuous, this integral must be dominated by the value of γ' that minimizes $1 + q\gamma' - h(\gamma')$. Thus we replace γ with $\gamma(q)$, defined by the conditions

$$\frac{d}{d\gamma'}\left[1 + q\gamma' - h(\gamma')\right]|_{\gamma'=\gamma(q)} = 0 \tag{24.11}$$

and

$$\frac{d^2}{d\gamma'^2}\left[1 + q\gamma' - h(\gamma')\right]|_{\gamma'=\gamma(q)} > 0. \tag{24.12}$$

It follows that

$$h'(\gamma) = q \tag{24.13}$$

$$h'(\gamma) < 0. \tag{24.14}$$

From comparison with (24.2) we find

$$q\alpha_q = 1 + q\gamma(q) - h(\gamma_q) \tag{24.15}$$

and

$$\gamma(q) = \frac{d}{dq}(q\alpha_q). \tag{24.16}$$

The last two relations give us the relationship between the exponents α_q, the local scaling exponents γ, and the fractal dimension $h(\gamma)$ of the points that share the same γ. Note that the points that share the same exponents are not randomly distributed, but form a fractal.

The above results were derived for one-dimensional interfaces, and generalization to higher dimensions is straightforward.

Suggested further reading:

[33, 160]

Exercises:

24.1 Calculate the multi-affine exponents α_q for the model illustrated in Fig. 24.3. What is the value of the roughness exponent?

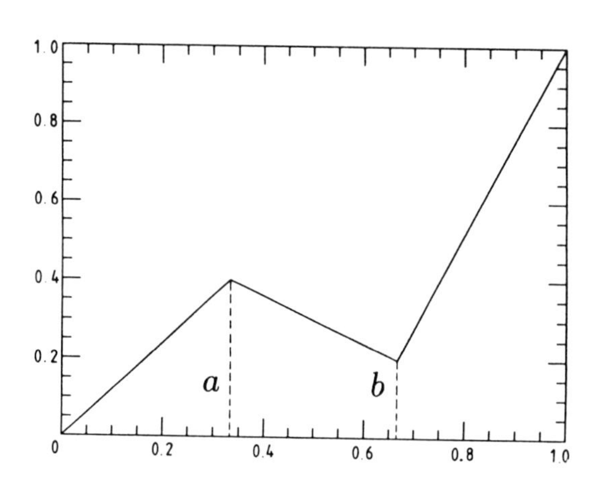

Figure 24.3 Construction of a deterministic multi-affine surface. In each step of the recursion, the intervals obtained in the previous step are replaced with the properly rescaled version of the initial configuration shown in the figure (see also Fig. 24.1).

25 Variants of the KPZ equation

It is a characteristic feature of modern statistical mechanics that if a given model does not apply to a given situation, a simple variant may. Sometimes the variant results in a change of universality class, and sometimes it does not. For example, to describe isotropically-interacting spins, Heisenberg proposed the variant of the Ising model that today bears his name and we now know the exponents are rather different for the two models.

In this chapter we discuss variants of the KPZ equation. In particular, we present some key results regarding the noiseless KPZ equation and the interesting properties that arise when the KPZ approach is extended to anisotropic systems.

25.1 Deterministic KPZ equation

Thus far, we have studied the interplay between the deterministic terms in the KPZ equation and the various possible sources of noise. In fact, the KPZ equation without any noise can be solved, and its solution may be useful in interpreting certain experimental systems.

In the absence of noise, Eq. (6.4) is called the deterministic KPZ equation; it can be solved exactly, subject to any initial condition [217]. If the interface at $t = 0$ is $h_0(\mathbf{x}) \equiv h(\mathbf{x}, 0)$, then the reader may confirm that its time evolution is given by†

$$h(\mathbf{x}, t) = \frac{2\nu}{\lambda} \ln \int_{-\infty}^{\infty} \frac{d^d\mathbf{y}}{(4\pi\nu t)^{d/2}} \exp\left[-\frac{(\mathbf{x}-\mathbf{y})^2}{4\nu t} + \frac{\lambda}{2\nu} h_0(\mathbf{y})\right]. \qquad (25.1)$$

An initially rough interface becomes smooth if growth is governed by

† The deterministic KPZ equation can be mapped into the diffusion equation (26.8) with $V = 0$, whose solution leads to (25.1).

the noiseless KPZ equation (Fig. 25.1), developing paraboloids of the form

$$h = A - (\mathbf{x} - \xi)^2/2\lambda t, \tag{25.2}$$

which are joined together by discontinuities in ∇h.

The deterministic KPZ equation 'asymptotically smooths' an initially rough interface. Relaxation proceeds by the growth and coalescence of the largest parabolas. The relaxation of an initially rough interface can be studied by following the time evolution of the density of surface steps [251]

$$\rho(t) \equiv L^{-d} \int d^d\mathbf{x} \langle |\nabla h(\mathbf{x}, t)| \rangle. \tag{25.3}$$

A typical slope of the interface is proportional to $w/\xi_\parallel$, where w is the interface width and $\xi_\parallel$ the parallel correlation length. Hence using $w \sim \xi_\parallel^\alpha$ and (2.13), we obtain for the step density

$$\rho(t) \sim \frac{w}{\xi_\parallel} \sim \xi_\parallel^{\alpha-1} \sim t^{-(1-\alpha)/z}, \tag{25.4}$$

which, for $\alpha < 1$, predicts a decay in the step density with time – i.e., a smoothing of the interface [9, 251].

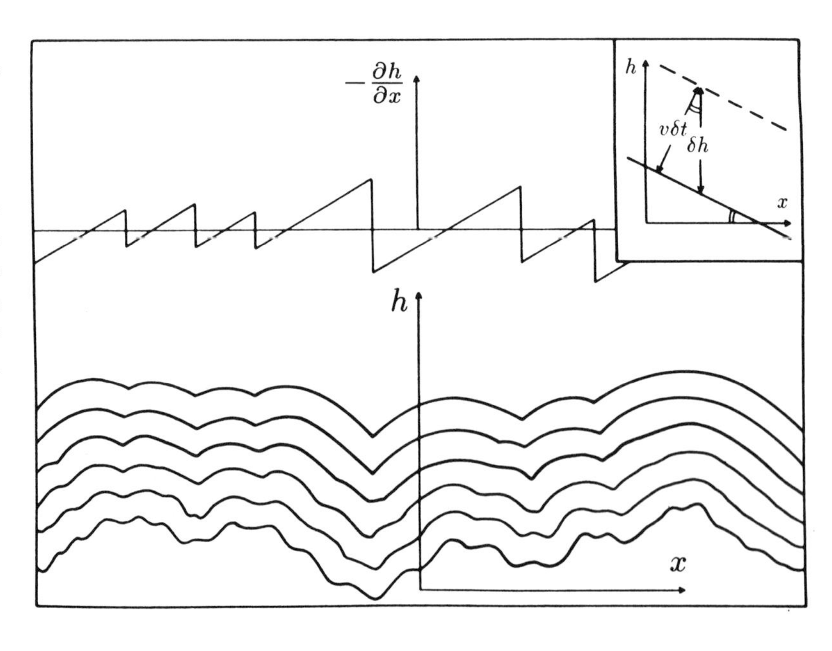

Figure 25.1 Successive profiles of an initially rough interface whose time evolution is described by the deterministic KPZ equation. The gradient of the profile develops shock waves. (After [217]).

25.2 Anisotropic KPZ equation

In Chapters 6 and 7, we discussed in detail the scaling behavior predicted by the KPZ equation. For two dimensions, we found that if λ is nonzero, the exponents are described by the strong coupling regime, while for $\lambda = 0$ the growth in L is described by the EW equation, the roughness being only logarithmic with $\alpha = \beta = 0$. There is no reason to suspect any other behavior, so it was a surprise when in 1991 Wolf found that the presence of anisotropy may drastically change this scenario [482].

As a physical example, consider a stepped surface, such as the one shown in Fig. 25.2. The existence of the steps selects a preferred direction for the growth process. To derive a continuum equation to describe the growth, we must consider that the two directions, parallel and perpendicular to the average step edges, are not equivalent. Suppose we choose our x coordinate along the step direction and the y perpendicular to it. We then see that while the system looks invariant under a reflection around the x axis, it is certainly not symmetric around the y – due to the step structure. Thus we conclude that the x and y directions are not equivalent.

This anisotropy is expected to show up in the growth equation as well. If we consider the basic symmetries of the problem (see §5.2), for a stepped surface, we find that the rotational symmetry (iv) is broken, leading to an *anisotropic* growth equation of the form (for $d = 2$)

$$\frac{\partial h}{\partial t} = \nu_x \partial_x^2 h + \nu_y \partial_y^2 h + \frac{1}{2}\lambda_x(\partial_x h)^2 + \frac{1}{2}\lambda_y(\partial_y h)^2 + \eta \qquad \text{[AKPZ]} \rightarrow (25.5)$$

Certainly, the presence of the anisotropy is expected to lead to surface tension and nonlinear terms that are different in the two directions, which has been incorporated in the growth equation by considering different values for the coefficients ν and λ. Due to the broken $x \rightarrow -x$ symmetry, a linear term $a\partial_y h$ is not excluded, but does

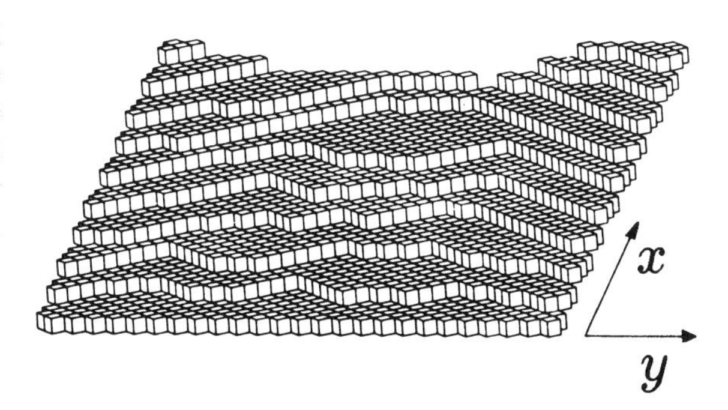

Figure 25.2 A typical stepped interface generated by the three-dimensional Toom model. (After [29]).

not affect the scaling exponents since it can be transformed away with a simple change of coordinates of the form $y \to y - at$. Equation (25.5) is called the anisotropic KPZ (AKPZ) equation. It was introduced by Villain, and its nontrivial properties were studied by Wolf [463, 482]. We note that if $\nu_x = \nu_y$ and $\lambda_x = \lambda_y$, (25.5) reduces to the KPZ equation (6.4).

25.2.1 Renormalization group analysis

The AKPZ equation can be studied using the dynamic RG [482]. The existence of anisotropy means that we expect to have different scaling exponents in the two scaling directions – i.e. we must rescale x and y differently: $x \to bx$, $y \to b^{\zeta} y$. Here ζ is the *anisotropy exponent*, reflecting the difference in the scaling of the correlation lengths in the two directions

$$\xi_x \sim (\xi_y)^{\zeta}. \tag{25.6}$$

The RG calculations show that λ_x and λ_y are not renormalized, resulting in the scaling relations

$$\alpha + z = 2, \qquad \zeta = 1. \tag{25.7}$$

The flow equations for the remaining coefficients are

$$\frac{\partial \nu_y}{\partial l} = \nu_y \left[z - 2 + \frac{g_y}{16\pi^2} \left(1 - \frac{r_\lambda}{r_\nu} \right) A(1, r_\nu) \right], \tag{25.8}$$

$$\frac{\partial r_\nu}{\partial l} = -r_\nu \frac{g_y}{16\pi^2} \left[1 - \left(\frac{r_\lambda}{r_\nu} \right)^2 \right] A(1, r_\nu), \tag{25.9}$$

$$\frac{\partial g_y}{\partial l} = \frac{g_y^2}{32\pi^2} \left[8\frac{r_\lambda}{r_\nu} - 3\left(1 - \left(\frac{r_\lambda}{r_\nu} \right)^2 \right) \right] A(1, r_\nu), \tag{25.10}$$

where $g_y \equiv D\lambda_y^2/\nu_y^3$, $r_\lambda \equiv \lambda_x/\lambda_y$, $r_\nu \equiv \nu_x/\nu_y$, and

$$A(\zeta, r_\nu) = \left[\tan^{-1}(r_\nu^{1/2}) + \zeta \ \tan^{-1}(1/r_\nu^{1/2}) \right] / r_\nu^{1/2}. \tag{25.11}$$

The nontrivial effect of the anisotropy can be seen from equations (25.9) and (25.10). Equation (25.9) has two fixed points, $r_\nu^* = r_\lambda$ and $r_\nu^* = -r_\lambda$. Choosing ν_x and ν_y positive ($r_\nu > 0$), for the first fixed point we find $r_\nu^* = r_\lambda$, λ_x and λ_y have the same sign. The physical properties of the interface corresponding to this fixed point are the same as for the isotropic KPZ equation (6.4), so there is algebraic roughness.

The situation is quite different in the case of the other fixed point, $r_\nu^* = -r_\lambda$, which corresponds to values of λ with opposite signs. In this case, any finite coupling constant g_y is renormalized to zero since the larger parenthesis in Eq. (25.10) has a negative sign. Thus if the signs of the two λs are opposite, the nonlinearity is irrelevant, and the scaling is described by the linear EW equation (5.6).

Thus we see that the AKPZ equation is quite different from the isotropic KPZ equation where, once we are convinced that the nonlinear term is present, we know the scaling exponents. In contrast, for the AKPZ equation we must inspect the sign of the nonlinear terms as well: if they have opposite signs, then the nonlinearity is irrelevant.

If by varying some external parameter, we would be able to tune the value of at least one of the λ coefficients, we could observe a phase transition from algebraic roughness ($\beta \approx 0.24$) to logarithmic roughness ($\beta = 0$) as r_λ changes sign. The nontrivial effect of the anisotropy can be observed in the case of stepped surfaces [331, 482] or in the Toom model [29, 201].

25.2.2 Integration of the AKPZ equation

The effect of anisotropy in the two-dimensional KPZ equation can also be studied by means of simulations, and there are several relevant results. Moser and Wolf integrated the AKPZ equation (25.5) [331]. The numerical method is the same as the one used for the integration of the *isotropic* KPZ equation. As shown in Fig. 25.3(a), for $\lambda_x = \lambda_y/2$ the interface roughens algebraically in time, consistent with an exponent $\beta = 0.23 \pm 0.01$, which within the limits of the error bars coincides with the exponent of the isotropic KPZ equation. However, for $\lambda_x = -\lambda_y/2$ the width increases as a logarithm of the time, as shown in Fig. 25.3(b), so if the signs of the nonlinear terms are opposite, then the scaling is described by the EW equation.

25.2.3 The Toom model

A much-investigated discrete model described by the AKPZ equation is the three-dimensional Toom model [29, 201]. In the Toom model, spins with values $s = \pm 1$ are simultaneously updated at every time step with the following simple rule: s becomes equal to 1 with probability p, -1 with probability q, and becomes aligned with the majority of itself and a specified set $\{S\}$ of neighboring spins with probability $1 - p - q$.

When $p = q = 0$, the Toom model is deterministic. For small enough p and q, the model for any dimension has two stable phases; one phase has most spins aligned up ($+1$) and the other phase has most spins aligned down (−1) [41, 102, 103, 142, 268, 450]. Figure 25.2 shows a snapshot of the interface separating the two phases. The origin of the surface anisotropy is the anisotropy of the set $\{S\}$ in the updating rule: the x and y directions are not equivalent. Numerical simulations indicate that the interface is described by the AKPZ equation in the strong-coupling or KPZ limit, the scaling exponents being $\beta = 0.21 \pm 0.03$, and $\alpha = 0.43 \pm 0.04$ [29]. Using a different updating rule Jeong *et al.* have demonstrated that the interface is still described by the AKPZ equation, but in the EW limit [201].

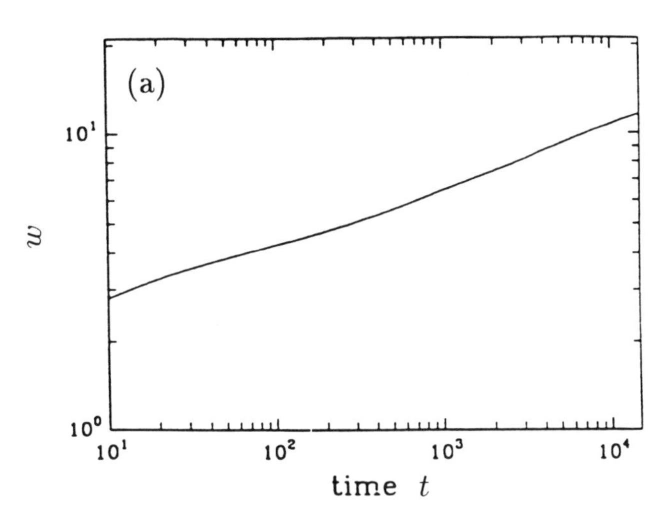

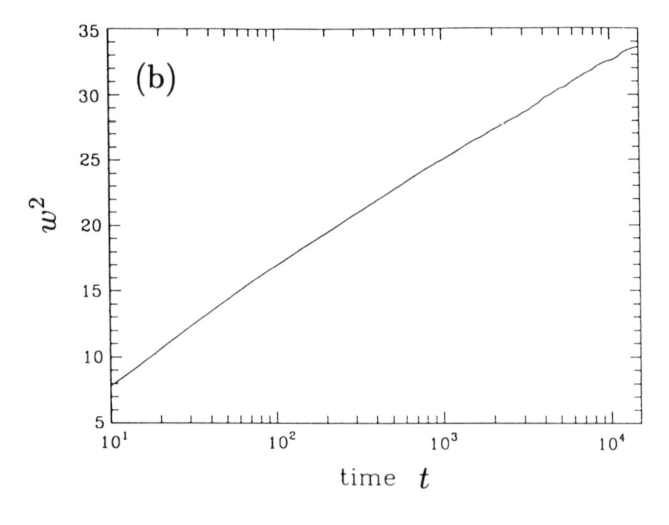

Figure 25.3 Time-dependent width obtained from numerical integration of the AKPZ equation (25.5). (a) If $\lambda_x = \lambda_y/2$ (same sign case), the interface roughens algebraically, the scaling being described by the isotropic KPZ equation. (b) If $\lambda_x = -\lambda_y/2$ (opposite sign case), the interface roughens logarithmically, the scaling being described by the EW equation. (After [331]).

25.3 Universal amplitudes

In previous chapters we mentioned that while various growth models may generate different interfaces by different growth mechanisms, they belong to the same universality class if they share the same set of growth exponents. In addition to the scaling exponents, the scaling function $f(u)$, defined in (2.8), is also universal. In this section we discuss the universal properties of $f(u)$ and an important consequence, the existence of universal amplitudes [7, 8, 186, 249].

Let us consider the one-dimensional KPZ equation (6.4). As we discussed in §6.4.2, in one dimension the height fluctuations become Gaussian and for $t \to \infty$, their probability distribution is given by (6.20). The variance of the height distribution in the *stationary* state is given by

$$\lim_{t\to\infty} \langle |h(k,t)|^2 \rangle = \frac{D}{\nu L k^2}, \tag{25.12}$$

where $h(k,t)$ is the discrete spatial Fourier transform of $h(x,t)$, with $k = 2\pi m/L$ and $m = -L/2, \ldots, -1, 1, \ldots, L/2$.

From (25.12) one obtains the height–height correlation function

$$C^2(\ell) = \lim_{t\to\infty} \langle [h(x+\ell,t) - h(x,t)]^2 \rangle = \frac{D}{\nu}\ell. \tag{25.13}$$

However, in the *transient* regime, (25.12) must be replaced with the scaling form

$$\langle |h(k,t)|^2 \rangle = \frac{D}{\nu L k^2} f\left[(\lambda^2 A)^{1/3} k t^{2/3} \right], \tag{25.14}$$

where $f(u)$ is a universal scaling function.

One of the consequences of the universality of $f(u)$ concerns the moments

$$a_n \equiv \lim_{t\to\infty} \lim_{L\to\infty} t^{-n/3} \langle [h(x,t) - \langle h(x,t) \rangle]^n \rangle, \tag{25.15}$$

which have the general form

$$a_n = (|\lambda| A^2)^{n/3}\, c_n. \tag{25.16}$$

To see that the quantities c_n are universal, we note that c_n depends only on the scaling function $f(u)$, which is universal. For example, we have

$$c_2 = \frac{1}{\pi} \int_0^\infty du \frac{f(u)}{u^2}. \tag{25.17}$$

The universality of the amplitudes has been confirmed by numerical simulations. For example, the values of c_2, measured for five different growth models, lead to $c_2 = 0.404 \pm 0.013$ [249]. The transient amplitudes, a_n, are not the only quantities that lead to universal, model independent, numbers. Fluctuations in the stationary regime, or corrections to the growth rate, can also be related to universal scaling functions, and lead to universal amplitudes.

25.4 Discussion

The sensitivity of the two dimensional KPZ equation to the anisotropy in the coefficients comes as a surprise. However, it turns out to be a very important property that one must consider when analyzing simulations or experimental data. We learn the lesson that measurement of the scaling exponents may not be a perfect indication concerning the growth equation if there is reason to believe that anisotropy is present in the system. For example, one can observe logarithmic roughening, which may be taken as an indication that the correct equation to describe the growth process is the EW equation, when in fact nonlinear terms of different sign may be present.

Suggested further reading:

[66, 251, 249, 482]

Exercises:

25.1 Consider an initially-rough interface, characterized by a roughness exponent α, whose time evolution is described by the deterministic KPZ equation. How does the total mass deposited per unit (up to a time t) scale with t for this system?

25.2 Consider the growth process described by the deterministic equation

$$\frac{\partial h}{\partial t} = \nu \nabla^2 h + \lambda |\nabla h|^{\gamma}, \tag{25.18}$$

where $\gamma \geq 1$. (a) Repeat the calculation of the previous problem for this system. (b) Find a scaling relation linking α (the roughness exponent of the interface at $t = 0$), γ, and the dynamic exponent z.

26 Equilibrium fluctuations and directed polymers

In Chapter 9 we discussed the dynamics of a *driven* interface in a porous medium. However, there are a number of problems in physics in which one is interested in the properties of an *equilibrium* interface, when there is no driving force pushing the interface in a selected direction. A closely related problem is the equilibrium fluctuations of an elastic line in a porous medium, which is a problem of interest in many branches of physics ranging from flux lines in a disordered superconductor (see Fig. 1.4) or the motion of stretched polymer in a gel. In this chapter we discuss the properties of a directed polymer (DP) in a two-dimensional random medium, and the relation between the directed polymer problem and the interface problem.

26.1 Discrete model

Consider a discrete lattice whose horizontal axis is x, and vertical axis is h (see Fig. 26.1). The polymer starts at $x = 0$ and $h = 0$, and moves along the x direction in discrete steps. It can go directly along x, or it can move via transverse jumps, such that $|h(x+1) - h(x)| = 0, 1$. There is an energy cost ('penalty') ϑ, for every transverse jump. This simulates a line tension, discouraging motion along the h axis. On every bond parallel to x, a random energy $\varepsilon(x, h)$ is assigned.

The total energy of the DP is the sum over the energies along the polymer length, which includes the sum over the random energies arising from motion along the x direction, and the energy penalties for motion along the h axis. Thus for every *path* we can associate an energy E. The 'partition function' of the model is given by

$$Z_x(h) = \sum_{\{h(x)\}} \exp[-E(h)/k_B T], \qquad (26.1)$$

where the sum goes over all possible paths starting at $(0, 0)$ and ending at (x, h).

The energy of a path is affected by two competing forces: the energy penalty for transverse fluctuations favors straight configurations, but the polymer wants to wander in order to take advantage of the randomly distributed low energy bonds. At low temperature, the partition function (26.1) is dominated by polymers with the lowest energy.

26.2 Scaling properties

Our goal is to calculate the fluctuations of $h(x)$, the h coordinates of the endpoint of the DP after x steps. Before treating the DP in a general random medium, we note that if we set $\vartheta = 0$ and $\varepsilon(x, h) = 0$, the DP becomes a classic lattice random walk, for which

$$\langle |h(x)| \rangle \sim x^{1/2}. \tag{26.2}$$

For the DP model with nonzero ϑ and random ε we define the exponent ν_{DP} by

$$\langle |h(x)| \rangle \sim x^{\nu_{\mathrm{DP}}}. \qquad \rightarrow (26.3)$$

A second exponent can be defined as follows. Starting from (0,0), after x horizontal steps, there are many possible paths, with many different energies. At $T = 0$ the partition function is dominated by the path which has the lowest energy. We are interested, naturally, not in the value of this energy, but rather in the fluctuation in the energies of various paths – this quantity grows as a power law, and so leads us to define the exponent ω_{DP} by

$$\Delta E \equiv \langle (E - \langle E \rangle)^2 \rangle^{1/2} \sim x^{\omega_{\mathrm{DP}}}. \qquad \rightarrow (26.4)$$

We can calculate ν_{DP} and ω_{DP} using Monte Carlo simulations. A

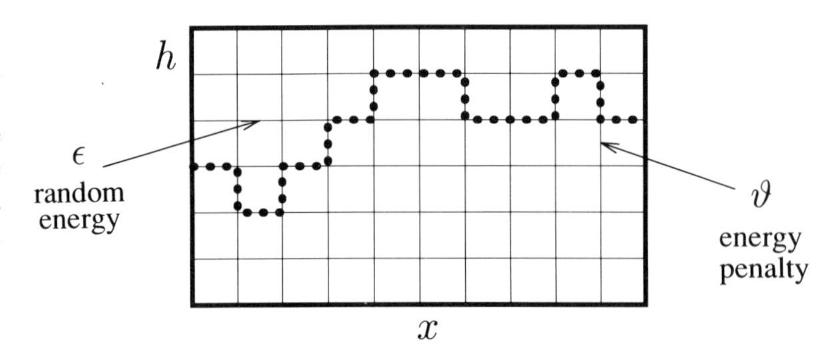

Figure 26.1 Schematic illustration of the directed polymer model. The dotted line corresponds to a particular realization of a directed walk.

more efficient method, the transfer matrix method, allows the calculation of *all* paths, enabling the exact determination of the partition function for a given realization of the randomness. One thereby finds the values [218, 232]

$$\nu_{\rm DP} \cong \frac{2}{3} \qquad \omega_{\rm DP} \cong \frac{1}{3}. \tag{26.5}$$

To develop our understanding of these exponents, we turn to the continuum description of the directed polymer problem.

26.3 Continuum description

In the continuum description the total energy of the DP path can be written as

$$\mathscr{H}_0 = \int dx \left[\frac{\nu}{2}(\nabla h)^2 + V(x,h)\right], \tag{26.6}$$

where $V(x,h)$ is the quenched randomness in the medium. Two limiting cases of (26.6) are easily discussed.

(i) If $V = 0$, then (26.6) reduces to the Hamiltonian (C.1) of the EW equation. The lowest energy according to (26.6) corresponds to a straight DP, for which $\nabla h = 0$ everywhere.

(ii) If $\nu = 0$, considering only the random potential $V(x,h)$ but ignoring the elastic term, the DP wanders to take advantage of the low-energy positions in the noise.

Thus the quenched noise wants to make the DP path *rough*, while the elastic energy acts to *smooth* it. The competition between these two effects determines the final morphology (see Fig. 26.2).

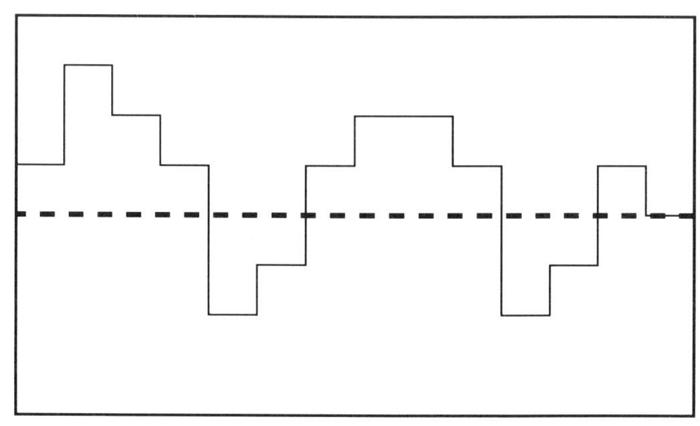

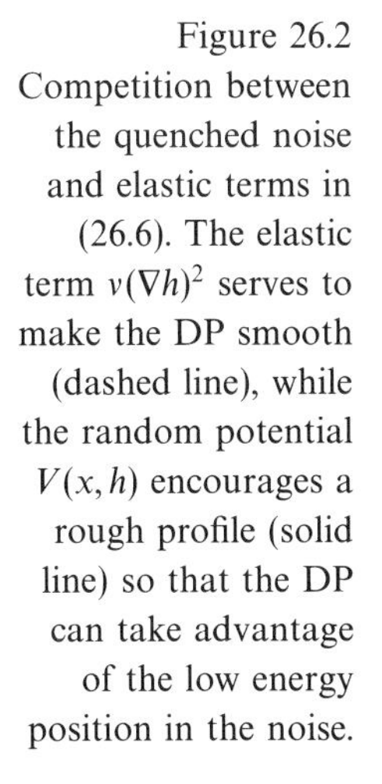

Figure 26.2 Competition between the quenched noise and elastic terms in (26.6). The elastic term $\nu(\nabla h)^2$ serves to make the DP smooth (dashed line), while the random potential $V(x,h)$ encourages a rough profile (solid line) so that the DP can take advantage of the low energy position in the noise.

Using (26.6), we have the partition function for a path starting at (0,0) and ending at (x, h)

$$Z_x(h) = \int_{(0,0)}^{(x,h)} \mathscr{D}h \exp\left(-\frac{1}{T}\int dx \left[\frac{\nu}{2}(\nabla h)^2 + V(x,h)\right]\right) \quad (26.7)$$

One can readily check that the partition function satisfies the differential equation with quenched multiplicative noise

$$\frac{\partial Z_x(h)}{\partial x} = \frac{1}{2}\frac{T}{\nu}\frac{\partial^2 Z}{\partial h^2} - \frac{1}{T}V(x,h)Z_x(h). \quad (26.8)$$

Fortunately, this equation has the form of the KPZ equation, since the change of variables $Z_x(h) = \exp[\text{const.} \times \tilde{h}(x,h)]$ transforms (26.8) to a KPZ equation, where $\tilde{h}(x,h)$ is the interface height, h is the transverse coordinate and x plays the role of time. Since for $d = 1$ we know the exact KPZ exponents, this mapping gives us the exponents for the DP problem

$$\nu_{DP} = \frac{1}{z_{KPZ}}, \qquad \omega_{DP} = \beta_{KPZ}. \qquad \text{[DP]} \quad (26.9)$$

The connection between the interface problem and the directed polymer problem is useful for many reasons. First, it expands the notion of universality, indicating that apparently different problems (interfaces, directed polymers) can be understood using the same analytical framework. Second, the possibility of exactly enumerating all directed paths in the DP problem provides good estimates for the $d = 1$ scaling exponents. Third, some analytical results can be obtained more easily in the language of one problem than the other.

Due to the connection between directed polymers and interfaces, both research areas benefit from progress achieved in either one. In particular, the DP problem has been studied using functional RG [157, 158] and real space RG methods [82, 83, 101].

The DP model is defined for finite temperatures, but its scaling properties are controlled by the zero temperature fixed point. For $T = 0$ the partition function is dominated by the path with the *lowest* energy, so the problem reduces to the identification of the lowest energy path in a system with quenched disorder (see Fig. 26.3). This is an example of a *global optimization problem*.

26.4 Equilibrium theory

In the previous section we discussed the properties of a DP in a two-dimensional random medium, which is a particular realization of the general problem of a d-dimensional equilibrium interface in a

$(d + 1)$-dimensional random medium. We will show that there is a more general approach to the problem, and that we can distinguish two types of disorder: random field and random bond. In the previous sections we considered a DP subject to random bond disorder.

Let us consider the interface shown in Fig. 9.2, with $F = 0$, defined by the single-valued function $h(\mathbf{x}, t)$. The equation of motion of the interface in a quenched environment has the form

$$\frac{\partial h}{\partial t} = \nu \nabla^2 h + \eta(\mathbf{x}, h) + \eta(\mathbf{x}, t), \tag{26.10}$$

which can be obtained from the Hamiltonian (26.6) using

$$\frac{\partial h}{\partial t} = -\frac{\delta H_0}{\delta h} + \eta(\mathbf{x}, t), \tag{26.11}$$

where $\eta(\mathbf{x}, t)$ is the thermal noise. The quenched random noise $\eta(\mathbf{x}, h)$ in (26.10) is linked to the random potential $V(\mathbf{x}, h)$ of (26.6) by

$$\eta(\mathbf{x}, h) = -\frac{\partial V(\mathbf{x}, h)}{\partial h}. \tag{26.12}$$

We assume that the quenched noise is given by (9.5).

26.4.1 Random field disorder

The two different classes of disorder, random-field and random-bond, reflect the form of the function $\Delta(u)$ in (9.5). What is termed the

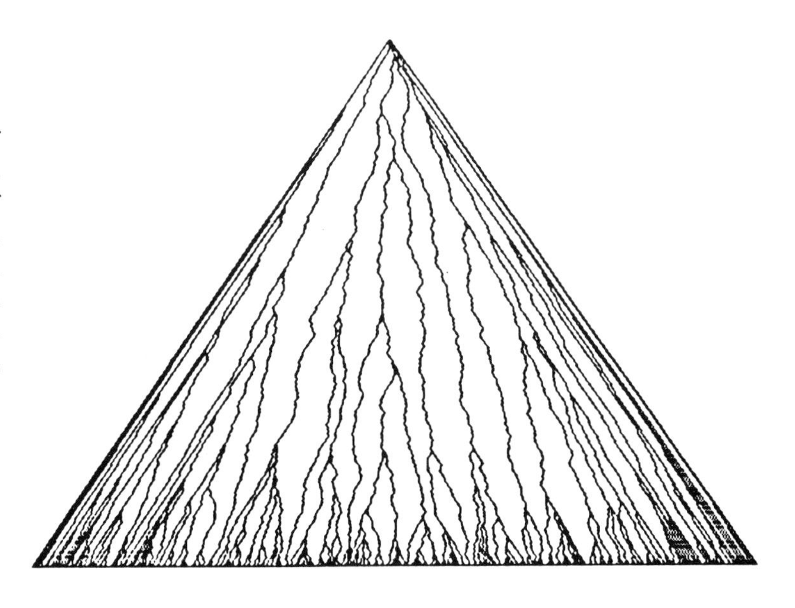

Figure 26.3 The DP problem in two dimensions. Shown is a set of one-dimensional directed polymers of very low energy on a square lattice with random bonds. Each DP has one end fixed. (After [218]).

random field model is obtained if, in (9.5), we choose

$$\Delta(h - h') = D_{\rm RF}\delta(h - h'). \tag{26.13}$$

The Hamiltonian corresponding to this type of noise [according to (26.6) and (26.13)] is

$$\mathscr{H}_{\rm RF} = \int_0^L d^d x \left[\frac{\nu}{2}(\nabla h)^2 + \int_{-\infty}^h dh' \; \eta(\mathbf{x}, h') \right], \tag{26.14}$$

which means that the pinning energy is affected by all the randomness that the interface has encountered during its previous motion (see Fig. 26.4). Random field disorder describes, for example, a fluid invading a porous medium, for which the final form of the interface is affected by all impurities in the region previously invaded by the fluid. The name 'random field' arises from the study of Ising systems: the model describes the equilibrium properties of an interface in the RFIM (see §10.3.2).

The *equilibrium* properties of the interface are defined by the Hamiltonian (26.14), while its *dynamical* properties can be obtained from the evolution equation (26.10). We can use scaling arguments to obtain the roughness exponent α. Let us concentrate on the Hamiltonian (26.14), having in mind that we could pursue similar arguments using the evolution equation (26.10).

Rescaling the Hamiltonian (8.7) using $\mathbf{x} \to b\mathbf{x}$ and $h \to b^\alpha h$, we find that the elastic term has a factor b^d arising from the integral over x^d, and a factor $b^{2\alpha-2}$ from the gradient squared which, when combined, results in $b^{d+2\alpha-2}$ as the scale factor. Similarly, we have $b^{d+\alpha}$ from the two integrals over the noise, and $b^{-d/2-\alpha/2}$ from the delta functions in the noise correlations, with $b^{(\alpha+d)/2}$ as the scaling factor. In equilibrium, the smoothing effect of the elastic term should balance the roughness created by the quenched randomness. Thus from $b^{d+2\alpha-2} \sim b^{(\alpha+d)/2}$, we obtain for the roughness exponent [153]

$$\alpha_{\rm RF}(d) = \frac{4-d}{3}. \tag{26.15}$$

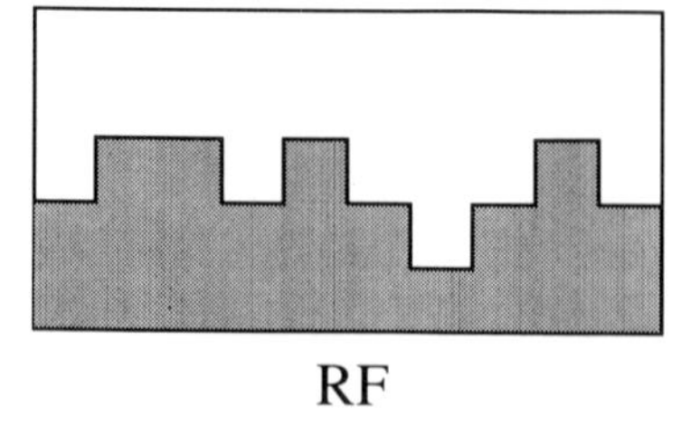

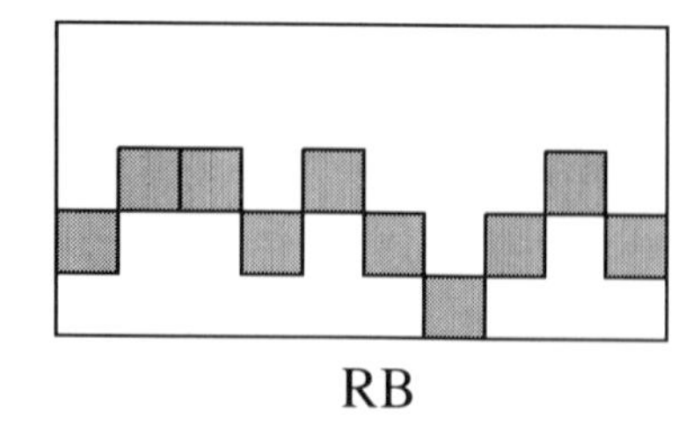

Figure 26.4 Difference between random field (RF) and random bond (RB) disorder. For RF disorder, the energy of the interface is determined by all disorder *under* the interface (gray area). For RB disorder the Hamiltonian (26.6) depends only on the noise *on* the interface (gray sites).

From (26.15), we identify $d = 4$ as the critical dimension of the model. For $d > 4$, the elastic term wins over the disorder. In this case the disorder becomes irrelevant, and the interface takes the lowest energy morphology determined only by the elastic term, i.e., it becomes a straight line. Functional renormalization group methods in $d = 4 - \epsilon$ dimensions suggest that the exponent (26.15) determined by these simple scaling arguments is actually exact [129].

Numerical simulations found exponents in good agreement with (26.15). Simulations on solid-on-solid interface representations of the random field Ising model in one dimension made possible the evaluation of the scaling exponent α, and good agreement was found with $\alpha_{\mathrm{RF}}(d = 1) = 1$ [128]. Similarly, the transfer matrix method extended to two dimension gave $\alpha_{\mathrm{RF}} = 0.59 \pm 0.07$, which is compatible with the prediction $\alpha = 2/3$ of (26.15) [219].

26.4.2 Random bond disorder

Interfaces with random bond disorder are described by the Hamiltonian (26.6) with random potential

$$\langle V(\mathbf{x}, h) V(\mathbf{x}', h') \rangle = D_{\mathrm{RB}} \delta^d(\mathbf{x} - \mathbf{x}') \delta(h - h'). \qquad (26.16)$$

In the language of the Ising systems, this corresponds to impurities that do not couple to the order parameter, but directly attract or repel the interface (i.e., nonmagnetic impurities). Formally the difference reduces to the absence of integration of the noise over h in the Hamiltonian for the random bond model. In other words, the total energy depends only on the disorder in the immediate neighborhood of the interface (see Fig. 26.4).

It is important to understand that in higher dimensions the DP and interface problems with RB disorder are *not* equivalent. For interfaces, we have a d-dimensional interface separating two $d + 1$ dimensional domains. The higher dimensional generalization of the DP problem deals with a one-dimensional line following a directed path in a $d + 1$ dimensional space. But the mapping between directed polymers and the KPZ equation is valid in *any* embedding dimension, the free energy of the polymer being the height of a d-dimensional interface driven by time-dependent noise.

To obtain the scaling exponents, we may use the same scaling arguments as for the RF system. The scaling factor for the elastic term remains unchanged from the RF model, but for the noise we obtain $b^{(d-\alpha)/2}$, giving the 'Flory exponent'

$$\alpha_{RB}^{F} = \frac{4-d}{5}. \tag{26.17}$$

An upper bound can be obtained by estimating the energy 'gain' due to random bond fluctuations to be $L^{d/2}$, since the surface covers L^d random bonds [337, 338]. Equilibrating with the elastic 'cost' $w^2 L^{d-2}$, where w is the width of the interface, leads to

$$\alpha_{RB} = \frac{4-d}{4}. \tag{26.18}$$

For $d = 1$, the roughness exponent can be determined exactly by mapping the problem to the KPZ equation (6.4). Since for $d = 1$ we know exactly the values of the KPZ scaling exponents, the mapping gives us the exponents for the quenched noise problem. The result, $\alpha_{RB} = 2/3$, is supported by numerical simulations [184, 185, 213]. This value is different from both $\alpha_{RB}^{F} = 3/5$ and $\alpha_{RB} = 3/4$, as predicted by (26.17) and (26.18) respectively. For higher dimensions, functional RG methods predict $\alpha_{RG} = 0.2083\epsilon$, where $\epsilon \equiv 4 - d$ [129]. The only two-dimensional numerical test of which we are aware is by Kardar and Zhang, who found $\alpha_{RB} = 0.50 \pm 0.08$ using a transfer matrix method [219].

26.5 Discussion

In this chapter we focused on the scaling properties of an equilibrium interface in a medium with quenched disorder. We found that the two types of disorder (RF and RB) result in different scaling exponents. For the RF model, simple power counting gives exponents that are in good agreement with the numerical results. The situation is more complicated for the RB problem. In 1+1 dimension, the problem is equivalent to directed polymers in a random environment and can be exactly mapped onto the KPZ equation.

Suggested further reading:

[130, 159, 338, 341]

27 Summary of the continuum growth equations

The universality class for a given growth process is determined by the symmetry properties of the underlying phenomena, and by the conservation laws. In this chapter we categorize the growth equations according to the conservation laws and symmetries they obey. We shall present a summary of the growth equations, pointing out the differences that lead to the various scaling exponents. Moreover, we shall briefly discuss some problems that can be studied using equations similar to those related to various growth processes, but that lead to new universality classes.

27.1 Universality classes

There are a number of questions that we must answer if we wish to classify the growth equations in terms of universality classes. For example, what exactly is a universality class? We shall use here the standard working definition, that two systems belong to the same universality class if they share the same set of scaling exponents. This often means that the two systems are described by the same growth equations as well.†

What are the building blocks we are interested in?‡ After neglecting

† However, this is not always true. A counterexample is the AKPZ equation, that – depending on the strength of the anisotropy – predicts the exponents of either the KPZ or the EW equation. Thus the AKPZ equation does not present a new universality class.

‡ As a general rule, we focus on the scaling properties of the system, i.e., we neglect

the irrelevant terms, the equations have the general form

$$\frac{\partial h}{\partial t} = G + \eta, \tag{27.1}$$

where in G we collected all the *deterministic* terms, while η is the noise. The explicit form of G is determined by the symmetries of the relaxation process *and* conservation laws acting during growth. Similarly, the noise can be conservative or nonconservative.

In the following we shall assume that the basic symmetries (i)–(iv) of §5.2 are valid. Then there are three parameters determining the universality class: (a) the conservation law obeyed by the deterministic part; (b) whether the system is linear or nonlinear, and (c) the nature of the noise.

Deterministic part – The different terms in the growth equations can be classified in two groups, whether they are conservative or nonconservative. A relaxation process is conservative if it does not change the number of particles in the system. For continuum equations, this translates into the statement that the integral of the term over the entire system is identically zero. The terms that have this property and occur in the growth equations are $\nabla^2 h$, $\nabla^4 h$, $\nabla^2(\nabla h)^2$. There is only one relevant nonconservative term: the KPZ nonlinearity $(\nabla h)^2$. The difference between these terms can be seen in Fig. 14.1, where we offer a geometrical interpretation: for the conservative terms the area above the x axis equals the area below it, which is not true for the nonconservative term. Among the conservative terms, two are linear, $\nabla^2 h$ and $\nabla^4 h$, while one is nonlinear, $\nabla^2(\nabla h)^2$.

Noise† – The noise can be conservative or nonconservative, the difference being reflected in the noise correlation functions. For nonconservative noise we have

$$\langle \eta(\mathbf{x}, t)\eta(\mathbf{x}', t')\rangle = 2D\delta^d(\mathbf{x} - \mathbf{x}')\delta(t - t'), \tag{27.2}$$

while for conservative noise,

$$\langle \eta_d(\mathbf{x}, t)\eta_d(\mathbf{x}', t')\rangle = (-2D_d\nabla^2 + D_d'\nabla^4)\delta^d(\mathbf{x} - \mathbf{x}')\delta(t - t'). \tag{27.3}$$

all terms in the growth equation that do not have a role in determining the scaling exponents (see §5.2 for a discussion). However, we must emphasize that the terms we neglect may be present in the growth equations. If we study the asymptotic properties of the growth process, they should be irrelevant, but they may determine the short length scale morphology, and also the short time scale behavior.

† Here we do not address the possibility that the noise η is quenched in the lattice, in which case $\eta = \eta(\mathbf{x}, h)$ (see Chapters 9–11).

27.2 Nomenclature

A simple nomenclature helps distinguish between the various growth equations. Let us denote every growth equation by three letters: the first tells us whether the growth equation is linear or nonlinear (L or N). The second letter (C or N) denotes whether the deterministic part is conservative or nonconservative. Finally, the third letter (C or N) denotes the conservation law obeyed by the noise. There is only one further problem with this nomenclature: there are two linear conservative terms, $\nabla^2 h$ and $\nabla^4 h$. To be able to distinguish between them, we shall put a 2 or 4 at the end of each symbol, denoting $\nabla^2 h$ and $\nabla^4 h$, respectively.

Why do we require such a classification? By following this nomenclature we shall be able to describe all the universality classes discussed in this book. Moreover, this simplification may reveal combinations that are missing a corresponding universality class. Their absence will help us to gain a deeper understanding on how the different terms influence the scaling behavior.

Next we discuss these growth equations systematically. We shall not reproduce every growth equation in the text. Table 27.1 presents a summary of all physically-relevant growth equations, with their names and the acronyms of the proposed classification. Table 27.1 also presents a summary of the scaling exponents for different dimensions, predicted by the growth equations or by numerical simulations.

- *Linear, conservative dynamics and conservative noise (LCC)*

There are two equations of this type, LCC2 and LCC4. Due to the linearity of the problem, the exponents can be calculated exactly. It is interesting that while LCC4 has been studied in detail, we do not know of any study focusing on LCC2.

- *Linear, conservative dynamics and nonconservative noise (LCN)*

There are again two equations belonging to this class, the Edwards–Wilkinson equation (LCN2) and the linear equation with surface diffusion (LCN4). Again the exponents can be calculated exactly using scaling arguments.

- *Linear, non-conservative dynamics and conservative or nonconservative noise (LNC or LNN)*

There are *no* such growth equations, since there is no nonconservative linear term.

- *Nonlinear, conservative dynamics and noise (NCC)*

There is only one growth equation of this type, studied in §21.4. This would correspond to NCC4. There is no NCC2 universality

Table 27.1 *Summary of the principal continuum growth equations and the predicted exponents. Nonconservative noise is denoted by η, and conservative noise by η_d.*

Symbol	Name	Equation	Exponents α	β	z
-	RD (4.7)	$\partial h/\partial t = \eta(\mathbf{x}, t)$	-	1/2	-
LCN2	EW (5.6)	$\partial h/\partial t = \nu\nabla^2 h + \eta(\mathbf{x}, t)$	$\frac{2-d}{2}$	$\frac{2-d}{4}$	2
NNN2	KPZ (6.4)	$\partial h/\partial t = \nu\nabla^2 h$ $+\frac{\lambda}{2}(\nabla h)^2 + \eta(\mathbf{x}, t)$	$\frac{1}{2}$	$d = 1$: $\frac{1}{3}$; $d > 1$: See Table 27.2	$\frac{3}{2}$
LCN4	— (13.9)	$\partial h/\partial t = -K\nabla^4 h$ $+\eta(\mathbf{x}, t)$	$\frac{4-d}{2}$	$\frac{4-d}{8}$	4
LCC2	— (21.8)	$\partial h/\partial t = \nu\nabla^2 h$ $+\eta_d(\mathbf{x}, t)$	$-\frac{d}{2}$	$-\frac{d}{4}$	2
LCC4	— (21.6)	$\partial h/\partial t = -K\nabla^4 h$ $+\eta_d(\mathbf{x}, t)$.	$\frac{2-d}{2}$	$\frac{2-d}{8}$	4
NCN4	— (14.2)	$\partial h/\partial t = -K\nabla^4 h$ $+\lambda_1\nabla^2(\nabla h)^2$ $+\eta(\mathbf{x}, t)$	$\frac{4-d}{3}$	$\frac{4-d}{8+d}$	$\frac{8+d}{3}$
NCC4	— (14.2)	$\partial h/\partial t = -K\nabla^4 h$ $+\lambda_1\nabla^2(\nabla h)^2 + \eta_d(\mathbf{x}, t)$	$d \leq 1$: $\frac{2-d}{3}$; $d > 1$: $\frac{2-d}{2}$	$d \leq 1$: $\frac{2-d}{10+d}$; $d > 1$: $\frac{2-d}{8}$	$d \leq 1$: $\frac{10+d}{3}$; $d > 1$: 4

class, because the nonlinear term $\nabla^2(\nabla h)^2$ is irrelevant if $\nabla^2 h$ is present in the growth equation.

• *Nonlinear, conservative dynamics and nonconservative noise (NCN)*
Again, for the same reason as for the NCC2 growth equation, NCN2 does not exist. The only universality class in NCN4.

• *Nonlinear, nonconservative dynamics and nonconservative noise (NNN)*
This is the KPZ equation.

Table 27.2 *Values of scaling exponents α, β and z, predicted by the continuum growth equations or obtained using simulations.*

Model	d=1			d=2			d=3		
	α	β	z	α	β	z	α	β	z
RD	-	$\frac{1}{2}$	-	-	$\frac{1}{2}$	-	-	$\frac{1}{2}$	-
EW (LCN2)	$\frac{1}{2}$	$\frac{1}{4}$	2	0	0	2	$-\frac{1}{2}$	$-\frac{1}{4}$	2
KPZ (NNN2)	$\frac{1}{2}$	$\frac{1}{3}$	$\frac{3}{2}$	0.38	0.24	1.58	0.30	0.18	1.66
(LCN4)	$\frac{3}{2}$	$\frac{3}{8}$	4	1	$\frac{1}{4}$	4	$\frac{1}{2}$	$\frac{1}{8}$	4
(LCC2)	$-\frac{1}{2}$	$-\frac{1}{4}$	2	-1	$-\frac{1}{2}$	2	$-\frac{3}{2}$	$-\frac{3}{4}$	2
(LCC4)	$\frac{1}{2}$	$\frac{1}{8}$	4	0	0	4	$-\frac{1}{2}$	$-\frac{1}{8}$	4
(NCN4)	1	$\frac{1}{3}$	3	$\frac{2}{3}$	$\frac{1}{5}$	$\frac{10}{3}$	$\frac{1}{3}$	$\frac{1}{11}$	$\frac{11}{3}$
(NCC4)	$\frac{1}{3}$	$\frac{1}{11}$	$\frac{11}{3}$	0	0	4	$-\frac{1}{2}$	$-\frac{1}{8}$	4

There are a number of observations regarding this classification. First, the nomenclature does not allow us to include the RD model, which lacks a deterministic part. Similarly, we assume that the systems are all isotropic. However, as we saw in §25.2, anisotropy can result in an interesting combination of universality classes: the exponents of the AKPZ equation depend on the isotropy of the system. Finally, for many physical phenomena more terms are present than those strictly determining the asymptotic scaling. However, in these cases we can decide which are the terms that actually determine the scaling exponents, and all other terms can be neglected. For example, if in a system $\nabla^2 h$ and $\nabla^4 h$ coexist, the former will play the leading role in determining the asymptotic scaling.

27.3 Related problems

In this book we are mainly concerned with the description of growth phenomena using discrete models and continuum growth equations. However, the methods developed here can be applied to other systems, not within the scope of this book. In this section, we shall give a brief review of some of these related problems. In particular we discuss three sets of problems: (a) the effect of coupling two nonequilibrium growth equations, (b) the continuum description of continuously driven systems related to self-organized criticality, and (c) driven diffusive systems.

27.3.1 Coupled nonequilibrium growth equations

To characterize the dynamics and morphology of interfaces, it is usually sufficient to consider a single parameter, the height of the interface $h(\mathbf{x}, t)$. However, there are some systems for which there is more than one relevant fluctuating quantity, or field, that must be dealt with in order to obtain a correct description of the scaling properties. For example, suppose that an interface with height $h_0(x, t)$ grows in the presence of a nonequilibrium field $h_1(x, t)$, and the interface and the field strongly affect each other. Such a perturbing field may be the density or pressure of the fluid during fluid displacement in porous media, the concentration of nutrient in bacterial growth, or the density of mesoscopic particles in imbibition experiments. Thus, in order to describe the coupled *nonequilibrium* dynamics of the two fields h_0 and h_1, we must consider a set of coupled growth equations. How do we obtain the relevant growth equations?

Let us consider that the time evolution of h_1 is described by the KPZ equation. If h_1 influences h_0, the growth equation for h_0 must include terms that depend on h_1. If both h_1 and h_0 obey the symmetries (i)–(iv) of §5.2, then the possible relevant coupling terms are $\nabla h_0 \nabla h_1$ and $(\nabla h_1)^2$. Higher order terms are irrelevant. Similarly, additional terms must be included in the growth equation of h_1 to account for the effect of h_0. The resulting equation represents the lowest order coupled equations that describe the coupling of two nonequilibrium fields [26]

$$\partial h_0/\partial t = v_0 \nabla^2 h_0 + \lambda_0 (\nabla h_0)^2 + \gamma_0 \nabla h_0 \nabla h_1 + \varphi_0 (\nabla h_1)^2 + \eta_0 \qquad (27.4)$$

$$\partial h_1/\partial t = v_1 \nabla^2 h_1 + \lambda_1 (\nabla h_1)^2 + \gamma_1 \nabla h_0 \nabla h_1 + \varphi_1 (\nabla h_0)^2 + \eta_1. \qquad (27.5)$$

If $\gamma_i = \varphi_i = 0$, then we are left with two independent KPZ equations. There are four scaling exponents characterizing the scaling properties of such a system: the roughness exponents α_0 and α_1, the dynamic exponents z_0 and z_1, for the two fields h_0 and h_1. However, if the system is fully coupled, one expects a single dynamic exponent $z = z_0 = z_1$.

This may not be the case if some of the terms are absent from the growth equations.

The coupled growth equations (27.4)–(27.5) can be studied using dynamic RG, and it is possible to construct the corresponding discrete models [26]. Regarding the universality class of the resulting growth process, in the fully coupled case (when all coefficients are nonzero), the exponents coincide with the KPZ exponents. However, this is not the case if some of the terms are missing, as we show next.

An interesting special case of (27.4)–(27.5) can be obtained by assuming that only h_1 has a broken up–down symmetry, while h_0 fluctuates in equilibrium. The $h_1 \to -h_1$ symmetry excludes γ_0 from (27.4), and λ_1 and φ_1 from (27.5). Thus we are left with the much simpler pair of coupled growth equations

$$\partial h_0/\partial t = \nu_0 \nabla^2 h_0 + \lambda_0 (\nabla h_0)^2 + \varphi_0 (\nabla h_1)^2 + \eta_0, \tag{27.6}$$

and

$$\partial h_1/\partial t = \nu_1 \nabla^2 h_1 + \gamma_1 \nabla h_0 \nabla h_1 + \eta_1. \tag{27.7}$$

This set of equations – proposed by Ertas and Kardar [117, 118] – describes the motion of a flux line in a three-dimensional superconductor driven in a preferred direction by an external field.

Next we describe briefly the scaling behavior of this system, as predicted by the dynamic RG calculations and direct integration of the equations.

(a) If $\gamma_1 \varphi_0 > 0$ the scaling exponents are the same as for the KPZ equation: $z = 3/2$, $\alpha_0 = \alpha_1 = 1/2$.

(b) If $\varphi_0 = 0$, (27.6) is the KPZ equation, with the corresponding scaling exponents. However, h_0 influences the fluctuations of h_1, acting as a correlated multiplicative noise. The dynamic RG predicts $z_1 = 3/2$ and $\alpha_1 = 3/4$ if $\gamma_1 > 0$, while if $\gamma_1 < 0$, γ_1 scales to zero.

(c) If $\gamma_1 = 0$, (27.7) reduces to the EW equation, with the corresponding exponents. The fluctuations in h_1 act as an additive noise for the growth equation in h_0. Since the predictions of the one-loop dynamic RG do not agree with the numerical results, we do not quote any of them.

(d) For $\gamma_1 < 0$ and $\varphi_0 > 0$ the dynamic RG indicates that $\gamma_1 \to 0$, so the system is not fully coupled. Simulations indicate the presence of instabilities.

(e) For $\gamma_1 > 0$ and $\varphi_0 \geq 0$ the dynamic RG indicates that the exponents should be $z_0 = z_1 = 3/4$ and $\alpha_0 = 1/2$, but α_1 is unknown.

The growth equations (27.4)–(27.7) have nonconservative dynamics and noise. As we saw in the previous section, the introduction of conservation laws, both in the deterministic part and in the noise, may change the universality class. Since now we have two equations, the number of possible universality classes is even larger than in the one-component case. However, at this point the possible coupled equations and the related universality classes are largely unexplored.

The small number of cases studied raises the possibility that the coupling of nonequilibrium fields may lead to unexpected phases and scaling behavior. Motivated by experimental studies on surfactant-mediated MBE growth [84, 85, 110], a minimal model describing the coupling between a conservative and a nonconservative field indicates the appearance of a new phase in which the roughness exponent is negative, thereby suggesting that such a coupling may stabilize the fluctuations and result in a smooth interface [27, 28]. Similarly, studies on two coupled conservative fields indicate the presence of a largely unexplored rich scaling behavior [32].

27.3.2 Self-organized criticality

Recent years have witnessed increasing interest in systems that are critical, but for which there is no external tuning parameter that must be adjusted to reach the critical state. This property of becoming critical without parameter tuning is referred to as *self-organized criticality* (SOC). The most prominent example of such a system is the sandpile model introduced by Bak, Tang and Wiesenfeld [21]. In this section we briefly introduce the main elements of the sandpile model. Next we discuss the continuum equation motivated by the sandpile model, as proposed by Hwa and Kardar [188], and the scaling exponents predicted by it.

(a) *The discrete model* – The original sandpile model considered deposition of particles on a flat surface. The one-dimensional version of the model is illustrated in Fig. 27.1, which illustrates the sandpile model for the case $d = 1$. The sandpile is characterized by the height of each column $H(x)$. A particle added to the system results in $H(x) \to H(x)+1$, and the column x for deposition is randomly selected. At each time step, we check the sites of the sandpile for stability. A particle is unstable if the local slope $m(x) = |H(x) - H(x+1)|$ exceeds a critical value m_c (here $m_c = 2$). If a particle is unstable, we remove 2 particles from the column x and add them to the column at $x + 1$ (algorithmically: if $m(x) > m_c$, then $H(x) \to H(x) - 2$, $H(x+1) \to H(x+1)+2$). If the particle reaches the boundary, it falls out.

Since our purpose is to arrive at the continuum equations describing the model, we shall consider here a slightly modified version [210]. Particles are deposited with a constant flux F. The original sandpile model corresponds to the $F \to 0$ limit – i.e., all unstable particles are relaxed before a new particle is deposited. As numerous numerical and analytical studies indicate, the distribution of the number of particles falling off the edge – and also the number of updating events between two depositions – follow simple power laws. It appears impossible to describe this limit with simple continuum equations. Thus we shall focus on the driven system, for which F is finite.

(b) *Continuum equations and exponents* – In order to introduce a growth equation describing this model we must consider a different system of coordinates. As illustrated in Fig. 27.1, we define the height $h(x,t)$ as the difference between the landscape and the straight 'average' profile,

$$h(x,t) \equiv H(x,t) - m(L-x), \tag{27.8}$$

where m is the average slope of the sandpile. In order to construct an appropriate growth equation, we must study the symmetries and

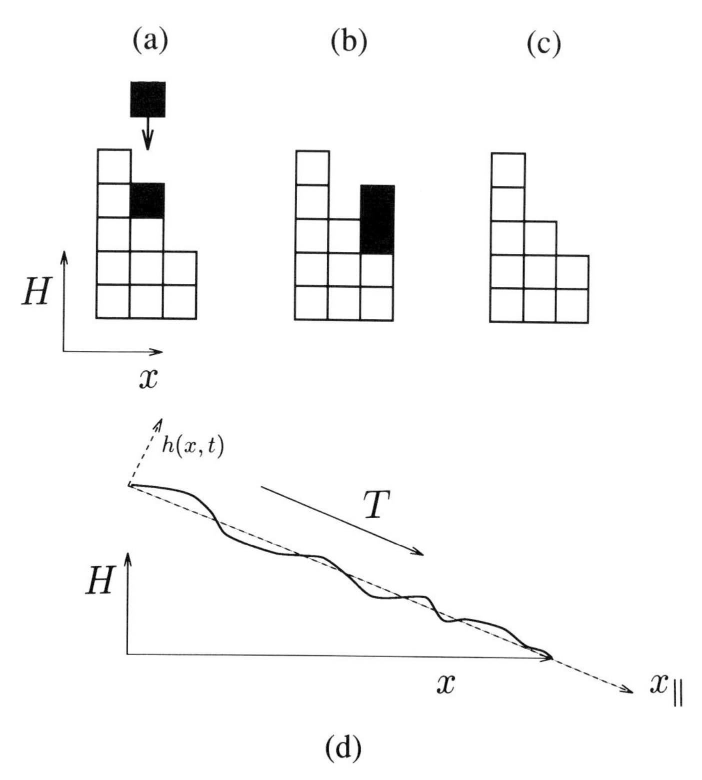

Figure 27.1 (a) The sandpile algorithm (see text) is illustrated on a small system. Depositing a particle on the second column, the sandpile becomes unstable ($m = 3 > m_c$). (b) The shaded particles will move to the right. (c) The third column becomes unstable, and two particles move to the right, leaving the system. (d) The transition from the coordinate system of the sandpile model to the local coordinates which are convenient for constructing a continuum growth equation.

conservation laws obeyed by the system. The grain motion is influenced by gravity; the component of the gravity parallel to the average interface profile leads to a preferred transport direction $\mathbf{T}$. Considering the coordinates $\mathbf{x}_\parallel \equiv (\mathbf{T} \cdot \mathbf{x})\mathbf{T}$, and $\mathbf{x}_\perp \equiv \mathbf{x} - \mathbf{x}_\parallel$, the system (i) has rotational invariance in $\mathbf{x}_\perp$ and translational invariance in $\mathbf{x}_\parallel$ and $\mathbf{x}_\perp$; (ii) lacks reflection symmetry in h and $\mathbf{x}_\parallel$ because of the preferred direction $\mathbf{T}$. (iii) There is no translational symmetry in h because of gravity.

Once the symmetry properties are elucidated, we must see whether there is any conservation law affecting the deterministic part of the noise. The local relaxation process is conservative, since the particle number is conserved except at the boundaries. However, the origin of the noise is the random deposition process that adds particles to the system, so the noise is nonconservative.

Thus the continuum equation describing the system should have the general form

$$\frac{\partial h(\mathbf{x},t)}{\partial t} = -\nabla \mathbf{j} + \eta(\mathbf{x},t). \tag{27.9}$$

Next we must find the appropriate current describing the dynamics of the reorganization process. We can proceed using symmetry arguments. For this, we first write down all possible terms for the current, and remove those that violate the listed symmetries. Assuming that the fluctuations in h are small, we find up to second order in h,

$$\mathbf{j} = a_1 \nabla h + \lambda_1 h \mathbf{T} + \lambda_2 h^2 \mathbf{T}. \tag{27.10}$$

Due to the anisotropy induced by the preferred direction $\mathbf{T}$, the surface tension may have different values for the two preferred spatial directions. From (27.9) and (27.10), we thereby obtain the equation of motion of the continuously driven sandpile model [188]

$$\frac{\partial h(\mathbf{x},t)}{\partial t} = \nu_\parallel \partial_\parallel^2 h + \nu_\perp \nabla_\perp^2 h - \frac{\lambda}{2}\partial_\parallel (h^2) + \eta(\mathbf{x},t). \tag{27.11}$$

For arbitrary dimensions the interface has only one preferred direction along $\mathbf{T}$, while it may have more than one transverse direction.

The growth equation (27.11) can be studied using dynamic RG. The analysis shows that $\nu_\perp$, λ and D do not renormalize, which provides three simultaneous algebraic equations for the three exponents, α, z and ζ, where the anisotropic exponent is defined by the scaling of $\mathbf{x}_\perp$. The solution is

$$\alpha = \frac{1-d}{7-d} \qquad z = \frac{6}{7-d} \qquad \zeta = \frac{3}{7-d}. \tag{27.12}$$

Further analysis reveals the existence of an upper critical dimension,

$d_c = 4$. For $d < d_c$ the nonlinear term is relevant, and the existing fixed point leads to the scaling exponents (27.12). For $d > d_c$ the nonlinear term is irrelevant, and the exponents reduce to those of the EW equation.

The exponents (27.12) define a new universality class, different from those discussed in the previous section. Since the goal of this chapter is to discuss the origin of universality classes, it is important to see what makes equation (27.11) different. We know that in order to have a critical system we must not have a term proportional to h in the growth equation, since that would lead to a characteristic length scale. In the growth equations such a term is forbidden by the translational symmetry in h. No such symmetry is present in this case, but such a term is excluded by the conservation law obeyed by the deterministic relaxation process. The absence of translational symmetry in h apparently leads to a different universality class. Another ingredient of Eq. (27.11) is the existence of the selected direction $\mathbf{T}$, which generates anisotropy in the system and leads to a nonlinear term.

Strictly speaking, the continuum equation (27.11) describes a continuous sandpile model [189], since it neglects the discreteness of the sand particles – i.e. the interface does not grow continuously, but in unit jumps. Such a lattice effect can be included by considering a periodic potential in the Hamiltonian, which reduces to a term periodic in h in the growth equation (see Chapter 18). However, any term included in the growth equation must conform to the conservative nature of the relaxation process. A simple equation useful for describing such a discrete system is [152]

$$\frac{\partial h(\mathbf{x},t)}{\partial t} = \nu_{\parallel}\nabla_{\parallel}^2 h + \nu_{\perp}\nabla_{\perp}^2 h - \lambda\nabla_{\parallel}\cos(2\pi h) + \eta(\mathbf{x},t). \tag{27.13}$$

On applying dynamic RG to (27.13) [152], we find the following results:

(a) For $d > 4$ the nonlinear term is irrelevant, and the model reduces to the linear equation.

(b) For $2 < d < 4$ the scaling behavior is controlled by the same fixed point that we found for (27.11), the exponents being given by (27.12).

(c) For $1 < d \leq 2$ both the Gaussian (obtained for $\lambda = 0$) and the nonlinear ($\lambda \neq 0$) fixed points are stable. The two fixed points correspond to different morphologies: rough for the Gaussian, smooth for the nonlinear case. There is a Kosterlitz–Thouless phase transition between these two phases in $d = 2$.

27.3.3 Driven diffusive systems

Another problem for which continuum equations are helpful in obtaining the scaling properties is the driven diffusive system (for a review see [390]). These are simple lattice gas models describing the diffusion of an ensemble of particles in the presence of an external driving field (that adds a drifting motion to the particles). A simple isotropic version of the problem, aimed to describe non-interacting particles, leads to the continuum equation

$$\frac{\partial h}{\partial t} = \nu \nabla^2 h + \frac{\lambda}{2} \nabla^2 (h^2) + \eta(\mathbf{x}, t). \tag{27.14}$$

where h is now the coarse grained particle density. Various versions of this equation, especially its anisotropic version, have been studied in detail in the literature [194]. The full anisotropic equations describing a driven lattice gas of interacting particles are more complicated, and lead to various equilibrium and nonequilibrium phase transitions.

27.4 Discussion

The existence of various universality classes is a powerful concept in modern statistical mechanics, since it allows us to understand the differences between growth processes. In this chapter we introduced a simple classification of various continuum growth equations and discussed the corresponding universality classes. We saw that most growth equations can be partitioned into universality classes based on the conservation laws obeyed by the deterministic part and the noise, as well as whether they are linear or nonlinear.

In practical situations, there are a large number of effects shaping the actual morphology of the interface. However, out of these there are only a few that actually determine the scaling exponents of the system. Identifying them allows us to reduce the problem to onc of a few growth equations. Once this is possible, we can predict the scaling exponents. In this fashion, a large number of seemingly unrelated phenomena may be seen to belong to the same universality class, even though there is no apparent connection among them.

We also showed that there may be, in some situations, other factors determining the universality class. For example, if there is an external field h_1 influencing the growth of the interface h_0, the coupling between the two fluctuating quantities may lead to new scaling exponents. Similarly, we have shown that there are a number of phenomena of current interest, including self-organized criticality and driven diffusive systems, that can be studied using the same analytical techniques that we used in the study of various roughening processes.

The discussion in this chapter focused entirely on roughening dominated by thermal noise. However, if we change the properties of the noise by considering power-law distributed amplitudes, correlated noise, or quenched noise, we may find new scaling exponents and new universality classes.

Suggested further reading:

[151, 169, 390]

28 Outlook

It is a truism to remark that no one – not even a theoretical physicist – can predict the future. Nonetheless, after asking the beleaguered reader to indulge in the rather extensive 'banquet' of the preceding 27 chapters, it seems only fair to offer a light 'dessert' that affords some outlook and perspective on this rapidly-evolving field.

What concepts loom above the details is a question worth addressing at the end of any large meal. Charles Kittel wrote his first edition of *Introduction to Solid State Physics* almost 50 years ago. He surely realized that solid state physics was a rapidly-evolving field, so his book ran the risk of becoming dated in short order. Therefore the first chapter systematically discusses the various crystal symmetries – and the group theory mathematics that describes these symmetries. The topics comprising solid state physics have changed rather dramatically, and most chapters of Kittel's 7th edition hardly resemble the chapters of the first edition. Nevertheless, the opening chapter of the first edition could serve as well today as an introduction to the essential underpinnings of the subject.

Inspired by Kittel's example, we have attempted in this short book to highlight where possible what seems to us to be the analog for disorderly surface growth of the various symmetries obeyed by crystalline materials. These newer 'symmetries', described using terms that may frighten the neophyte – such as scale invariance and self-affinity – are as straightforward to describe as translation, rotation, and inversion. Just as group theory expresses the implications of the symmetries of a periodic structure, so fractal theory enables one to decipher the implications of these newer symmetries. The analog of the 32 point groups are the various universality classes, each characterized by its set of critical exponents.

We have also attempted to illustrate the ubiquity of fractal concepts.

On the one hand, there are a wide range of relatively simple surface phenomena, such as interface motion treated in the early chapters. On the other hand, there are rather complex phenomena treated in later chapters – such as deposition followed by diffusion of the deposited species and eventual aggregation into 'submonolayer nanostructures.'

What, then, of the future? Is it possible that the new concepts of scale invariance and self-affinity will lead to revolutionary discoveries? It has often been said of group theory that one learns nothing new. Nonetheless, generations of students have mastered basic group theoretical concepts because they know that the principles of group theory facilitate classification, nomenclature, and quantification, without which understanding and eventual new discoveries are improbable. Group theory is not the end goal of ones study of solid state physics, but rather the first step. So also, the principles of abstract scale invariance are not the end of our study, but are rather stepping stones to viewing experimental facts in a rather different light.

When one undertakes the study of a new system, one does not even know the correct questions to ask. If we find that the exponents characterizing the new system are the same as those for a known class of systems, then we can form hypotheses and exclude other possibilities, just as when we know the positions of spectroscopic lines we can eliminate various possible structures of a crystalline system. Is it possible that the classifications, nomenclature, and quantification of disorderly growth phenomena achieved by defining growth exponents – and using the measured or calculated values of these exponents to partition phenomena into distinct universality classes – will help provide the 'infrastructure' needed to eventually understand the underlying phenomena and harness this understanding for the betterment of mankind?

Another utility of building a suitable infrastructure that permits classification, nomenclature, and quantification is to prepare for future advances in disparate fields of science. For example, as time goes on, scientists and engineers are likely to discover newer and more useful materials for which surface phenomena are relevant. This book hopefully will provide the reader with the background needed to navigate among the zoo of new materials that surely will dominate the 21st century.

There is a second class of investigations for which the basic symmetries discussed in this book may be relevant, namely when one formally maps some property of interest onto a 'disorderly surface'. An example was mentioned in the introduction: the DNA walk, which forms a easily-visualizable surface in 1+1 dimensions. The DNA walk is an abstract surface that has nothing to do with the surface of DNA itself. Rather, it is constructed by focusing on an aspect of DNA that is unrelated to surface physics, the sequence of base pairs that carries the genetic code, and constructing an artificial surface by means of a one-to-one mapping of this sequence of base pairs and a self-affine surface.

The number of such abstract surfaces is limited only by the imaginations of those who construct them, so one may ask 'why bother?' One possible answer to this question is that by representing abstract information in the graphic form of a surface, the eye (and eventually the computer) can recognize features that would have passed unnoticed in the original data. An example of such a feature is the long-range correlation in the base pair sequence of DNA: The 'raw data' of 666 000 base pairs (the number in a single recently-sequenced chromosome of yeast) would fill as many pages of paper as there are pages of this book. No one is likely to spot a pattern given such a wealth of information unless he or she knows what to look for in advance. For example, if one uses a digital computer to search for a pattern, then one must know at the outset what sort of a pattern to search for in order to write the appropriate search program. Nonetheless, when the same information is displayed as a 'self-affine' surface, one recognizes features that were never searched for using the analog computer known as the eye/brain complex. The eye/brain complex is of course also programmed in a sense, but the patterns searched for represent the years of experience of the brain that is connected to the eye – something the digital computer cannot be readily programmed to do.

Are there additional quantities that can be mapped onto such abstract surfaces and examined by the eye/brain complex? Can the analysis of such surface representations lead to uncovering basic facts about nature that have hitherto remained obscured? As noted at the outset of this chapter, 'no one can predict the future.'

APPENDIX A

Numerical recipes

One of the main methods used to study the roughening of nonequilibrium interfaces is to construct discrete models and study them using computer simulations. To simulate the models on a computer and to study their scaling properties, we must use a number of numerical methods. If the model is simple, the corresponding simulation is also simple, e.g., a program to simulate RD or BD requires only a few lines. However, the *analysis* of the simulation results is far from being simple, requiring special care due to finite size effects, slow crossover behavior, and other complications. In this chapter we collect some additional methods that are useful in obtaining a thorough analysis of the scaling properties.

A.1 Measuring exponents for self-affine interfaces

As noted, α and β are universal exponents, which do not depend on the particular details of the model. Thus obtaining these scaling exponents helps to identify the universality class of the growth process. The method used for the determination of α and β for a particular interface depends on the information available. If the interface is obtained from numerical simulation, every detail of the interface and its dynamics can be extracted from the computer. However, experiments may not be able to follow every detail of the time evolution of the growth, and one can analyze only the final interface.

We briefly present five methods for the numerical analysis of a rough interface. In the ideal case, α does not depend on the method we use. But in practice, α sometimes varies from method to method, and it is useful to use as many methods as possible in order to get a feeling for the range of values of α that are consistent with the data.

• *Method 1*

Assume that we have calculated the saturation width $w_{sat}(L)$ for different system sizes L. Then the roughness exponent α can be estimated using (2.5) by plotting the data on a log-log plot, and measuring the slope of the straight line.

To use this method, we must be able to obtain $w_{sat}(L)$ for various system sizes L. In most experimental situations we do not have results for different systems sizes, and in addition there may be strong boundary effects influencing $w_{sat}(L)$. Thus Method 1 is more useful for numerical simulations, where we can simulate systems of arbitrary sizes, and we can require periodic boundary conditions to reduce the boundary effects. Method 1 cannot always be used, since for sufficiently large systems it is not possible to saturate the interface. According to (2.6), the time necessary for saturation increases with the system size. If the dynamic exponent z is large, this increase can be quite dramatic, requiring a prohibitive computation time.

• *Method 2*

In many situations, we do not have any information on the dynamics of the growth, nor do we have the possibility of producing interfaces with different system sizes. Suppose the only data we have are collected at the final stage of an experiment, consisting of the value of the height at different points. In this situation, we can study the scaling of the *local* width $w_L(\ell, t)$ defined by

$$w_L^2(\ell, t) \equiv \langle [h(\mathbf{x}, t) - h_\ell(\mathbf{x}, t)]^2 \rangle_{\mathbf{x}}. \tag{A.1}$$

The subscript ℓ on the rhs means that we select a portion ('window') of length ℓ on the interface and measure the width and average height $h_\ell(\mathbf{x}, t)$ in this window. The brackets $\langle \ldots \rangle_{\mathbf{x}}$ denote spatial (over $\mathbf{x}$) and ensemble averages – we choose many different windows along the surface and average over the obtained results. For small ℓ

$$w_L(\ell, t) \sim \ell^\alpha \qquad [\ell \ll \xi_\parallel], \tag{A.2}$$

where α is the same roughness exponent defined in (2.5).

One can verify that (A.2) reduces to (2.5) in the long time limit, by replacing ℓ with the system size L. For a saturated system, this scaling would last all the way up to $\ell = L$, but for a nonsaturated interface the scaling behavior is observable only for $\ell \ll \xi_\parallel(t)$. Thus if we measure $w_L(\ell, t)$, the scaling region increases with time as predicted by (2.13). The local width $w_L(\ell, t)$ reaches its saturation value at $\ell = \xi_\parallel(t)$, which is in fact $\xi_\perp(t)$,

$$\xi_\perp(t) = w_L(\xi_\parallel, t). \tag{A.3}$$

Method 2 is useful for the determination of α in numerical simulations, and is complementary to Method 1.

- *Method 3*

If, in addition to w_{sat}, we have the full time evolution of the width, as shown in Fig. 2.4, we can use (2.8) to determine both α and z. This can be done by attempting to rescale the data as we did in Chapter 2, and is illustrated in Fig. 2.6 for BD. If we use the correct exponents for the rescaling, we should obtain good data collapse. Any significant deviation would lead to the breakdown of data collapse. To illustrate this, Fig. A.1 shows the data of Fig. 2.4 rescaled with *incorrect* exponents α and z.

- *Method 4*

Another quantity that scales in the same way as the interface width is the height–height correlation function,

$$C(\ell) \equiv \left[\langle\left(h(\mathbf{x}) - h(\mathbf{x}')\right)^2\rangle_{\mathbf{x}}\right]^{1/2} \qquad [|\mathbf{x} - \mathbf{x}'| = \ell]. \tag{A.4}$$

The interface heights are considered at the same moment t, so we do not write out time explicitly. The roughness exponent can be determined from the relation

$$C(\ell) \sim \ell^{\alpha} \qquad [\ell \ll \xi_{\parallel}]. \tag{A.5}$$

A finite correlation length has a similar effect on the scaling of $C(\ell)$ as on interface width: scaling extends for values of ℓ up to the parallel correlation length $\xi_{\parallel}$. Although the equivalence of the scaling exponents determined from (A.5) with those obtained by studying the local width w is not *a priori* transparent, various numerical simulations underline their interchangeable use. The height–height correlation

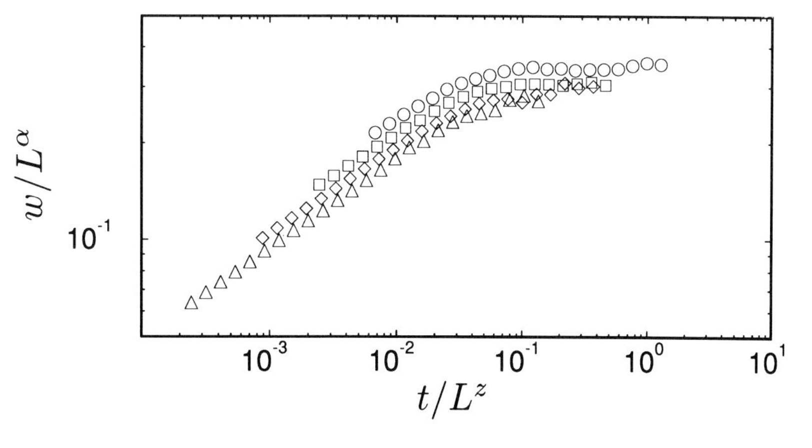

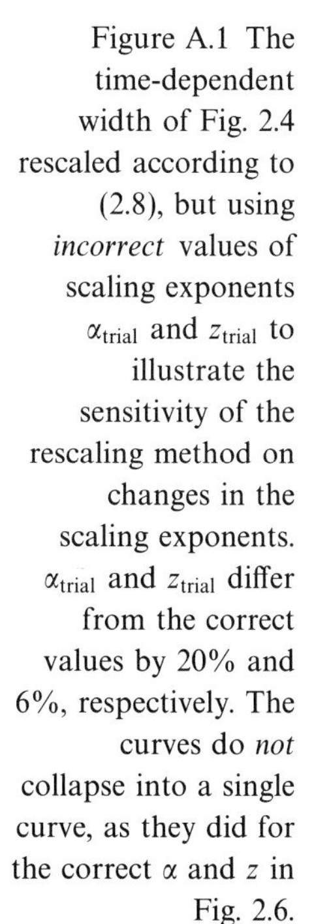
Figure A.1 The time-dependent width of Fig. 2.4 rescaled according to (2.8), but using *incorrect* values of scaling exponents α_{trial} and z_{trial} to illustrate the sensitivity of the rescaling method on changes in the scaling exponents. α_{trial} and z_{trial} differ from the correct values by 20% and 6%, respectively. The curves do *not* collapse into a single curve, as they did for the correct α and z in Fig. 2.6.

function is especially useful when we calculate analytically the roughness exponent, since, as we shall see in Appendix B, it can be directly related to the continuum equations.

The correlation function $C(\ell)$ defined in (A.4) is sensitive to bias in the data, i.e., if the interface has an overall slope, the scaling is affected. If the interface is obtained from numerical simulations with periodic boundary conditions, there should be no bias in the data, so (A.4) gives the correct scaling. However, interfaces obtained from experimental data may have an overall slope. To overcome the effect of the bias, we can use a slightly different height variable $\tilde{h}(\mathbf{x},t)$ from which the overall average slope has been subtracted, and define

$$C(\ell,t) \equiv [\langle(\tilde{h}(\mathbf{x},t') - \tilde{h}(\mathbf{x}',t'+t))^2\rangle_{\mathbf{x}',t'}]^{1/2}. \qquad \text{(A.6)}$$

If the surface is inhomogeneous, consisting of regions with differing overall average slopes, one can separately analyze each region (the "min-max method" [365]) or one can use a method that does not require one to partition the surface into separate regions, such as the method called "detrended fluctuation analysis" [366].

If we are interested only in the roughness of the interface at the same time t, we take $t = 0$ (i.e., there is no time delay), and

$$C(\ell,0) \sim \ell^{\alpha} \qquad [\ell \ll \xi_{\|}]. \qquad \text{(A.7)}$$

If we wish to determine β, we can measure the correlations between interfaces with a time delay. Provided we study times shorter than the crossover time $t_{\times}$ where the width saturates, we find

$$C(0,t) \sim t^{\beta} \qquad [t \ll t_{\times}]. \qquad \text{(A.8)}$$

We shall frequently refer to the local width (A.1) and correlation function (A.4). Since the correlation function $C(\ell,t)$ and the local width $w_L(\ell,t)$ scale in the same way, in most cases we shall not distinguish between them.

• *Method 5*

Some experiments measure the power spectrum of the interface, and not the height $h(\mathbf{x},t)$. Thus we must examine how the dynamic scaling relation (2.8) is modified in Fourier space. Consider the structure factor

$$S(\mathbf{k},t) \equiv \langle h(\mathbf{k},t)h(-\mathbf{k},t)\rangle, \qquad \text{(A.9)}$$

where

$$h(\mathbf{k},t) \equiv \frac{1}{L^{d/2}}\sum_{\mathbf{x}}[h(\mathbf{x},t) - \bar{h}]\exp(i\mathbf{k}\cdot\mathbf{x}). \qquad \text{(A.10)}$$

The dynamic scaling hypothesis (2.8) can be translated to the structure factor, with result

$$S(\mathbf{k}, t) = k^{-d-2\alpha} g(t/k^{-z}). \tag{A.11}$$

Here $g(u) \sim u^{(2\alpha+d)/z}$ for $u \ll 1$ and $g(u) \to$ const for $u \gg 1$.

Note that $w^2(L,t) = (1/L^d)\sum_{\mathbf{k}} S(\mathbf{k},t)$ gives a simple relation between the structure factor and the width of the interface. The calculation of the structure factor provides an alternative way to estimate the scaling exponents. It has the advantage over the real space methods that only the long wavelength modes contribute to its scaling. The scaling of the real space form (2.8) is determined by all modes, including the short wavelength modes. Thus (2.8) is expected to have stronger finite size effects than (A.11) [276, 372].

Before concluding this section, we briefly discuss several points that may help the newcomer.

• *Upper and lower cutoffs*

Both experiments and models have upper and lower cutoffs which limit scaling regimes. Sometimes it is easy to identify the origin of the cutoff (such as lattice spacing, or the system size), but this is not easy for many experiments. In such situations it is useful to study *both* the local width and correlation function, since they have the same scaling properties, but may behave somewhat differently at the cutoffs, thus providing information on the position of the scaling regime.

• *Consecutive slopes*

A useful technique for the determination of the scaling region and the scaling exponent α or β is the use of *consecutive slopes.* Let us consider the scaling of the width $w(L,t)$ with t. The consecutive slope $\beta_s(L,t)$ is the local slope of the $\log w(L,t)$ curve plotted as a function of $\log t$ *fit over a region of size s.* Methodologically, we plot $\log w(L,t)$ against $\log t$ and fit a straight line to the data between t and $(t+s)$; then we plot the set of slopes so obtained as a function of $\log t$ (Fig. A.2). If there were no upper and lower cutoffs and no corrections to the scaling, then the consecutive slopes would scatter around a horizontal line at ordinate β. With the cutoffs, the curves deviate from the straight line outside the scaling region.

From Fig. A.2, we see that a 'blind' regression fit to the data will result in misleading values of the exponent. In general, there is no universally agreed-upon procedure for estimating the exponent value, much less the error bars. While the plateau indicated by the consecutive slopes in Fig. A.2 is quite well defined, in practice it is often rather small, and the data can be quite noisy. However, plotting consecutive slopes for systems with different sizes L may be helpful in identifying the correct scaling region.

- *Averaging vs. self-averaging*

An important part of any numerical analysis is related to averaging over independent sets of data. Fig. A.3 shows the time evolution of the width $w(L,t)$ for a *single* run for two different system sizes for BD. The first thing we observe is the noisiness of the data, particularly for the smaller system. Noise is a general property of all statistical systems. The noise may be substantially reduced by either (i) averaging over different *independent* measurements or (ii) by increasing the system size, whereupon one often finds 'self-averaging' of the measurable quantities. For numerical simulations (i) is accomplished by repeatedly re-running the simulation with a different seed for the random number generator, and averaging the results obtained from different runs. Experimentally one repeats the experiment, and averages the results obtained for different interfaces. It is very important that the results to be averaged are for systems that evolved under identical conditions –

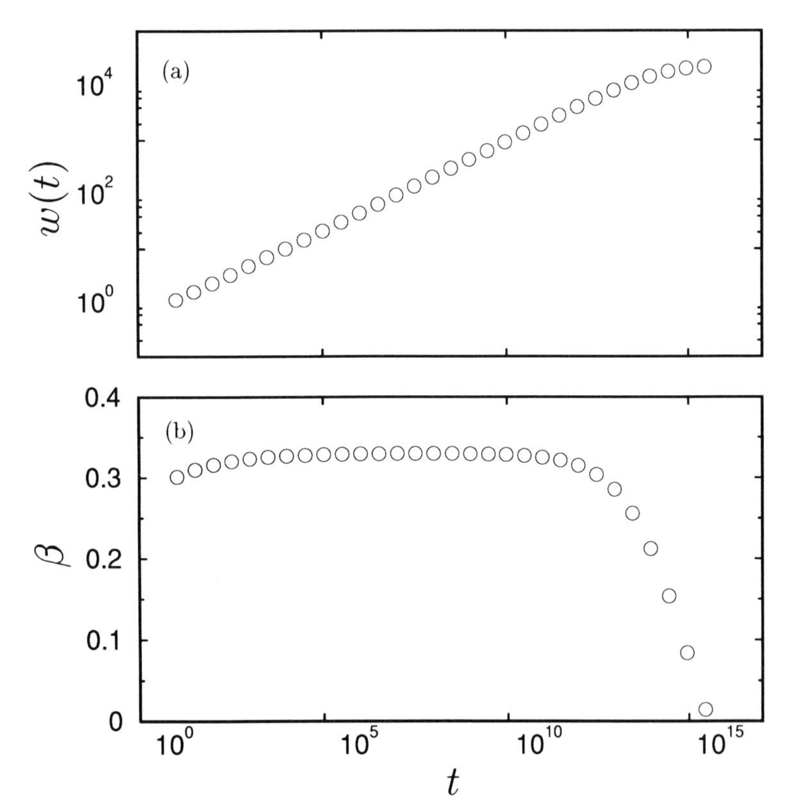

Figure A.2 Use of consecutive slopes for the determination of the optimal scaling region. (a) Scaling of the interface width with time. (b) Consecutive slopes displaying a plateau at an exponent $\beta \approx 0.33$ between the lower and upper cutoffs, t_ℓ and t_u.

e.g., we must not average over the results of two interfaces obtained at different times (unless the system is saturated), or between the scaling obtained for interfaces with different system size L. The larger the number of averages we form, the more reliable is the scaling obtained. The results presented in Fig. 2.4 are obtained after averaging typically 1000 independent runs.

A.2 The coefficient λ of the nonlinear term

When studying a growing interface, whether generated by experiment or by a model, in order to obtain the proper understanding of the growth process, we must identify the universality class to which it belongs. The different universality classes predict different scaling exponents, so studying the scaling of the interface width and determining the roughness exponents should allow us to distinguish between them. In many situations, however, it is difficult to obtain reliable scaling exponents, due to complicated crossover and finite-size effects. These effects become negligible in the limit of infinite system size, but this limit is often not accessible by computer simulations or experiments. Another problem arises in one dimension, where the roughness exponents predicted by the EW and the KPZ equations are the same ($\alpha = 1/2$), so based on calculations of α alone we cannot distinguish the two possibilities.

An alternative method for identifying the universality class is to obtain direct evidence for the presence of different terms in the growth equation. Among these is the nonlinear term $\lambda(\nabla h)^2$. The determination of the coefficient λ is of special interest since, if present, λ controls the scaling properties of the interface. Recently, a number of methods have been proposed to identify this nonlinear term [183, 253, 261, 487].

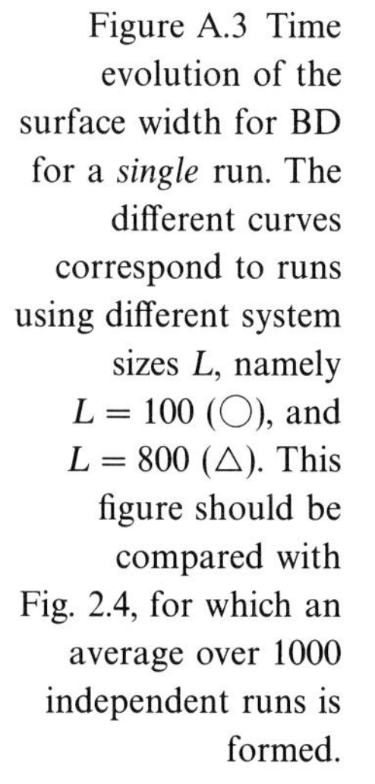

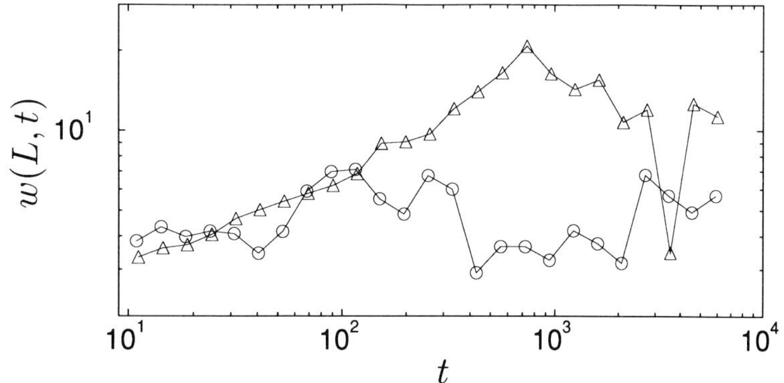

Figure A.3 Time evolution of the surface width for BD for a *single* run. The different curves correspond to runs using different system sizes L, namely $L = 100$ (○), and $L = 800$ (△). This figure should be compared with Fig. 2.4, for which an average over 1000 independent runs is formed.

The simplest method of calculating λ is based on the fact that the average interface velocity depends on the interface tilt, as is expressed by (6.6) [253]. Let us consider an interface which grows according to the KPZ equation with an average velocity

$$v = v_o + \frac{\lambda}{2} \int_0^L d^d\mathbf{x}\ (\nabla h)^2. \tag{A.12}$$

Here v denotes the average velocity for an interface with periodic boundary conditions, and v_o the drift velocity, the contribution of an external force acting on the interface. The interface has a zero average slope. For model calculations, we can generate an overall slope $m \equiv \langle(\nabla h)\rangle$ of the interface by tilting the surface. Operationally, this can be accomplished by fixing the boundary conditions such that $h(L,t) = h(1,t) - m(L-1)$.

According to (A.12), the nonzero tilt changes the velocity of the interface, with

$$v(m) = v(0) + \frac{\lambda}{2} m^2. \tag{A.13}$$

If we measure the tilt-dependent velocity, we expect to find a parabola which, when fit with (A.13), provides the coefficient of the nonlinear term λ (Fig. A.4).

This method provides the value of λ for a given growth model. However, in certain situations we may be interested in ν and D as well. These parameters can be calculated using the *inverse method* [261]. The basic idea behind the inverse method is that we can attempt to fit a continuum growth equation to a given discrete model.

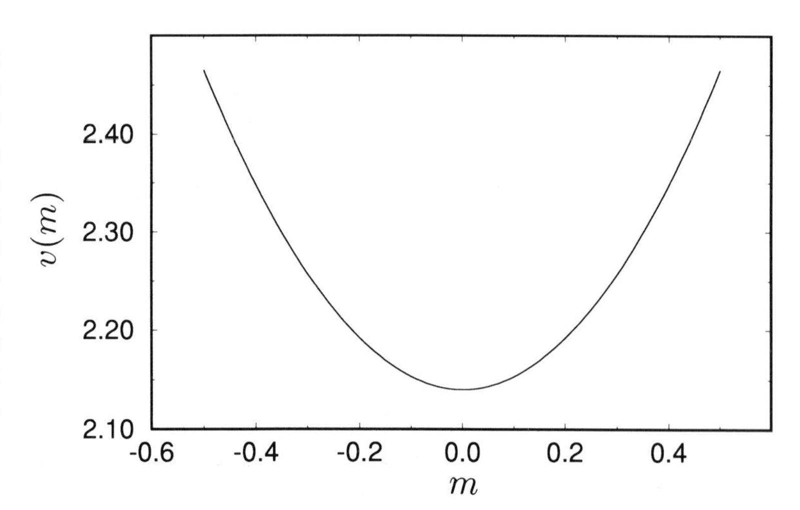

Figure A.4 Tilt dependence of the interface velocity for a growth process described by the KPZ equation. If, on the other hand, the model is described by the EW equation, we have $\lambda = 0$, and the velocity does not depend on the tilt.

Assume that the growth model is described by the growth equation

$$\frac{\partial h}{\partial t} = F + \nu \partial^2 h + \frac{\lambda}{2}(\partial h)^2 + K\partial^4 h + \lambda_4 \partial^2 (\partial h)^2 + ... + \eta, \qquad (A.14)$$

which can be written in the discretized form

$$\frac{\Delta h}{\Delta t} = \mathbf{a} \cdot \mathbf{H}(x,t) + \eta. \qquad (A.15)$$

Here $\mathbf{a} \equiv (F, \nu, \lambda/2, K, ...)$ and $\mathbf{H}(x,t) \equiv (1, \partial^2 h, (\partial h)^2, \partial^4 h, ...)$ are vectors.

Our goal is to take an interface generated by the growth model, and calculate **a** by fitting the variation in h with (A.15). To this end, we Fourier transform the interface $h(x,t)$ generated by the growth model and coarse grain the Fourier components for wavelengths smaller than l, after which we obtain **H**, which now depends on l. Now we continue to grow the surface, obtaining $h(x, t+\Delta t)$, so we can calculate Δh and **H** in (A.15). We do not know the value of the noise η, but if we average (A.15) over $\mathscr{R}$ realizations of $h(x, t+\Delta t)$ – by starting from the same $h(x,t)$ – the noise averages to zero. Hence we can calculate **a** using the least square fitting procedure.

The parameters we obtain depend on the coarse graining length scale l. During the coarse graining procedure, we neglect the short length scale fluctuations, doing exactly what one does in an RG calculation. Thus plotting the obtained parameters as a function of l provides the *flow* of that parameter under successive renormalizations of the interface. Thus the method provides not only the bare values of the parameters in the growth equation, but also their flow under renormalization.

A.3 Intrinsic width

For many models, the roughening of the interface is hampered by strong finite size and crossover effects, resulting in slow convergence to the asymptotic scaling behavior. One major source of the slow convergence is the existence of the *intrinsic width*, w_i, which represents an additive term in the expression of scaling [225, 486]. For models with a nonzero intrinsic width w_i, the scaling law (2.8) becomes

$$w^2(L,t) - w_i^2 \sim L^{2\alpha} f\left(\frac{t}{L^z}\right). \qquad (A.16)$$

Unlike the width $w(L,t)$, the intrinsic width is independent of the system size L. Its value depends on the parameters of the model, and in some models w_i can be reduced by using noise reduction techniques [225].

For small system sizes, the intrinsic width may be comparable to the final width, and so strongly affects the scaling behavior. Figure A.5 shows the scaling of the time-dependent width in the presence of a nonzero intrinsic width. At early times we observe a crossover, with slow convergence to the asymptotic scaling behavior. According to (A.16) we have

$$w^2(L,t) - w_i^2 \sim t^{2\beta}. \tag{A.17}$$

Plotting $w^2(L,t) - w_i^2$ as a function of time, we should obtain improved scaling behavior. As is shown in Fig. A.5, by choosing $w_i = 1.1$ we indeed find that the data conform to a straight line, facilitating a more accurate estimate of the exponent β.

What is the origin of the intrinsic width? There is no single answer to this question, since w_i is *not* a universal quantity and its presence and magnitude depend on the microscopic details of the model. The existence of overhangs, or the development of large local slopes can all contribute to w_i.

A.4 Measuring surface diffusion currents

For nonconservative growth models, measuring the interface velocity and its dependence on the overall interface tilt is a useful method to obtain information on the presence of the KPZ or higher order nonlinear terms, and thereby to establish the universality class. Models with surface diffusion, which conserve the number of particles and do not allow overhangs, have a velocity which depends only on the deposition rate, so a tilt in the interface does not affect the growth velocity of the interface.

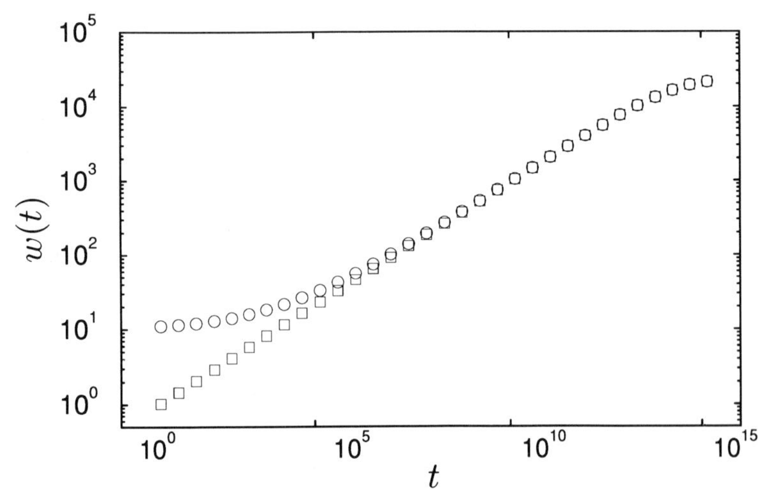

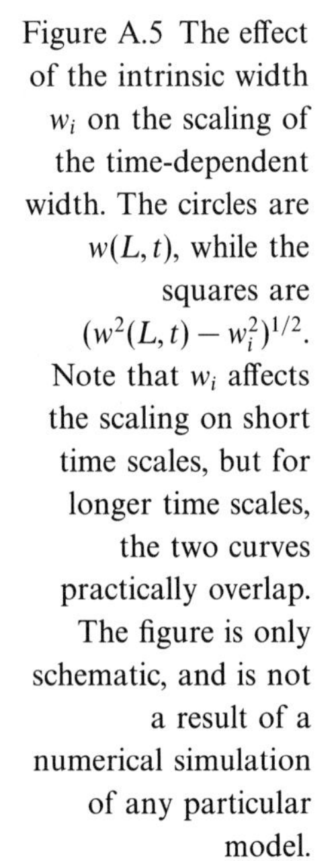
Figure A.5 The effect of the intrinsic width w_i on the scaling of the time-dependent width. The circles are $w(L,t)$, while the squares are $(w^2(L,t) - w_i^2)^{1/2}$. Note that w_i affects the scaling on short time scales, but for longer time scales, the two curves practically overlap. The figure is only schematic, and is not a result of a numerical simulation of any particular model.

However, Krug *et al.* [250] have introduced a method that measures the surface current as a function of the tilt, and which can give us very important indications on the relevant process and the universality class characterizing the growth model. The method is based on the continuity equation (13.2), indicating that the surface height can change due to the surface current $\mathbf{j}(\mathbf{x}, t)$. For a one dimensional interface this current can be measured by choosing a point on the interface and measuring the number of atoms moving to the right or to the left. Evidently, for an interface with periodic boundary conditions this surface current will be zero. But this is not always the case, if we *tilt* the interface, to have an average slope m.

For a tilted interface, a nonzero uphill or downhill current is possible, depending on the growth mechanism. Consider, e.g., the EW equation, for which $\mathbf{j} \sim -\nu \nabla h = -\nu m$, corresponding to an average downhill current. We can expand the current as a function of the tilt $j(m) = j(0) + j'(0)\, m + \frac{1}{2} j''(0)\, m^2 + ...$ Since the average current is zero if the interface is not tilted, $j(0) = 0$. The expansion of the current should contain only odd order terms in the tilt; if an even order term were present, that would correspond to a term with an odd number of derivatives in x, which is excluded by the $x \to -x$ symmetry of the interface. Thus the relevant expansion is

$$j(m) = j'(0)\, m + \frac{1}{3!} j'''(0)\, m^3 + \ldots \tag{A.18}$$

By imposing different tilts m, $j(m)$ can be calculated. The behavior in the vicinity of zero tilt provides the coefficient $\nu = -j'(0)$. A positive ν indicates the existence of the EW linear terms, leading to a stable interface. However, a negative ν signals that the interface is linearly unstable at long wavelengths.

We shall illustrate the method on the Wolf–Villain (WV) model, introduced in §15.1. Surprisingly, simulations show that the one dimensional WV model has a very small *downhill* current, $j(m = 2) \sim -0.0012$ [250]. Thus the scaling behavior should cross over to EW scaling for very long time scales. The crossover time is estimated from (13.14), giving $t_\times \sim \nu^{-2} \sim 10^6$. Fig. A.6 shows the tilt-dependent current for the two dimensional WV model, and one estimates $\nu \sim 0.0075$ corresponding to a crossover time $t_\times \sim 2 \times 10^4$.

A.5 Generating noise in simulations

A.5.1 Correlated noise

To study the effect of long-range correlations on the roughness exponent, the most difficult part of a simulation is the generation of noise with the power-law correlations of (22.1) or (22.2). There are a number of methods for generating long-range correlated noise, ranging from Fourier methods to using Lévy walks or the 'fast fractional Gaussian noise generator' [165, 262, 285, 291, 364, 375, 469]. Most of these methods suffer from the finiteness of the scaling region over which the noise is actually correlated. Here we present the key ingredients of the Fourier method. We also discuss recent work that allows correlated noise without a finite cutoff to be generated [281, 282]. We have limited ourselves to one dimension, but generalizing to higher dimensions is straightforward.

Consider a sequence of N *uncorrelated* random numbers $\{u_j\}_{j=1,..,N}$. The goal is to generate a new sequence of N *correlated* numbers, $\{\eta_j\}_{j=1,..,N}$, with the correlation function

$$C(\ell) \equiv \langle \eta_j \, \eta_{j+\ell} \rangle \sim \ell^{2\varphi-1}. \tag{A.19}$$

We can relate u_j and η_i using a 'response' function, ϕ_j,

$$\eta_j \equiv \sum_{i=1}^{N} \phi_{j-i} \, u_i. \tag{A.20}$$

By Fourier transforming (A.19) and (A.20), one finds

$$\hat{\eta}_q = \hat{\Phi}_q \hat{U}_q = \sqrt{\hat{C}_q} \hat{U}_q, \tag{A.21}$$

where $\hat{\eta}_q$, $\hat{\Phi}_q$, $\hat{U}_q$ and $\hat{C}_q$ are the Fourier transforms of η_j, ϕ_j, u_j and $C(j)$, respectively. Thus knowing the desired function $C(\ell)$ and the

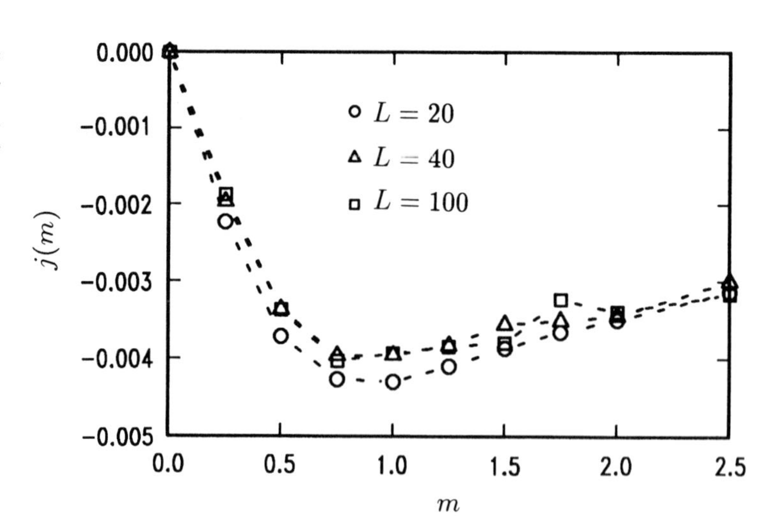

Figure A.6 Surface diffusion current for the two-dimensional WV model. (After [250]).

original noise u_j, we obtain the Fourier transform of the correlated variables $\hat{\eta}_q$ response function using (A.21).

The algorithm consists of the following steps:

(a) Generate the sequence, $\{u_j\}$, of uncorrelated random numbers and obtain its Fourier transform, $\hat{U}_q$.

(b) Generate the Fourier transform, $\hat{C}_q$, of the desired correlation function $C(\ell)$.

(c) Multiply $\hat{U}_q$ by $\sqrt{\hat{C}_q}$ to get $\hat{\eta}_q$, according to Eq. (A.21).

(d) Calculate the inverse Fourier transform of $\hat{\eta}_q$ to obtain $\{\eta_q\}$, the sequence of correlated numbers in real space.

The method discussed so far is the general procedure to generate long-range correlations using Fourier transforms. However, the range over which the numbers η_j are *actually* correlated depends sensitively upon the algorithm used. In fact, a straightforward implementation of the steps (a)–(d) generates a correlated sequence with a correlation length as small as 1% of the system size.

Recently Makse *et al.* introduced an algorithm to increase the range over which the number have long-range correlation [281]. The first problem is that the correlation function $C(\ell) \sim \ell^{2\psi-1}$ does not have a well-defined Fourier transform due to its singularity at $\ell = 0$. To overcome this difficulty, Makse *et al.* use

$$C(\ell) \equiv (1 + \ell^2)^{(2\psi-1)/2}, \tag{A.22}$$

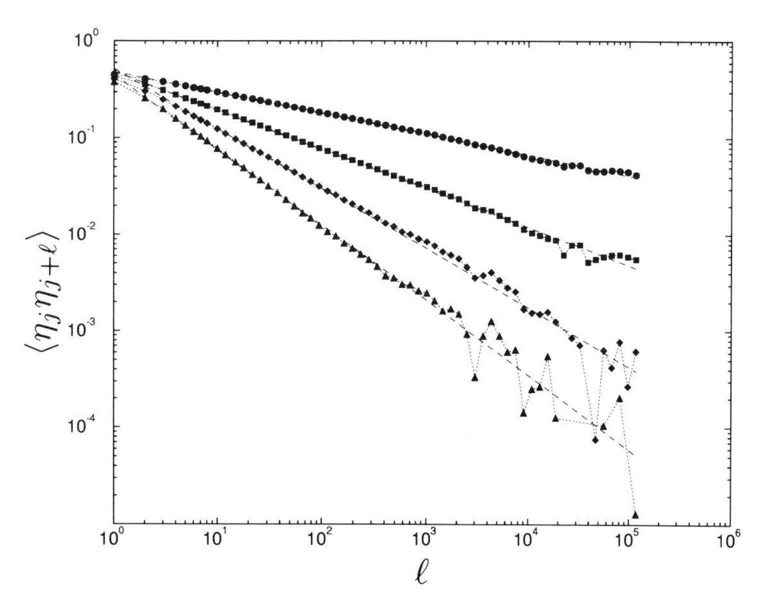

Figure A.7 Log-log plot of the average correlation function $\langle \eta_j \, \eta_{j+\ell} \rangle$ as a function of the distance ℓ between the sites for $N = 131\,072$. The average is taken over 20 realizations. Shown are results for different values of ψ, which range from 0.4, 0.3, 0.2, and 0.1 (from the top to the bottom). The slopes of the straight segments are -0.21 ± 0.01, -0.40 ± 0.01, -0.61 ± 0.01, and -0.70 ± 0.01, respectively, which are consistent with the expectation $2\psi - 1$ of Eq. (A.22). (After [281]).

which has the desired power law behavior for large ℓ. The Fourier-transformed form of this function,

$$\hat{C}_q = \frac{2\pi^{1/2}}{\Gamma[(1-2\varphi)/4]} (\pi q)^{-\varphi} K_{-\varphi}(2\pi q), \tag{A.23}$$

can be used directly in (A.21) to calculate $\hat{\eta}_q$. Here $K_{-\varphi} = K_\varphi$ is the modified Bessel function of order φ and Γ is the gamma function.

The second problem is related to the boundary conditions used during Fourier transformation. The condition (A.21) relies on the validity of the convolution theorem and therefore the correlation function (A.22) must satisfy the proper periodic boundary conditions. The function $C(\ell)$ is extended to negative values by defining it in the interval $[-N/2, N/2]$ with periodic boundary condition $C(\ell) = C(\ell+N)$ imposed. Thus the method actually generates $N/2$ correlated numbers from the initial sequence of N uncorrelated numbers. Thus to have N correlated numbers as an output, we must start with a sequence of $2N$ uncorrelated numbers. Indeed, the method incorporating the previous observations generates a sequence of numbers with long-range correlations, as is shown in Fig. A.7.

A.5.2 Power-law distributed noise

Numerical methods used to simulate roughening dominated by rare events have been studied in detail (e.g., [50]). Here we discuss briefly how to generate uncorrelated noise with power-law distribution (23.1).

A random number obeying the distribution (23.1) can be obtained by choosing a number y_i *uniformly* distributed on the interval (0,1), and then computing

$$\eta \equiv y^{-1/\mu}. \tag{A.24}$$

It is the small values of y close to 0 that produce the large values of η. A cutoff in the low-end of the $(0, 1)$ distribution of y leads to a cutoff of the tail of the corresponding power law. The effect of such a cutoff can be effectively decreased by increasing the precision in the random number y – e.g., by using double-precision or quadruple-precision representations for y, instead of single-precision.

Suggested further reading:

[250, 456]

APPENDIX B

Dynamic renormalization group

In this appendix we discuss the application of the renormalization group (RG) method to nonequilibrium systems. The dynamic RG was developed by Forster, Nelson and Stephen, who studied the dynamical properties of a problem equivalent to the KPZ equation, the noisy Burgers equation [132]. Their approach was transposed into the language of interfaces by Kardar, Parisi and Zhang [217]. A detailed description of the method for the more involved situation of correlated noise is given by Medina *et al.* [314]. Here we develop the dynamic RG method and apply it to the KPZ equation with uncorrelated noise. In particular, we shall illustrate the RG procedure by offering a derivation of the flow equations (7.12)–(7.14), whose predictions are discussed in Chapter 7.

B.1 Introduction

The KPZ equation can be studied using dimensional analysis. This method turned out to be successful in the case of the EW equation, providing the exact value of the scaling exponents. After a change of scale $\mathbf{x} \to b\mathbf{x}$, $h \to b^\alpha h$, and $t \to b^z t$, the KPZ equation becomes

$$\frac{\partial h}{\partial t} = \nu b^{z-2}\nabla^2 h + \frac{\lambda}{2} b^{z+\alpha-2}(\nabla h)^2 + b^{-d/2+z/2-\alpha}\eta. \tag{B.1}$$

Under this scale transformation the parameters of the equation change to

$$\nu \to b^{z-2}\nu, \tag{B.2}$$

$$D \to b^{z-d-2\alpha}D, \tag{B.3}$$

$$\lambda \to b^{z+\alpha-2}\lambda. \tag{B.4}$$

If the nonlinearity is absent, the exponents are given by (5.16). According to (B.4), a small nonlinearity added to the linear equation renormalizes λ to zero for $d > d_c = 2$ and to infinity for $d < d_c$.† This is a first indication that the nonlinear term may generate nontrivial exponents for $d < d_c$.

In order to consider systematically the effect of the nonlinearities, we shall try to do a series expansion around the exact solution of the EW equation in powers of λ. In §B.2 we present the details of such a series expansion using a diagrammatic technique. The integrals involved are calculated in §B.4. We will find that the terms obtained from the perturbation expansion diverge for $d < d_c$. To handle this divergence, we must renormalize the system. The renormalization procedure is presented in §B.3. After renormalization, we arrive at the flow equations (7.12)-(7.14) analyzed in §7.3.

B.2 Perturbation expansion

B.2.1 Fourier space

The RG can be done most conveniently using the Fourier components of the height

$$h(\mathbf{x}, t) = \int_{-\infty}^{\infty} \frac{d\omega}{2\pi} \int^{k<\Lambda} \frac{d^d\mathbf{k}}{(2\pi)^d} h(\mathbf{k}, \omega) e^{i(\mathbf{k}\cdot\mathbf{x}-\omega t)}, \tag{B.5}$$

where the momentum integrals are subject to an upper cutoff Λ, corresponding to the 'lattice spacing' in the real space. In the following, this cutoff is understood implicitly.

The KPZ equation in Fourier space becomes

$$-i\omega h(\mathbf{k}, \omega) = -\nu k^2 h(\mathbf{k}, \omega)$$
$$-\frac{\lambda}{2}\int\int \frac{d^d\mathbf{q}d\Omega}{(2\pi)^{d+1}}\ \mathbf{q}\cdot(\mathbf{k}-\mathbf{q})\ h(\mathbf{q}, \Omega) h(\mathbf{k}-\mathbf{q}, \omega-\Omega) + \eta(\mathbf{k}, \omega). \tag{B.6}$$

The noise is assumed to be uncorrelated in space:

$$\langle \eta(\mathbf{k}, \omega) \rangle = 0 \tag{B.7}$$

and

$$\langle \eta(\mathbf{k}, \omega)\eta(\mathbf{k}', \omega') \rangle = 2D\ (2\pi)^d\ \delta^d(\mathbf{k}+\mathbf{k}')\delta(\omega+\omega'). \tag{B.8}$$

Equation (B.6) can be rewritten as

$$h(\mathbf{k}, \omega) = G_0(\mathbf{k}, \omega)\eta(\mathbf{k}, \omega)$$
$$-\frac{\lambda}{2}G_0(\mathbf{k}, \omega)\int\int \frac{d^d\mathbf{q}d\Omega}{(2\pi)^{d+1}}\ \mathbf{q}\cdot(\mathbf{k}-\mathbf{q})\ h(\mathbf{q}, \Omega) h(\mathbf{k}-\mathbf{q}, \omega-\Omega), \tag{B.9}$$

† To obtain d_c, substitute the exponents of the linear theory in (B.4).

where $G_0(\mathbf{k}, \omega)$ is the bare propagator defined by

$$G_0(\mathbf{k}, \omega) \equiv \frac{1}{\nu k^2 - i\omega}. \tag{B.10}$$

Equation (B.9) allows us to calculate $h(\mathbf{k}, \omega)$ perturbatively in powers of λ.

B.2.2 Diagrammatic expansion

To simplify the calculations, it is convenient to use the diagrammatic representation of (B.9) as indicated in Fig. B.1. Those familiar with diagrammatic expansions may proceed directly to (B.13). For those who are not, we offer a brief 'hands-on' explanation of how to use diagrammatic expansions.

Diagrammatic expansion is an efficient method to escape from manipulating long and complicated expressions through boring calculations. Instead we replace every term in the equation with a symbol and establish the rules for handling the symbols. With properly-defined rules, the perturbation expansion can be carried out merely by combining the diagrams. In the final step, the diagrams are translated back into formulas.

Figure B.1(a) illustrates a method of replacing the starting equation (B.9) of the perturbation expansion with symbols. We have for $\lambda = 0$

$$h(\mathbf{k}, \omega) = G_0(\mathbf{k}, \omega)\eta(\mathbf{k}, \omega). \tag{B.11}$$

The expansion is done in powers of λ around this zeroth order solution

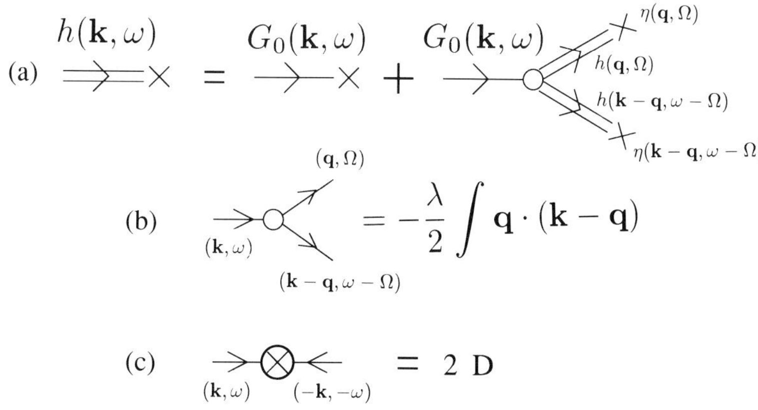

Figure B.1 (a) Diagrammatic representation of the integral equation (B.9). (b) The vertex λ and (c) the contracted noise D. In (b) the $\mathbf{q} \cdot (\mathbf{k} - \mathbf{q})$ is associated with the outgoing momenta.

for h. In general we define

$$h(\mathbf{k}, \omega) \equiv G(\mathbf{k}, \omega)\eta(\mathbf{k}, \omega). \tag{B.12}$$

The major steps are the following:

(i) $G_0(\mathbf{k}, \omega)$ is replaced with a simple arrow, carrying a momentum $(\mathbf{k}, \omega)$.
(ii) $G(\mathbf{k}, \omega)$ is replaced with a double arrow, carrying a momentum $(\mathbf{k}, \omega)$.
(iii) According to (B.11) h is represented by a double line ending in a noise term η $(\times)$.
(iv) The integral over $\mathbf{q}$ and Ω is represented by the vertex shown in Fig. B.1(b). It is a property of the Fourier transform that the arguments of the h's under the integral in (B.9) sum up to $(\mathbf{k}, \omega)$. This leads to a conservation rule, shown in Fig. B.1(c): The momenta $(\mathbf{k}, \omega)$ going into the vertex are equal to the sum of the momenta coming out from the vertex.

B.2.3 Propagator

To perform a perturbation expansion in powers of λ, we must replace the h terms under the integral in (B.9) with the full form of $h(\mathbf{k}, \omega)$ given by (B.9). The newly-appearing h terms can again be replaced by (B.9). This procedure can be truncated by replacing h by (B.11) instead of (B.9). Using the diagrams we replace the double line corresponding to $h(\mathbf{k}, \omega)$ in the rhs of Fig. B.1(a) with an expansion of Fig. B.1(a). The essential steps are shown in Fig. B.2. To second order in λ we obtain four contributions (A–D) to $h(\mathbf{k}, \omega)$.

Our goal is to calculate diagrammatically the contributions to $G(\mathbf{k}, \omega)$. According to (B.12), $G(\mathbf{k}, \omega)$ can be obtained from $h(\mathbf{k}, \omega)$ by eliminating the noise term. This means that we 'simplify' with one noise term $(\times)$ from every element of the perturbation expansion,† meaning that we multiply by $\eta(-\mathbf{k}, -\omega)$ and average over the noise, where we assume (B.12) holds.

Average over the noise — Next we must average over the stochastic noise – i.e., we must perform the $\langle \ldots \rangle$ average for every diagram. According to (B.7), if there is only one noise term in the diagram, that diagram vanishes, since the noise has zero average. If there are two noise terms (which within a single diagram multiply each other),

† If we wish to carry out the perturbation calculation without the help of the diagrammatic representation, we must formally introduce an additional 'external field' added to the noise. This field will be set to zero only in the final step of the calculation.

according to (B.8) the average over the product of the two noise terms generates a delta function. Thus two noise terms in the same diagram can be paired up ('contracted') and replaced with a delta function. The delta function leads to an another useful property of the average over the noise: denoting by a bubble the contraction of the two noise terms (as shown in Fig. B.1(c)), the sum of the momenta going into the bubble must be zero. In general, if there is an even number of noise terms in a diagram, they can be paired up, and the diagram may be nonzero. However, if there is an odd number of noise terms, there is always one unpaired noise, whose average is zero, so the whole diagram gives zero contribution.

Diagrams — Next we average over the noise in all terms of the series expansion shown in Fig. B.2(b). Diagram A has only one noise term, but we 'simplify' all diagrams with one noise, since our goal is to calculate the contributions to $G(\mathbf{k}, \omega)$. After the simplification A contains no noise, and so survives the average over the noise, as is

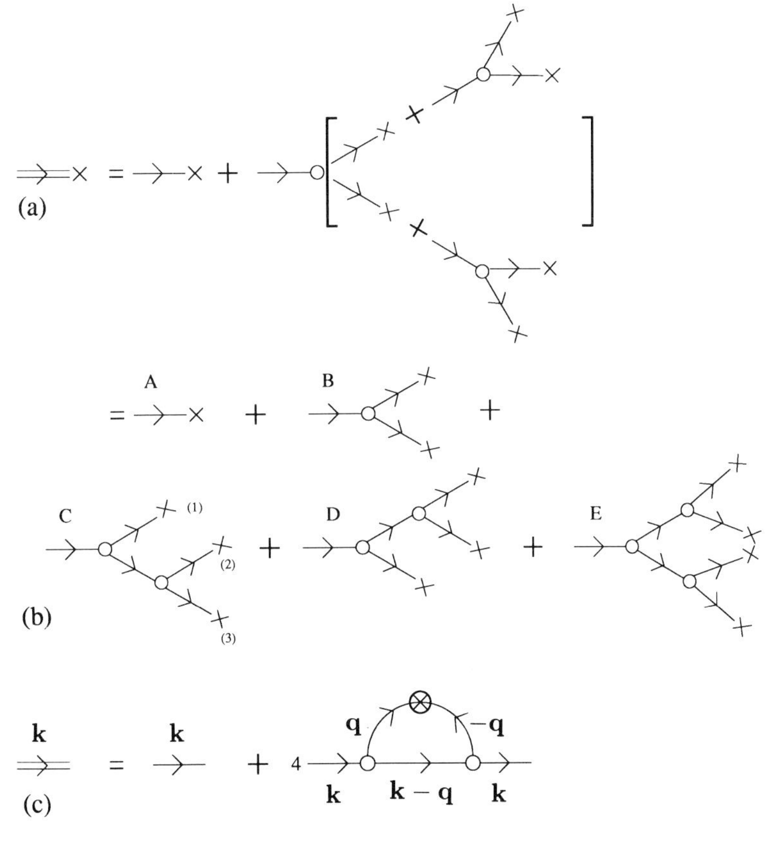

Figure B.2 Perturbation expansion for the propagator $G(\mathbf{k}, \omega)$. (a) Replace the double arrows in Fig. B.1(a) with the full perturbation expansion. (b) The corresponding five terms obtained after reordering (a). Note that diagram E is already one of the contributions to third order in λ. (c) The effective propagator obtained by averaging over the noise. We isolate the essential variables by showing only $\mathbf{k}$ and $\mathbf{q}$, which represent the full variables $(\mathbf{k}, \omega)$ and $(\mathbf{q}, \Omega)$, respectively.

shown in Fig. B.2(c). However, B is left with only one noise term after simplifying with one $\times$, so it vanishes. C and D do not vanish: we can contract the two remaining noise terms. Moreover, the two diagrams give exactly the same contribution. The resulting diagram is shown on Fig. B.2(c).

There are different ways of pairing up the noise, generating identical diagrams. For C there are two such combinations (corresponding to the pairing (1–2) and (1–3)), and there are two for D as well. Thus the 'symmetry factor' of the diagrams (how many identical diagrams we can generate) is four. Note that the pairing (2–3) gives a diagram which is identically zero. And finally, E vanishes again, because after eliminating one noise term, there will always be one dangling noise.

Momenta — The final diagrams, the one-loop corrections† to G_0, are shown in Fig. B.2(c). Concerning the momentum on the legs of the diagrams, we note that following three simple rules, one can find all the variables on the legs. These rules have been mentioned above, but it is useful to summarize them here:

(i) The incoming and outgoing variables are $(\mathbf{k}, \omega)$, set by the variables of the expansion $h(\mathbf{k}, \omega)$.
(ii) The sum of the momenta going into a vertex equals the sum of the momenta going out from the vertex.
(iii) The noise term has two identical momenta both going into the 'bubble'.

Perturbation expansion — Using these rules systematically, we can derive all the variables on the legs of the diagrams. Having the diagrams, we can go back to the formulas and identify every element of the diagram with the corresponding term. In particular, every simple arrow corresponds to G_0, the bubble to a noise term, and a vertex to an integral (see Fig. B.1). We integrate over the internal coordinates $(\mathbf{q}, \Omega)$. The variables can be read off from the legs of the diagrams of Fig. B.1(c).

Finally, we obtain $G(\mathbf{k}, \omega)$ as a perturbation expansion in λ:

$$\begin{aligned} G(\mathbf{k}, \omega) &= G_0(\mathbf{k}, \omega) + 4\left(-\frac{\lambda}{2}\right)^2 2D\; G_0^2(\mathbf{k}, \omega) \\ &\times \int\!\!\int \frac{d\Omega d^d\mathbf{q}}{(2\pi)^{d+1}}\, [\mathbf{q}(\mathbf{k}-\mathbf{q})]\; [-\mathbf{q}\mathbf{k}] \\ &\quad G_0(\mathbf{k}-\mathbf{q}, \omega-\Omega) G_0(\mathbf{q}, \Omega) G_0(-\mathbf{q}, -\Omega) + O(\lambda^4). \end{aligned} \quad \text{(B.13)}$$

† The diagrams contain only one loop, that is why this is called one-loop calculation. To obtain a two loop expansion, we must replace once more the double arrows on Fig. B.2 with the full form of h.

This result could have been obtained by following the same program without diagrams, doing the integrals and averaging over the noise directly. However, the diagrammatic expansion, once understood, is much faster and simpler.

The details on the calculation of the integral (B.13) are given in §B.4.1, the result being:

$$G(\mathbf{k},0) = G_0(\mathbf{k},0) + \frac{\lambda^2 D}{v^2} G_0^2(\mathbf{k},0) \frac{d-2}{4d} k^2 K_d \int dq\, q^{d-3}. \tag{B.14}$$

Since $G_0(\mathbf{k},0) = 1/(vk^2)$, the previous equation allows us to determine an effective surface tension $\tilde{v}$ from $G(\mathbf{k},0) = 1/(\tilde{v}k^2)$

$$\frac{1}{\tilde{v}} = \frac{1}{v}\left[1 + \frac{\lambda^2 D}{v^3}\frac{d-2}{4d} K_d \int dq\, q^{d-3}\right]. \tag{B.15}$$

Since λ is the perturbation parameter, λ is small so we can expand, obtaining

$$\tilde{v} = v\left[1 - \frac{\lambda^2 D}{v^3}\frac{d-2}{4d} K_d \int dq\, q^{d-3}\right]. \tag{B.16}$$

This result is the one-loop correction to v. In order to reach the flow equation (7.12), we must apply the renormalization procedure. But

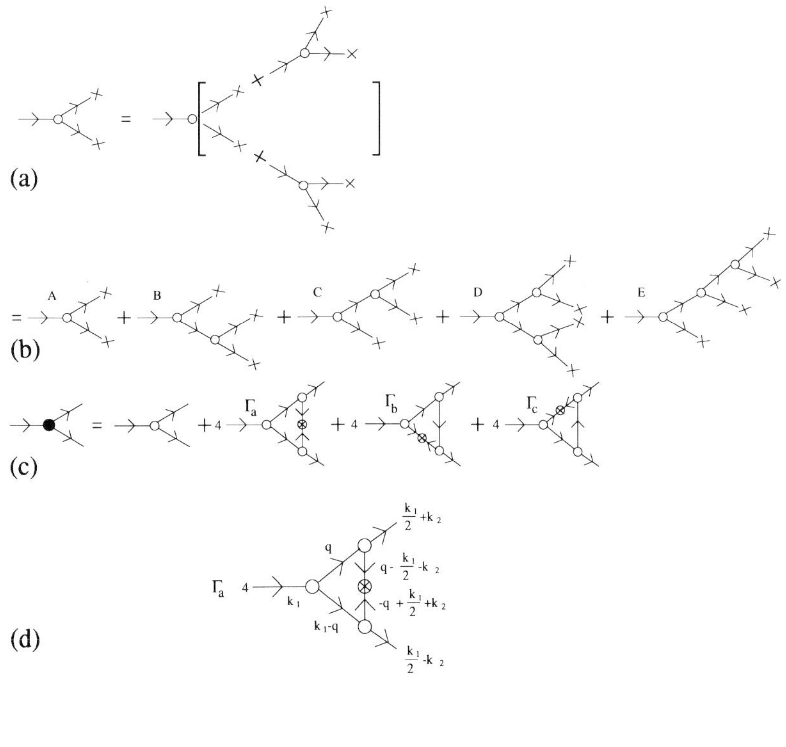

Figure B.3 Perturbation expansion for the vertex λ. (a) Replace the legs of the vertex with the full expansion shown in Fig. B.1. However, this time we must iterate the perturbation expansion twice in order to obtain all one-loop diagrams. In (a) we show the first step only; the second step is done by replacing the legs again with the full expansion. (b) The five terms so obtained. Note that E comes from the second iteration of the perturbation expansion. (c) The effective vertex function obtained after averaging over the noise. The three nonzero diagrams all have a symmetry factor of four. (d) The variables for the calculation of Γ_a; we mark only the spatial components.

before that we continue by calculating the perturbation expansion for λ and D.

B.2.4 Vertex

The goal here is to calculate perturbatively the vertex shown in Fig. B.1(b). The general method is to replace the two legs of the vertex with the expression for $h(\mathbf{k}, \omega)$, as shown in Fig. B.3. Now we iterate the perturbation expansion twice in order to obtain all one-loop diagrams. We simplify the diagrams with two noise terms, and average over all remaining noise terms. As a result contributions may arise from diagrams A, D and E since B and C vanish, having only one noise term remaining. There are three possible noise contractions that can be carried out on the diagrams D and E, giving the perturbation expansion shown in Fig. B.3(c). The three diagrams are calculated in §B.4.2. They all give nonzero contributions, but their sum vanishes. This is the consequence of the Galilean invariance of the system (see §6.4). While this calculation is done up to one-loop, due to the Galilean invariance the contributions to λ vanish up to all orders in the perturbation expansion. Thus we have for the effective λ the simple expression

$$\tilde{\lambda} = \lambda. \tag{B.17}$$

B.2.5 Spectral function

We next calculate $\tilde{D}$, the renormalization of the noise term D, defined as

$$\langle h^*(\mathbf{k}, \omega) h(\mathbf{k}, \omega) \rangle = 2\tilde{D} G(\mathbf{k}, \omega) G(-\mathbf{k}, -\omega) \tag{B.18}$$

which can be obtained using (B.8). The diagrammatic representation of the perturbation series is given in Fig. B.4. After noise contraction the diagrams B and C vanish, nonzero contributions arising from A and D as shown in Fig. B.4(c).

The resulting expansion is

$$2\tilde{D} = 2D + 8D^2 \left(-\frac{\lambda}{2}\right)^2 \int\int \frac{d\Omega}{2\pi} \frac{d^d\mathbf{q}}{(2\pi)^d} [\mathbf{q} \cdot (\mathbf{k} - \mathbf{q})]^2$$
$$\times G_0(\mathbf{q}, \Omega) G_0(-\mathbf{q}, -\Omega) G_0(\mathbf{k} - \mathbf{q}, \omega - \Omega) G_0(-\mathbf{k} + \mathbf{q}, -\omega + \Omega). \tag{B.19}$$

This expression is calculated in §B.4.3, resulting in the simpler form

$$\tilde{D} = D \left[1 + \frac{\lambda^2 D}{\nu^3} \frac{K_d}{4} \int dq \, q^{d-3} \right]. \tag{B.20}$$

B.3 Renormalization procedure

So far we have calculated the one loop corrections to ν, λ and D, with the results (B.16), (B.17), and (B.20), respectively. Since the integrals over q are truncated by Λ for large q, divergences can only arise from the $q \to 0$ limit. The corrections are well behaved for $d > 2$, but diverge for $d < 2$. The divergence appearing at $d_c = 2$ indicates that two is a critical dimension for the model, and that nontrivial scaling behavior might be expected. The RG procedure resumes the perturbation series to avoid this singularity. This procedure has the following general steps (see §7.2):

(1) Integration over momenta k in the range $\Lambda e^{-l} \leq k \leq \Lambda$; this step is called the 'elimination of the fast modes,' since small wavelength components are eliminated, while the long wavelength part, $0 \leq k^{<} \leq \Lambda e^{-l}$ is left untouched. This last part contains the

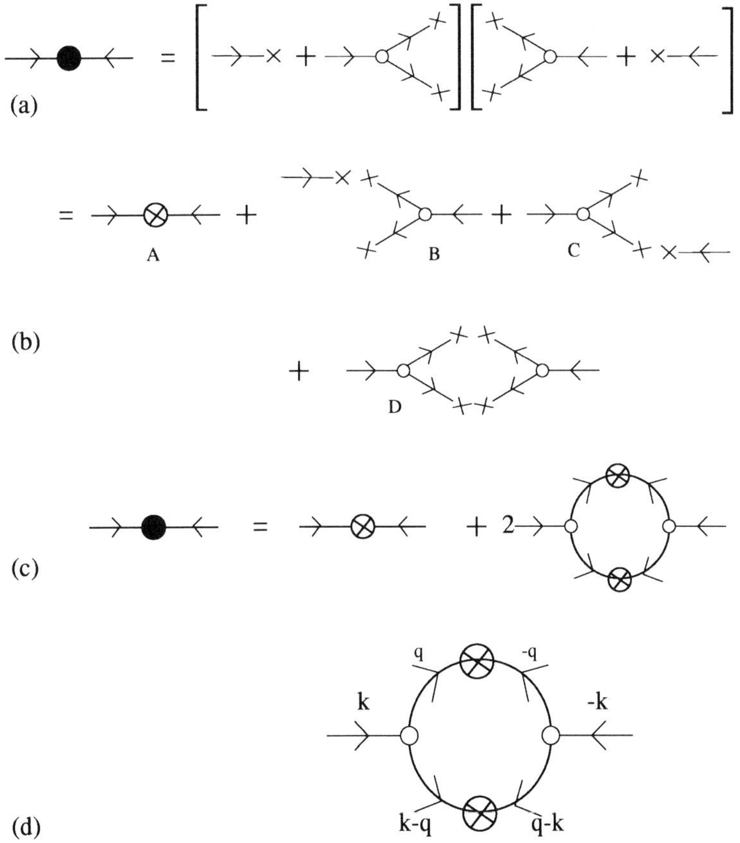

Figure B.4 Perturbation expansion for the spectral function D. (a) The noise term can be expanded in series using (B.4). The terms in (b) must be averaged over the noise, resulting in the nonzero diagrams shown in (c). The essential variables used for calculation of the diagram are given in (d).

singularity, thus the fast modes can be integrated out without any problem, since the integrals are performed over a momentum shell only.

(2) The resulting equations have a smaller cutoff, namely Λe^{-l}, where $b \equiv e^l$. This difference from the original model is removed by rescaling the momenta $\mathbf{k} \to e^{-l}\mathbf{k}$. After this step, we obtain the flow equations (7.12)–(7.14).

(3) The exponents are calculated by requiring that the flow equations are invariant under the rescaling, i.e., looking for the fixed points of the flow equations.

In this section we perform the program outlined in steps (1) and (2); step (3) is carried out in §7.3.

B.3.1 Flow equation for ν

Step 1 – We must perform the integral (B.16) over the domain $\Lambda(1-\delta l) < q < \Lambda$, where we have expanded $e^{-\delta l} = 1 - \delta l + ...$ We do all calculations up to first order in δl. To this end, we transform the integral as follows:

$$\int \frac{d^d\mathbf{q}}{(2\pi)^d} \to K_d \int_{\Lambda(1-\delta l)}^{\Lambda} dq\, q^{d-1}. \qquad \text{(B.21)}$$

Without loss of generality, Λ is set to unity. After some manipulation, we find

$$\nu^< = \nu \left[1 - \delta l\, K_d \frac{\lambda^2 D}{\nu^3} \frac{d-2}{4d} \right], \qquad \text{(B.22)}$$

where with the '<' index we denote that we are left with the long wavelengths only.

Step 2 – Now we must rescale the variables using $\mathbf{x} \to b\mathbf{x}$, which in $\mathbf{k}$ space corresponds to $\mathbf{k} \to (1-\delta l)\mathbf{k}$. To this end we rescale the frequency (corresponding to time in the real space) as $\omega \to (1 - z\delta l)\omega$, and the height $h \to (1 + \alpha\delta l)h$. The renormalized surface tension is related to $\nu^<$ by

$$\tilde{\nu} = b^{z-2}\nu^< = \nu^<[1 + \delta l(z-2)]. \qquad \text{(B.23)}$$

Replacing $\nu^<$ with (B.22), we obtain the flow equation for ν,

$$\frac{d\nu}{dl} = \nu \left[z - 2 + K_d \frac{\lambda^2 D}{\nu^3} \frac{2-d}{4d} \right], \qquad \text{(B.24)}$$

which coincides with the expression Eq. (7.12).

B.3.2 Flow equation for D

Step 1 – After performing the integral over the fast modes in (B.20) we obtain

$$D^{<} = D\left[1 + \delta l \frac{K_d}{4}\frac{\lambda^2 D}{\nu^3}\right]. \tag{B.25}$$

Step 2 – Rescaling the cutoff, we obtain for the flow equation

$$\frac{dD}{dl} = D\left[z - d - 2\alpha + \frac{K_d}{4}\frac{\lambda^2 D}{\nu^3}\right], \tag{B.26}$$

which coincides with (7.13).

B.3.3 Flow equation for λ

Since there is no correction to λ, we must perform only Step 2, resulting in

$$\frac{d\lambda}{dl} = \lambda[\alpha + z - 2], \tag{B.27}$$

which is the flow equation (7.14).

B.4 Calculation of the integrals

In this section we give the essential details for the calculation of the integrals resulting from the perturbation expansion.

B.4.1 Propagator renormalization

First we calculate the integral (B.13). A change of variables $\mathbf{q} \to \mathbf{k}/2 + \mathbf{q}$, $\Omega \to \omega/2 + \Omega$ leads to the symmetrised version of (B.13)

$$G(\mathbf{k},\omega) = G_0(\mathbf{k},\omega) + 4\left(-\frac{\lambda}{2}\right)^2 2D\; G_0^2(\mathbf{k},\omega)$$
$$\times \int \frac{d^d\mathbf{q}}{(2\pi)^d}\int_{-\infty}^{+\infty}\frac{d\Omega}{2\pi}\left[\left(\mathbf{q}+\frac{\mathbf{k}}{2}\right)\cdot\left(\frac{\mathbf{k}}{2}-\mathbf{q}\right)\right]\left[-\left(\mathbf{q}+\frac{\mathbf{k}}{2}\right)\cdot\mathbf{k}\right]$$
$$\times G_0\left[\left|\frac{\mathbf{k}}{2}-\mathbf{q}\right|,\frac{\omega}{2}-\Omega\right] G_0\left[\left|\mathbf{q}+\frac{\mathbf{k}}{2}\right|,\Omega+\frac{\omega}{2}\right] G_0\left[\left|\mathbf{q}+\frac{\mathbf{k}}{2}\right|,-\Omega-\frac{\omega}{2}\right].$$

The $\mathbf{q}$-integral is calculated in spherical coordinates. We use $\mathbf{k}\cdot\mathbf{q} = kq\cos\Theta$, where Θ is the angle between $\mathbf{q}$ and $\mathbf{k}$. Replacing $G_0(\mathbf{k},\omega)$ with its form (B.10), we obtain

$$G(\mathbf{k},\omega) = G_0(\mathbf{k},\omega) + 4\left(-\frac{\lambda}{2}\right)^2 2D\; G_0^2(\mathbf{k},\omega)$$

$$\times \frac{S_{d-1}}{(2\pi)^d} \int dq\, q^{d-1} \int_0^\pi d\Theta \sin^{d-2}\Theta \int_{-\infty}^{\infty} \frac{d\Omega}{2\pi}$$
$$\times \frac{\left[q^2 - \frac{k^2}{4}\right]\left[kq\cos\Theta + \frac{k^2}{2}\right]}{\left[\nu\left(q^2 - kq\cos\Theta + \frac{k^2}{4}\right) - i\left(\frac{\omega}{2} - \Omega\right)\right]}$$
$$\times \left[\nu^2\left(q^2 + kq\cos\Theta + \frac{k^2}{4}\right)^2 + \left(\frac{\omega}{2} + \Omega\right)^2\right],$$

where S_d is the surface area of a d-dimensional sphere.

We are interested in the hydrodynamic behavior of the system, which gives the scaling behavior in the large system size – long time regime. In Fourier space this corresponds to the $\mathbf{k} \to 0$ and $\omega \to 0$ limit. The $\omega \to 0$ limit can be taken right away in the previous equation. With the $\mathbf{k} \to 0$ limit we must be more careful, keeping the leading order contributions in k. It is convenient to use the dimensionless variables $x \equiv k/q$ and $z \equiv \Omega/\nu q^2$, whereupon

$$G(\mathbf{k}, 0) = G_0(\mathbf{k}, 0) + G_0^2(\mathbf{k}, 0) \int_0^\Lambda dq\, q^{d-1}$$
$$\times \left(\frac{\lambda^2 D}{\nu^2}\right) \frac{S_{d-1}}{(2\pi)^d} \int_{-\infty}^{\infty} \frac{dz}{\pi} \int_0^\pi d\Theta \sin^{d-2}\Theta$$
$$\times \frac{\left[1 - \frac{x^2}{4}\right]\left[x\cos\Theta + \frac{x^2}{2}\right]}{\left[1 + iz - x\cos\Theta + O(x^2)\right]\left[1 + z^2 + 2x\cos\Theta + O(x^2)\right]}. \tag{B.28}$$

Let us define $I(x, q)$ to be the integral in the second and third line. Expanding the denominator around $(1 + iz)$ and $(1 + z^2)$, respectively, we obtain to second order in x

$$I(x, q) = \left(\frac{\lambda^2 D}{\nu^2}\right) \frac{S_{d-1}}{(2\pi)^d} \int_{-\infty}^{\infty} \frac{dz}{\pi} \frac{1}{(1 + iz)(1 + z^2)} \int_0^\pi d\Theta \sin^{d-2}\Theta$$
$$\times \left[1 - \frac{x^2}{4}\right]\left[x\cos\Theta + \frac{x^2}{2}\right]\left[1 + \frac{x\cos\Theta}{1 + iz}\right]\left[1 - \frac{2x\cos\Theta}{1 + z^2}\right]$$
$$= \left(\frac{\lambda^2 D}{\nu^2}\right) \frac{S_{d-1}}{(2\pi)^d} \int_{-\infty}^{\infty} \frac{dz}{\pi} \frac{1}{(1 + iz)(1 + z^2)} \int_0^\pi d\Theta \sin^{d-2}\Theta$$
$$\times \left[x\cos\Theta + \frac{1}{2}x^2 + x^2\cos^2\Theta\left(\frac{1}{1 + iz} - \frac{2}{1 + z^2}\right) + O(x^3)\right].$$

After the angular integration, the $\cos\Theta$ term vanishes. Using

$$K_d \equiv \frac{S_{d-1}}{(2\pi)^d} \int_0^\pi d\Theta \sin^{d-2}\Theta = \frac{S_d}{(2\pi)^d} \tag{B.29}$$

and from integration by parts

$$\int_0^\pi d\Theta \sin^{d-2}\Theta \cos^2\Theta = \frac{1}{d} \int_0^\pi d\Theta \sin^{d-2}\Theta, \tag{B.30}$$

we obtain

$$I(x,q) = x^2 K_d \left(\frac{\lambda^2 D}{v^2}\right) \int_{-\infty}^{\infty} \frac{dz}{\pi} \left[\frac{1}{2}\frac{1}{(1+iz)(1+z^2)} - \frac{1}{d}\frac{1}{(1+z^2)^2}\right]. \quad \text{(B.31)}$$

Calculating the integrals over z, we obtain the final form for $I(x,q)$

$$I(x,q) = x^2 K_d \left(\frac{\lambda^2 D}{v^2}\right) \frac{d-2}{4d}. \quad \text{(B.32)}$$

Substituting (B.32) in (B.28) and changing back variables from x to k/q, we obtain for $G(\mathbf{k},0)$ the form given in Eq. (B.14).

B.4.2 Vertex renormalization

There are three vertices contributing to the renormalization of λ. According to Fig. B.3, we have the full expression

$$-\frac{\tilde{\lambda}}{2}\left(\frac{\mathbf{k}_1}{2}+\mathbf{k}_2\right)\cdot\left(\frac{\mathbf{k}_1}{2}-\mathbf{k}_2\right) G_0\left(\hat{\mathbf{k}}_1\right) G_0\left(\frac{\hat{\mathbf{k}}_1}{2}+\hat{\mathbf{k}}_2\right) G_0\left(\frac{\hat{\mathbf{k}}_1}{2}-\hat{\mathbf{k}}_2\right) =$$

$$-\frac{\lambda}{2}\left(\frac{\mathbf{k}_1}{2}+\mathbf{k}_2\right)\cdot\left(\frac{\mathbf{k}_1}{2}-\mathbf{k}_2\right) G_0\left(\hat{\mathbf{k}}_1\right) G_0\left(\frac{\hat{\mathbf{k}}_1}{2}+\hat{\mathbf{k}}_2\right) G_0\left(\frac{\hat{\mathbf{k}}_1}{2}-\hat{\mathbf{k}}_2\right)$$

$$\times\left[1+\Gamma_a+\Gamma_b+\Gamma_c\right].$$

Here Γ_a, Γ_b, and Γ_c are the bare vertices corresponding to the diagrams shown on Fig. B.3(c). To simplify the notation, we defined $\hat{\mathbf{k}}_i \equiv (\mathbf{k}_i, \omega_i)$ and $\hat{\mathbf{q}} \equiv (\mathbf{q}, \Omega)$.

In this section we present the details for the calculation of Γ_a only, since for calculating Γ_b, Γ_c we proceed along similar lines. Using the variables indicated in Fig. B.3(d), we have

$$\Gamma_a = 4\left(-\frac{\lambda}{2}\right)^2 \frac{2D}{\left(\frac{k_1^2}{4}-k_2^2\right)} \int \frac{d^d\mathbf{q}}{(2\pi)^d} \int_{-\infty}^{+\infty} \frac{d\Omega}{2\pi} \left[\mathbf{q}\cdot(\mathbf{k}_1-\mathbf{q})\right]$$

$$\times\left[\left(\frac{\mathbf{k}_1}{2}+\mathbf{k}_2\right)\cdot\left(\mathbf{q}-\frac{\mathbf{k}_1}{2}-\mathbf{k}_2\right)\right]\left[\left(\frac{\mathbf{k}_1}{2}-\mathbf{k}_2\right)\cdot\left(-\mathbf{q}+\frac{\mathbf{k}_1}{2}+\mathbf{k}_2\right)\right]$$

$$\times G_0(\hat{\mathbf{q}})\, G_0\left(\hat{\mathbf{k}}_1-\hat{\mathbf{q}}\right) G_0\left(\hat{\mathbf{q}}-\frac{\hat{\mathbf{k}}_1}{2}-\hat{\mathbf{k}}_2\right) G_0\left(-\hat{\mathbf{q}}+\frac{\hat{\mathbf{k}}_1}{2}+\hat{\mathbf{k}}_2\right).$$

We perform the change of variables $\hat{\mathbf{q}} \to \hat{\mathbf{q}} + \hat{\mathbf{k}}_1/2$ and use $\mathbf{k_1}\cdot\mathbf{q} = k_1 q\cos\Theta_1$ and $\mathbf{k_2}\cdot\mathbf{q} = k_2 q\cos\Theta_2$. Replacing $G_0(\mathbf{k},\omega)$ with its form given by (B.10), we obtain

$$\Gamma_a = \frac{2\lambda^2 D}{\left(\frac{k_1^2}{4}-k_2^2\right)} \int \frac{d^d\mathbf{q}}{(2\pi)^d} \int_{-\infty}^{+\infty} \frac{d\Omega}{2\pi}$$

$$\times \frac{\left[\frac{k_1^2}{4} - q^2\right] \left[q \frac{k_1}{2} \cos\Theta_1 + q k_2 \cos\Theta_2\right]}{\left[\nu \left(\mathbf{q} + \frac{\mathbf{k}_1}{2}\right)^2 - i\left(\Omega + \frac{\omega_1}{2}\right)\right] \left[\nu \left(-\mathbf{q} + \frac{\mathbf{k}_1}{2}\right)^2 - i\left(-\Omega + \frac{\omega_1}{2}\right)\right]}$$

$$\times \frac{\left[-q \frac{k_1}{2} \cos\Theta_1 + q k_2 \cos\Theta_2\right]}{\left[\nu^2 (\mathbf{q} - \mathbf{k}_2)^4 + (\Omega - \omega_2)^2\right]},$$

where in the numerator we systematically neglect terms of order k^2 compared with terms of order k. Performing the change of variables $z \equiv \Omega/\nu q^2$ and setting $\omega_1 = \omega_2 = 0$, we obtain

$$\Gamma_a = \frac{\lambda^2 D}{\left(\frac{k_1^2}{4} - k_2^2\right)} \frac{S_{d-1}}{(2\pi)^d} \int_0^\Lambda dq\, q^{d-1} \int_{-\infty}^{\infty} \frac{dz}{\pi} \int_0^\pi d\Theta \sin^{d-2}\Theta$$

$$\times \frac{\nu q^2 (-) q^2 \left[1 - \frac{k_1^2}{4q^2}\right] q \left[\frac{k_1}{2} \cos\Theta_1 + k_2 \cos\Theta_2\right]}{\nu q^2 \left[1 - iz + \frac{k_1}{q} \cos\Theta_1\right] \nu q^2 \left[1 + iz - \frac{k_1}{q} \cos\Theta_1\right]}$$

$$\times \frac{(-q) \left[\frac{k_1}{2} \cos\Theta_1 - k_2 \cos\Theta_2\right]}{\nu^2 q^4 \left[1 + z^2 - 4\frac{k_2}{q} \cos\Theta_2\right]}.$$

After expanding the numerator in series and keeping terms of leading order in k_1 and k_2, we have

$$\Gamma_a = \frac{\lambda^2 D}{\nu^3 \left(\frac{k_1^2}{4} - k_2^2\right)} \frac{S_{d-1}}{(2\pi)^d} \int_0^\Lambda dq\, q^{d-3} \int_{-\infty}^{\infty} \frac{dz}{\pi} \int_0^\pi d\Theta \sin^{d-2}\Theta$$

$$\times \frac{1}{(1+z^2)^2} \left[1 - \frac{k_1^2}{4q^2}\right] \left[\frac{k_1^2}{4} \cos^2\Theta_1 - k_2^2 \cos^2\Theta_2\right]$$

$$\times \left[1 - \frac{k_1}{q} \frac{\cos\Theta_1}{1 - iz}\right] \left[1 + \frac{k_1}{q} \frac{\cos\Theta_1}{1 + iz}\right] \left[1 + 4 \frac{k_2}{q} \frac{\cos\Theta_2}{1 + z^2}\right]$$

$$= \frac{\lambda^2 D}{\nu^3} \frac{S_{d-1}}{(2\pi)^d} \int_0^\Lambda dq\, q^{d-3} \int_{-\infty}^{\infty} \frac{dz}{\pi} \int_0^\pi d\Theta \sin^{d-2}\Theta$$

$$\times \frac{1}{(1+z^2)^2} \frac{\left[\frac{k_1^2}{4} \cos^2\Theta_1 - k_2^2 \cos^2\Theta_2 + O(k^3)\right]}{\left(\frac{k_1^2}{4} - k_2^2\right)}.$$

The two angular integrals are equal, both giving a contribution $1/d$ according to (B.30). Hence we obtain the final result

$$\Gamma_a = \left(\frac{\lambda^2 D}{\nu^3}\right) \frac{K_d}{2\, d} \int dq\, q^{d-3}. \tag{B.33}$$

The two remaining integrals, Γ_b and Γ_c are equal, and can be calculated similarly, giving

$$\Gamma_b = \Gamma_c = -\left(\frac{\lambda^2 D}{\nu^3}\right) \frac{K_d}{4d} \int dq\, q^{d-3}. \tag{B.34}$$

The sum of the three integrals is zero, giving no correction to $\tilde{\lambda}$, and resulting in (B.17).

B.4.3 Noise renormalization

The calculations for (B.19) follow the same general lines as presented for the propagator or the vertex discussed in the previous sections. Using the symmetrized momenta $\hat{\mathbf{q}} \to \hat{\mathbf{q}} + \hat{\mathbf{k}}/2$, we perform the $\omega \to 0$ limit, and change variables to $z \equiv \Omega/vq^2$. We find

$$2\tilde{D} = 2D + \lambda^2 D^2 \frac{S_{d-1}}{(2\pi)^d} \int_0^{\Lambda} dq\ q^{d-1} \int_{-\infty}^{\infty} \frac{dz}{\pi} \int_0^{\pi} d\Theta \sin^{d-2}\Theta$$
$$\times \frac{vq^2\ q^4 \left[1 - \frac{k^2}{4q^2}\right]^4}{v^2 q^4 \left[1 + z^2 + 2\frac{k}{q}\cos\Theta\right] v^2 q^4 \left[1 + z^2 - 2\frac{k}{q}\cos\Theta\right]}. \quad \text{(B.35)}$$

Expanding the denominator in a series, and systematically neglecting terms of order k^2, we obtain

$$2\tilde{D} = 2D + \frac{\lambda^2 D^2}{v^3} K_d \int_0^{\Lambda} dq\ q^{d-3} \int_{-\infty}^{\infty} \frac{dz}{\pi} \frac{1}{(1+z^2)^2} \quad \text{(B.36)}$$

which, after performing the integral over z, results in (B.20).

Suggested further reading:

[179, 277, 278, 314]

APPENDIX C

Hamiltonian description

As a manifestation of the equilibrium character of the EW growth process, the equation of motion (5.6) can be obtained from the Hamiltonian

$$\mathscr{H}_0 \equiv \nu \int d^d\mathbf{x} \frac{1}{2}(\nabla h)^2 \tag{C.1}$$

using

$$\frac{\partial h}{\partial t} = -\frac{\delta \mathscr{H}_0}{\delta h} + \eta. \tag{C.2}$$

The Hamiltonian (C.1) describes in general the equilibrium motion of an elastic line, or elastic interface, that is smoothed by surface tension ν.

The lowest energy configuration corresponds to a straight line, for which $\nabla h \equiv 0$ everywhere, so $\mathscr{H}_0 = 0$ from (C.2). Fluctuations generate roughening, which is penalized by the increase in the interface energy $\mathscr{H}$. The elastic term in the Hamiltonian acts to reduce this energy, i.e., to decrease the roughness. The dynamics of the interface is described by (C.2), which is equivalent to the EW equation. In fact, (C.2) can be used to associate a stochastic dynamics to any Hamiltonian.

Further linear terms can be generated. As a first example, let us consider the standard surface tension driven interface free energy described by the generalization of (C.1) [94]

$$\mathscr{H}_1 \equiv \nu \int d^d\mathbf{x}\sqrt{1 + (\nabla h)^2} = \nu \int d^d\mathbf{x} \left[1 + \frac{1}{2}(\nabla h)^2 - \frac{1}{8}(\nabla h)^4 + \ldots\right]. \tag{C.3}$$

The last equality follows from expanding the square root term. Using (C.2), we find the more general growth equation

$$\frac{\partial h}{\partial t} = \nu\nabla^2 h - \frac{\nu}{2}\nabla \cdot (\nabla h)^3 + \eta(\mathbf{x}, t), \tag{C.4}$$

which describes an equilibrium interface, since the $h \to -h$ symmetry is not violated. The first term of (C.3) generates the surface tension term $\nabla^2 h$ in the growth equation, while the contribution from the second term is $\nabla \cdot (\nabla h)^3$, so this term acts as a higher order correction to the surface tension.

While the $\nabla(\nabla h)^3$ term is irrelevant compared with the Laplacian term $\nabla^2 h$, it is relevant if it is compared to a term $\nabla^4 h$ [258]. The fourth order linear term $-K\nabla^4 h$ describing relaxation by surface diffusion can be generated from a curvature dependent Hamiltonian

$$\mathscr{H}_2 = K \int d^d\mathbf{x} \frac{1}{2}(\nabla^2 h)^2, \qquad \text{(C.5)}$$

and plays a major role in describing MBE. Such a Hamiltonian is extensively studied in the context of membranes, where the curvature-dependent bending energy dominates the scaling behavior [344].

Suggested further reading:

[94]

Bibliography

[1] A. W. Adamson, *Physical Chemistry of Surfaces* (John Wiley & Sons, New York, 1982).

[2] T. Ala-Nissila, T. Hjelt, J. M. Kosterlitz and O. Venölöinen, 'Scaling exponents for kinetic roughening in higher dimensions,' *J. Stat. Phys.* **72**, 207–225 (1993).

[3] S. Alexander, 'Fractal surfaces,' in *Transport and Relaxation in Random Materials*, edited by J. R. Klafter, J. Rubin and M. F. Shlesinger (World Scientific, Singapore, 1986), pp. 59–71.

[4] J. G. Amar and F. Family, 'Numerical solution of a continuum equation for interface growth in 2+1 dimensions,' *Phys. Rev. A* **41**, 3399–3402 (1990).

[5] J. G. Amar and F. Family, 'Phase transition in a restricted solid-on-solid surface growth model in 2+1 dimensions,' *Phys. Rev. Lett.* **64**, 543–546 (1990).

[6] J. G. Amar and F. Family, 'Scaling of surface fluctuations and dynamics of surface growth models with power-law noise,' *J. Phys. A* **24**, L79–L86 (1991).

[7] J. G. Amar and F. Family, 'Universal scaling functions and amplitude ratios in surface growth,' *Phys. Rev. A* **45**, R3373–R3376 (1992).

[8] J. G. Amar and F. Family, 'Universality in surface growth: Scaling functions and amplitude ratios,' *Phys. Rev. A* **45**, 5378–5393 (1992).

[9] J. G. Amar and F. Family, 'Deterministic and stochastic growth with generalized nonlinearity,' *Phys. Rev. E* **47**, 1595–1603 (1993)

[10] J. G. Amar, F. Family and P.-M. Lam, 'Dynamical scaling of the island-size distribution and percolation in a model of sub-monolayer molecular beam epitaxy,' *Phys. Rev. B* **50**, 8781–8797 (1994).

[11] J. G. Amar, P.-M. Lam and F. Family, 'Surface growth with long-range correlated noise,' *Phys. Rev. A* **43**, 4548–4550 (1991).

[12] J. G. Amar, P.-M. Lam and F. Family, 'Groove instabilities in surface growth with diffusion,' *Phys. Rev. E* **47**, 3242–3245 (1993).

[13] L. A. N. Amaral, A.-L. Barabási, S. V. Buldyrev, S. T. Harrington, S. Havlin, R. Sadr-Lahijany and H. E. Stanley, 'Avalanches and the directed percolation depinning model: Experiments, simulations and theory,' *Phys. Rev. E* **51**, xx–xx (1995).

[14] L. A. N. Amaral, A.-L. Barabási, S. V. Buldyrev, S. Havlin and H. E. Stanley, 'Anomalous interface roughening: The role of a gradient in the density of pinning sites,' *Fractals* **1**, 818–826 (1993).

[15] L. A. N. Amaral, A.-L. Barabási, S. V. Buldyrev, S. Havlin and H. E. Stanley, 'New exponent characterizing the effect of evaporation on imbibition experiments,' *Phys. Rev. Lett.* **72**, 641–644 (1994).

[16] L. A. N. Amaral, A.-L. Barabási and H. E. Stanley, 'Universality classes for interface growth with quenched disorder,' *Phys. Rev. Lett.* **73**, 62–65 (1994).

[17] F. Anselmet, Y. Gagne, E. J. Hopfinger and R. A. Antonia, 'High-order velocity structure functions in turbulent shear flows,' *J. Fluid Mech.* **140**, 63–89 (1984).

[18] J. R. Arthur, 'Interaction of Ga and As_2 molecular beams with GaAs surfaces,' *J. Appl. Phys.* **39**, 4032–4033 (1968).

[19] D. Avnir, ed., *The Fractal Approach to Heterogeneous Chemistry: Surfaces, Colloids, Polymers* (John Wiley & Sons Ltd., Chichester, 1989).

[20] R. Baiod, D. Kessler, P. Ramanlal, L. Sander and R. Savit, 'Dynamical scaling of the surface of finite-density ballistic aggregation,' *Phys. Rev. A* **38**, 3672–3678 (1988).

[21] P. Bak, C. Tang and K. Wiesenfeld, 'Self-organized criticality: An explanation of $1/f$ noise,' *Phys. Rev. Lett.* **59**, 381–384 (1987).

[22] G. S. Bales, R. Bruinsma, E. A. Eklund, R. P. U. Karunasiri, J. Rudnick and A. Zangwill, 'Growth and erosion of thin solid films,' *Science* **249**, 264–268 (1990).

[23] G. S. Bales and A. Zangwill, 'Growth dynamics of sputter deposition,' *Phys. Rev. Lett.* **63**, 692–692 (1989).

[24] S. Balibar and J. P. Bouchaud, 'Kardar–Parisi–Zhang equation and the dynamic roughening of the crystal surfaces,' *Phys. Rev. Lett.* **69**, 862–862 (1992).

[25] A.-L. Barabási, 'A model for the temporal fluctuations of the surface width: A stochastic one-dimensional map,' *J. Phys. A* **24**, L1013–L1019 (1991).

[26] A.-L. Barabási, 'Dynamic scaling of coupled nonequilibrium interfaces,' *Phys. Rev. A* **46**, R2977–R2980 (1992).

[27] A.-L. Barabási, 'Surfactant-mediated growth of nonequilibrium interfaces,' *Phys. Rev. Lett.* **70**, 4102–4105 (1993).

[28] A.-L. Barabási, 'Surfactant-mediated surface growth: Nonequilibrium theory,' in *Fractals in natural sciences*, edited by M. Matsushita, M. Shlesinger and T. Vicsek, *Fractals* **1**, 846–859 (1993).

[29] A.-L. Barabási, M. Araujo and H. E. Stanley, 'Three-dimensional Toom model: Connection to the Kardar–Parisi–Zhang equation,' *Phys. Rev. Lett.* **68**, 3729–3732 (1992).

[30] A.-L. Barabási, R. Bourbonnais, J. Kertész, M. H. Jensen, T. Vicsek and Y. C. Zhang, 'Multifractality in surface growth,' *Phys. Rev. A* **45**, R6951–R6954 (1992).

[31] A.-L. Barabási, S.V. Buldyrev, S. Havlin, G. Huber, H. E. Stanley and T. Vicsek, 'Imbibition in porous media: Experiment and theory,' in *Surface disordering: Growth, roughening and phase transitions*, edited by

R. Jullien, J. Kertész, P. Meakin and D. E. Wolf (Nova Science, New York, 1992), pp. 193–204.

[32] A.-L. Barabási and E. Kaxiras, 'Dynamics scaling of coupled conserved systems,' preprint.

[33] A.-L. Barabási, P. Szépfalusy and T. Vicsek, 'Multifractal spectra of multi-affine functions,' *Physica A* **178**, 17–28 (1991).

[34] A.-L. Barabási and T. Vicsek, 'Multifractality of self-affine fractals,' *Phys. Rev. A* **44**, 2730–2733 (1991).

[35] M. C. Bartelt and J. W. Evans, 'Scaling analysis of diffusion-mediated island growth in surface adsorption processes,' *Phys. Rev. B* **46**, 12675–12687 (1992).

[36] M. C. Bartelt and J. W. Evans, 'Dendritic islands in metal-on-metal epitaxy: I. Shape transitions and diffusion at island edges.' *Surf. Sci. Lett.* **314**, L829–L834 (1994).

[37] M. C. Bartelt and J. W. Evans, 'Dendritic islands in metal-on-metal epitaxy: II. Coalescence and multilayer growth,' *Surf. Sci. Lett.* **314**, L835–L842 (1994).

[38] H. van Beijeren, 'Exactly solvable model for the roughening transition of a crystal surface,' *Phys. Rev. Lett.* **38**, 993–995 (1977).

[39] E. Ben-Jacob, O. Shochet, A. Tenenbaum, I. Cohen, A. Czirók and T. Vicsek, 'Communication, regulation and control during complex patterning of bacterial colonies,' *Fractals* **2**, 15–44 (1994).

[40] E. Ben-Jacob, O. Shochet, A. Tenenbaum, I. Cohen, A. Czirók and T. Vicsek, 'Communicating walkers model for cooperative patterning of bacterial colonies,' *Nature* **368**, 46–49 (1994).

[41] C. H. Bennett and G. Grinstein, 'Role of irreversibility in stabilizing complex and nonergodic behavior in locally interacting discrete systems,' *Phys. Rev. Lett.* **55**, 657–660 (1985).

[42] M. Benoit and R. Jullien, 'Phase transition in the 2D ballistic growth model with quenched noise,' *Physica A* **207**, 500–516 (1994).

[43] K. R. Bhaskar, B. S. Turner, P. Garik, J. D. Bradley, R. Bansil, H. E. Stanley and J. T. LaMont, 'Viscous fingering of HCl through gastric mucin,' *Nature* **360**, 458–461 (1992).

[44] R. Bidaux and R. B. Pandey, 'Driven diffusion of particles, first-passage front and interface growth,' *Phys. Rev. E* **48**, 2382–2385 (1993).

[45] A. Birovljev, L. Furuberg, J. Feder, T. Jøssang, K. J. Måløy and A. Aharony, 'Gravity invasion percolation in two dimensions: Experiment and simulation,' *Phys. Rev. Lett.* **67**, 584–587 (1991).

[46] G. Blatter, M. V. Fiegel'man, V. B. Geshkenbein, A. I. Larkin and V. M. Vinokur, 'Vortices in high temperature superconductors,' *Rev. Mod. Phys.* **66**, 1125–1388 (1994).

[47] E. Bouchaud, G. Lapasset and J. Planés, 'Fractal dimension of fractured surfaces: A universal value?' *Europhys. Lett.* **13**, 73–79 (1990).

[48] J. P. Bouchaud and M. E. Cates, 'Self-consistent approach to the KPZ equation,' *Phys. Rev. E* **47**, R1455–R1458 (1993).

[49] J. P. Bouchaud and A. Georges, 'Anomalous diffusion in disordered media: Statistical mechanisms, models and physical applications,' *Phys. Rep.* **195**, 127–293 (1990).

[50] R. Bourbonnais, H. J. Herrmann and T. Vicsek, 'Simulations of kinetic roughening with power-law noise on the Connection Machine,' *Int. J. Mod. Phys. C* **2**, 719–733 (1991).

[51] R. Bourbonnais, J. Kertész and D. Wolf, 'Surface growth with power law noise in 2+1 dimensions,' *J. Physique II* **1**, 493–500 (1991).

[52] G. Bracco, C. Malo, C. J. Moses and R. Tatarek, 'On the primary mechanism of surface roughening: The Ag(110) case,' *Surf. Sci.* **287/288**, 871–875 (1993).

[53] R. M. Bradley and J. M. E. Harper, 'Theory of ripple topography induced by ion bombardment.' *J. Vac. Sci. Technol. A* **6**, 2390–2395 (1988).

[54] G. Brocks, P. J. Kelly and R. Car, 'Binding and diffusion of a Si adatom on the Si(100) surface,' *Phys. Rev. Lett.* **66**, 1729–1732 (1991).

[55] B. Bruinsma, 'The KPZ model and sputter erosion,' in *Surface Disordering: Growth, Roughening and Phase Transitions*, edited by R. Jullien, J. Kertész, P. Meakin and D. E. Wolf (Nova Science, New York, 1992).

[56] R. Bruinsma and G. Aeppli, 'Interface motion and nonequilibrium properties of the random-field Ising model,' *Phys. Rev. Lett.* **52**, 1547–1550 (1984).

[57] H. Brune, C. Romainczyk, H. Röder and K. Kern, 'Fractal and dendritic growth patterns in two dimensions,' *Nature* **369**, 469–471 (1994).

[58] S. V. Buldyrev, A.-L. Barabási, F. Caserta, S. Havlin, H. E. Stanley and T. Vicsek, 'Anomalous interface roughening in porous media: Experiment and model,' *Phys. Rev. A* **45**, R8313–R8316 (1992).

[59] S. V. Buldyrev, A.-L. Barabási, S. Havlin, J. Kertész, H. E. Stanley and H. S. Xenias, 'Anomalous roughening of interfaces in porous media: Experiment and model,' *Physica A* **191**, 220–226 (1992).

[60] S. V. Buldyrev, A.-L. Goldberger, S. Havlin, C.-K. Peng and H. E. Stanley, 'Fractals in biology and medicine: From DNA to the heartbeat,' in *Fractals in Science*, edited by A. Bunde and S. Havlin (Springer-Verlag, Berlin, 1994).

[61] S. V. Buldyrev, S. Havlin, J. Kertész, A. Shehter and H. E. Stanley, 'Surface roughening with quenched disorder in *d*-dimensions,' *Fractals* **1**, 827–839 (1993).

[62] S. V. Buldyrev, S. Havlin, J. Kertész, H. E. Stanley and T. Vicsek, 'Ballistic deposition with power-law noise: A variant of the Zhang model,' *Phys. Rev. A* **43**, 7113–7116 (1991).

[63] S. V. Buldyrev, S. Havlin and H. E. Stanley, 'Anisotropic percolation and the *d*-dimensional surface roughening problem,' *Physica A* **200**, 200–211 (1993).

[64] A. Bunde and S. Havlin, eds., *Fractals and Disordered Systems* (Springer-Verlag, Berlin, 1991).

[65] A. Bunde and S. Havlin, eds., *Fractals in Science* (Springer-Verlag, Berlin, 1994).

[66] J. M. Burgers, *The Nonlinear Diffusion Equation* (Riedel, Boston 1974).

[67] T. W. Burkhardt and J. M. J. van Leeuwen, eds., *Real-Space Renormalization* (Springer-Verlag, Berlin, 1982).

[68] W. K. Burton, N. Cabrera and F. C. Frank, 'The growth of crystals and the equilibrium structure of their surfaces,' *Phil. Trans. R. Soc.* [London] **243A**, 299–358 (1951).
[69] Y. C. Cao and E. H. Conrad, 'Approach to thermal roughening of Ni(110): A study by high-resolution low-energy electron diffraction,' *Phys. Rev. Lett.* **64**, 447–450 (1990).
[70] F. Caserta, H. E. Stanley, W. Eldred, G. Daccord, R. Hausman and J. Nittmann, 'Physical mechanisms underlying neurite outgrowth: A quantitative analysis of neuronal shape,' *Phys. Rev. Lett.* **64**, 95–98 (1990).
[71] A. Chakrabarki and R. Toral, 'Numerical study of a model for interface growth,' *Phys. Rev. B* **40**, 11419–11421 (1989).
[72] E. Chason and T. M. Mayer, 'Low energy ion bombardment induced roughening and smoothing of SiO_2 surfaces,' *Appl. Phys. Lett.* **62**, 363–365 (1993).
[73] E. Chason, T. M. Mayer, B. K. Kellerman, D. T. McIlroy and A. J. Howard, 'Roughening instability and evolution of the Ge(001) surface during ion sputtering,' *Phys. Rev. Lett.* **72**, 3040–3043 (1994).
[74] E. Chason, T. M. Mayer and A. Payne, 'In-situ energy dispersive X-ray reflectivity measurements of H ion bombardment on SiO_2/Si and Si,' *Appl. Phys. Lett.* **60**, 2353–2355 (1992).
[75] J. Chevrier, V. Le Thanh, R. Buys and J. Derrien, 'A RHEED study of epitaxial growth of iron on a silicon surface: Experimental evidence for kinetic roughening,' *Europhys. Lett.* **16**, 737–742 (1991).
[76] R. Chiarello, V. Panella, J. Krim and C. Thompson, 'X-ray reflectivity and adsorption isotherm study of fractal scaling in vapor-deposited films,' *Phys. Rev. Lett.* **67**, 3408–3411 (1991).
[77] S. T. Chui and J. D. Weeks, 'Dynamics of the roughening transition,' *Phys. Rev. Lett.* **40**, 733–736 (1978).
[78] M. Cieplak, A. Maritan and J. R. Banavar, 'Interfacial geometry and overhanging configurations,' *J. Phys. A* **27**, L765–L769 (1994).
[79] M. Cieplak and M. O. Robbins, 'Dynamical transition in quasistatic fluid invasion in porous media,' *Phys. Rev. Lett.* **60**, 2042–2045 (1988).
[80] M. Cieplak and M. O. Robbins, 'Influence of contact angle on quasistatic fluid invasion of porous media,' *Phys. Rev. B* **41**, 11508–11521 (1990).
[81] S. Clarke and D. Vvedensky, 'Origin of reflection high-energy electron-diffraction intensity oscillations during molecular-beam epitaxy: A computational modeling approach,' *Phys. Rev. Lett.* **58**, 2235–2238 (1987).
[82] J. Cook and B. Derrida, 'Directed polymers in a random medium: $1/d$ expansion,' *Europhys. Lett.* **10**, 195–199 (1989).
[83] J. Cook and B. Derrida, 'Directed polymers in a random medium: $1/d$ expansion and the n-tree approximation,' *J. Phys. A* **23**, 1523–1554 (1990).
[84] M. Copel, M. C. Reuter, M. Horn von Hoegen and R. M. Tromp, 'Influence of surfactants in Ge and Si epitaxy on Si(001),' *Phys. Rev. B* **42**, 11682–11689 (1990).
[85] M. Copel, M. C. Reuter, E. Kaxiras and R. M. Tromp, 'Surfactants in epitaxial growth,' *Phys. Rev. Lett.* **63**, 632–635 (1989).

[86] M. A. Cotta, R. A. Hamm, T. W. Staley, S. N. G. Chu, L. R. Harriott, M. B. Panish and H. Temkin, 'Kinetic surface roughening in molecular beam epitaxy of InP,' *Phys. Rev. Lett.* **70**, 4106–4109 (1993).
[87] Z. Csahók, K. Honda, E. Somfai, M. Vicsek and T. Vicsek, 'Dynamics of surface roughening in disordered media,' *Physica A* **200**, 136–154 (1993).
[88] Z. Csahók, K. Honda and T. Vicsek, 'Dynamics of surface roughening in disordered media' *J. Phys. A* **26**, L171–L178 (1993).
[89] Z. Csahók and T. Vicsek, 'Kinetic roughening in a model of segmentation of granular materials,' *Phys. Rev. A* **46**, 4577–4581 (1992).
[90] R. Cuerno and A.-L. Barabási, 'Dynamic scaling of ion-sputtered surfaces,' preprint 1994.
[91] G. Daccord, 'Chemical dissolution of a porous medium by a reactive fluid,' *Phys. Rev. Lett.* **58**, 479–482 (1987).
[92] G. Daccord and R. Lenormand, 'Fractal patterns from chemical dissolution,' *Nature* **325**, 41–43 (1987).
[93] S. Das Sarma, 'MBE growth as a self-organized critical phenomenon,' *J. Vac. Sci. Tech.* **B10**, 1695–1703 (1992).
[94] S. Das Sarma, 'Kinetic surface roughening and molecular beam epitaxy,' *Fractals* **1**, 784–794 (1993).
[95] S. Das Sarma and S. V. Ghaisas, 'Solid-on-solid rules and models for nonequilibrium growth in 2+1 dimensions,' *Phys. Rev. Lett.* **69**, 3762–3765 (1992).
[96] S. Das Sarma, S. V. Ghaisas and J. M. Kim, 'Kinetic super-roughening and anomalous dynamic scaling in nonequilibrium growth models,' *Phys. Rev. E* **49**, 122–125 (1994).
[97] S. Das Sarma and R. Kotlyar, 'Dynamical Renormalization Group Analysis of Fourth-Order Conserved Growth Nonlinearities,' *Phys. Rev. E* **50**, R4275–R4278 (1994).
[98] S. Das Sarma, C. J. Lanczycki, S. V. Ghaisas and J. M. Kim, 'Defect formation and crossover behavior in the dynamic scaling properties of molecular-beam epitaxy,' *Phys. Rev. B* **49**, 10693–10698 (1994).
[99] S. Das Sarma and P. Tamborenea, 'A new universality class for kinetic growth: One-dimensional molecular-beam epitaxy,' *Phys. Rev. Lett.* **66**, 325–328 (1991).
[100] R. H. Daushardt, F. Haubensak and R. O. Ritchie, 'On the interpretation of the fractal character of fracture surfaces,' *Acta Metall. Mater.* **38**, 143–159 (1990).
[101] B. Derrida and R. B. Griffiths, 'Directed polymers on disordered hierarchical lattices,' *Europhys. Lett.* **8**, 111–116 (1989).
[102] B. Derrida, J. L. Lebowitz, E. R. Speer and H. Spohn, 'Fluctuations of a stationary nonequilibrium interface,' *Phys. Rev. Lett.* **67**, 165–168 (1991).
[103] B. Derrida, J. L. Lebowitz, E. R. Speer and H. Spohn, 'Dynamics of an anchored Toom interface,' *J. Phys. A* **24**, 4805–4834 (1991).
[104] B. Derrida and H. Spohn, 'Polymers on disordered trees, spin glasses, and traveling waves,' *J. Stat. Phys.* **51**, 817–840 (1988).
[105] P. Devillard and H. Spohn, 'Kinetic shape of Ising clusters,' *Europhys. Lett.* **17**, 113–118 (1992).

[106] P. Devillard and H. E. Stanley, 'Scaling properties of Eden clusters in three and four dimensions,' *Physica A* **160**, 298–309 (1989).

[107] C. R. Doering and D. Ben-Avraham, 'Interparticle distribution functions and rate equations for diffusion-limited reactions,' *Phys. Rev. A* **38**, 3035–3042 (1988).

[108] M. Dong, M. C. Marchetti, A. A. Middleton and V. Vinokur, 'Elastic string in a random potential,' *Phys. Rev. Lett.* **70**, 662–665 (1993).

[109] D. J. Eaglesham and G. H. Gilmer, 'Roughening during Si deposition at low temperatures,' in *Surface disordering: Growth, roughening and phase transitions*, edited by R. Jullien, J. Kertész, P. Meakin and D. E. Wolf (Nova Science, New York, 1992), pp. 69–76.

[110] D. J. Eaglesham, F. C. Underwald and D. C. Jacobson, 'Growth morphology and the equilibrium shape: The role of surfactants' in Ge/Si island formation,' *Phys. Rev. Lett.* **70**, 966–969 (1993).

[111] M. Eden, in *Proceedings of the Fourth Berkeley Symposium on Mathematical Statistics and Probability, Volume IV: Biology and Problems of Health*, edited by J. Neyman (University of California Press, Berkeley, 1961), pp. 223–239.

[112] S. F. Edwards and D. R. Wilkinson, 'The Surface Statistics of a Granular Aggregate,' *Proc. R. Soc. London A* **381**, 17–31 (1982).

[113] G. Ehrlich and F. G. Hudda, 'Atomic view of surface self-diffusion: Tungsten on tungsten,' *J. Chem. Phys.* **44**, 1039–1099 (1966).

[114] E. A. Eklund, R. Bruinsma, J. Rudnick and R. S. Williams, 'Submicron-scale surface roughening induced by ion bombardment,' *Phys. Rev. Lett.* **67**, 1759–1762 (1991).

[115] H.-J. Ernst, F. Fabre, R. Folkerts and J. Lapujoulade, 'Observation of a growth instability during low temperature molecular beam epitaxy,' *Phys. Rev. Lett.* **72**, 112–115 (1994).

[116] H.-J. Ernst, F. Fabre and J. Lapujoulade, 'Nucleation and diffusion of Cu Adatoms on Cu(100): A helium-atom-beam scattering study,' *Phys. Rev. B* **46**, 1929–1932 (1992).

[117] D. Ertas and M. Kardar, 'Dynamic roughening of directed lines,' *Phys. Rev. Lett.* **69**, 929–932 (1992).

[118] D. Ertas and M. Kardar, 'Dynamic relaxation of drifting polymers: A phenomenological approach,' *Phys. Rev. E* **48**, 1228–1245 (1993).

[119] F. Family, 'Scaling of rough surfaces: Effects of surface diffusion,' *J. Phys. A* **19**, L441–L446 (1986).

[120] F. Family, 'Dynamic scaling and phase transitions in interface growth,' *Physica A* **168**, 561–581 (1990).

[121] F. Family and J. G. Amar, 'The morphology and evolution of the surface in epitaxial and thin film growth: A continuum model with surface diffusion,' *Fractals* **1**, 753–766 (1993).

[122] F. Family, K. C. B. Chan and J. Amar, 'Dynamics of interface roughening in imbibition,' in *Surface Disordering: Growth, Roughening and Phase Transitions*, edited by R. Jullien, J. Kertész, P. Meakin and D. E. Wolf (Nova Science, New York, 1992), pp. 205–212.

[123] F. Family and P. M. Lam, 'Renormalization-group analysis and simulational studies of groove instability in surface growth,' *Physica A* **205**, 272–283 (1994).

[124] F. Family and T. Vicsek, 'Scaling of the active zone in the Eden process on percolation networks and the ballistic deposition model,' *J. Phys. A* **18**, L75–L81 (1985).

[125] F. Family and T. Vicsek, eds. *Dynamics of Fractal Surfaces* (World Scientific, Singapore, 1991).

[126] J. Feder, *Fractals* (Plenum Press, New York, 1988).

[127] M. V. Feigel'man, 'Propagation of a plane front in an inhomogeneous medium,' *Sov. Phys. JETP* **58**, 1076–1077 (1983).

[128] J. F. Fernandez, G. Grinstein, Y. Imry and S. Kirpatrick, 'Numerical evidence for $d_c = 2$ in the random-field Ising model,' *Phys. Rev. Lett.* **51**, 203–206 (1983).

[129] D. S. Fisher, 'Interface fluctuations in disordered systems: 5-ϵ expansion and failure of dimensional reduction,' *Phys. Rev. Lett.* **56**, 1964–1967 (1986).

[130] D. S. Fisher and D. A. Huse, 'Directed paths in a random potential,' *Phys. Rev. B* **43** 10728–10742 (1991).

[131] B. M. Forrest and L. Tang, 'Hypercube-stackings: A Potts-spin model for surface growth,' *J. Stat. Phys.* **60**, 181–202 (1990).

[132] D. Forster, D. Nelson and M. Stephen, 'Large-distance and long-time properties of a randomly stirred fluid,' *Phys. Rev. A* **16**, 732–749 (1977).

[133] A. D. Fowler, H. E. Stanley and G. Daccord, 'Disequilibrium silicate mineral textures: Fractal and non-fractal features,' *Nature* **341**, 134–138 (1989).

[134] P. Freche, D. Stauffer and H. E. Stanley, 'Surface structure and anisotropy of Eden clusters,' *J. Phys. A* **18**, L1163–L1168 (1985).

[135] E. Frey and U. C. Täuber, 'Two-loop renormalization group analysis of the Burgers–Kardar–Parisi–Zhang equation,' *Phys. Rev. E* **50**, 1024–1045 (1994).

[136] U. Frisch, *Turbulence: The Legacy of A. N. Kolmogorov* (Cambridge University Press, Cambridge, 1995).

[137] U. Frisch and G. Parisi, in *Turbulence and Predictability in Geophysical Fluid Dynamics and Climate Dynamics*, edited by M. Ghil, R. Benzi and G. Parisi (North-Holland, Amsterdam, 1985).

[138] H. Fujikawa and M. Matsushita, 'Fractal growth of bacillus subtilis on agar plates,' *J. Phys. Soc. Japan* **58**, 2117–2120 (1989).

[139] L. Furuberg, J. Feder, A. Aharony and T. Jøssang, 'Dynamics of invasion percolation,' *Phys. Rev. Lett.* **61**, 2117–2120 (1988).

[140] L. Furuberg, A. Hansen, E. Hinrichsen, J. Feder and T. Jøssang, 'Scaling of overhang distribution of invasion percolation fronts,' *Physica Scripta.* **T38**, 91–94 (1991).

[141] D. Futer, A.-L. Barabási, S. V. Buldyrev, S. Havlin and H. Makse, 'Rough surfaces,' in *Fractals in Science*, edited by Boston University Education Group (Springer-Verlag, New York, 1994).

[142] P. Gács, 'A Toom rule that increases the thickness of sets,' *J. Stat. Phys.* **59**, 171–193 (1990).

[143] F. Gallet, S. Balibar and E. Rolley, 'The roughening transition of crystal surfaces. II. Experiments on static and dynamic properties near the first roughening transition of hcp ^{4}He,' *J. Physique* **48**, 369–377 (1987).

[144] C. W. Gardiner, *Handbook of Stochastic Methods* (Springer-Verlag, Berlin, 1985).

[145] S. V. Ghaisas and S. Das Sarma, 'Surface diffusion length under kinetic growth conditions,' *Phys. Rev. B* **46**, 7308–7311 (1992).

[146] S. C. Glotzer, M. F. Gyure, F. Sciortino, A. Coniglio and H. E. Stanley, 'Pinning in phase-separating systems,' *Phys. Rev. E* **49**, 247–258 (1994).

[147] L. Golubović and R. Bruinsma, 'Surface diffusion and fluctuations of growing interfaces,' *Phys. Rev. Lett.* **66**, 321–324 (1991).

[148] J. M. Gómez-Rodriguez, A. M. Baró and R. C. Salvarezza, 'Fractal characterization of gold deposits by scanning tunneling microscopy,' *J. Vac. Sci. Technol. B* **9**, 495–499 (1991).

[149] J.-F. Gouyet, M. Rosso and B. Sapoval, 'Fractal surfaces and interfaces,' in *Fractals and Disordered Systems*, edited by A. Bunde and S. Havlin (Springer-Verlag, Berlin, 1991).

[150] D. Grier, E. Ben-Jacob, R. Clarke and L. M. Sander, 'Morphology and microstructure in electrochemical deposition of zinc,' *Phys. Rev. Lett.* **56**, 1264–1267 (1986).

[151] G. Grinstein, 'Generic scale invariance and self-organized criticality,' in *Scale Invariance, Interfaces and Non-Equilibrium Dynamics*, Proc. 1994 NATO Adv. Study Inst. (Newton Institute, Cambridge [UK], 1994).

[152] G. Grinstein and D.-H. Lee, 'Generic scale invariance and roughening in noisy model sandpiles and other driven interfaces,' *Phys. Rev. Lett.* **66**, 177–180 (1991).

[153] G. Grinstein and S.-K. Ma, 'Surface tension, roughening and lower critical dimension in the random-field Ising model,' *Phys. Rev. B* **28**, 2588–2601 (1983).

[154] G. Grüner, 'The dynamics of charge-density waves,' *Rev. Mod. Phys.* **60**, 1129–1181 (1988).

[155] H. Guo, B. Grossmann and M. Grant, 'Kinetics of interface growth in driven systems,' *Phys. Rev. Lett.* **64**, 1262–1265 (1990).

[156] E. Guyon and H. E. Stanley, *Les Formes Fractales* (Palais de la Decouverte, Paris, 1991). [English Translation: *Fractal Forms* (Elsevier, Amsterdam, 1991).]

[157] T. Halpin-Healy, 'Diverse manifolds in random media,' *Phys. Rev. Lett.* **62**, 442–445 (1990).

[158] T. Halpin-Healy, 'Disorder-induced roughening of diverse manifolds,' *Phys. Rev. A* **42**, 711–722 (1990).

[159] T. Halpin-Healy and Y.-C. Zhang, 'Surface growth, directed polymers and all that,' *Phys. Rep.* **254**, 215–362 (1995).

[160] T. C. Halsey, M. H. Jensen, L. P. Kadanoff, I. Procaccia and B. Shraiman, 'Fractal measures and their singularities,' *Phys. Rev. A* **33**, 1141–1149 (1986).

[161] S. Havlin, L. A. N. Amaral, S. V. Buldyrev, S. T. Harrington and H. E. Stanley, 'Dynamics of surface roughening with quenched disorder,' 1994 preprint.

[162] S. Havlin, A.-L. Barabási, S. V. Buldyrev, C. K. Peng, M. Scwartz, H. E. Stanley and T. Vicsek, 'Anomalous interface roughening: Experiment and models,' in *Growth Patterns in Physical Sciences and Biology*, edited by E. Louis, L. Sander and P. Meakin, Proc. 1991 NATO

Advanced Research Workshop, Granada, Spain. (Plenum, New York, 1992), pp. 85–98.

[163] S. Havlin and D. Ben-Avraham, 'Diffusion in disordered media,' *Adv. Phys.* **36**, 695–798 (1987).

[164] S. Havlin, S. V. Buldyrev, H. E. Stanley and G. H. Weiss. 'Probability distribution of the interface width in surface roughening: Analogy with a Lévy flight,' *J. Phys. A* **24**, L925–L931 (1991).

[165] S. Havlin, R. Selinger, M. Schwartz, H.E. Stanley and A. Bunde, 'Random multiplicative processes and transport in structures with correlated spatial disorder,' *Phys. Rev. Lett.* **61**, 1438–1441 (1988).

[166] S. He, G. L. M. K. S. Kahanda and P.-z. Wong, 'Roughness of wetting fluid invasion fronts in porous media,' *Phys. Rev. Lett.* **69**, 3731–3734 (1992).

[167] Y.-L. He, H.-N. Yang, T.-M. Lu and G.-C. Wang, 'Measurements of dynamic scaling from epitaxial growth front: Fe film on Fe(001),' *Phys. Rev. Lett.* **69**, 3770–3773 (1992).

[168] P. E. Hegeman, H. J. W. Zandvliet, G. A. M. Kip and A. van Silfhout, 'Kinetic roughening of vicinal Si(001),' *Surf. Sci.* **311**, L655–L660 (1994).

[169] H. G. E. Hentschel, 'Shift invariance and surface growth,' *J. Phys. A* **27**, 2269–2275 (1994).

[170] H. G. E. Hentschel and F. Family, 'Scaling in open dissipative systems,' *Phys. Rev. Lett.* **66**, 1982–1985 (1991).

[171] H. G. E. Hentschel and I. Procaccia, 'The infinite number of generalized dimensions of fractals and strange attractors,' *Physica D* **8**, 435–444 (1983).

[172] P. Herrasti, P. Ocón, L. Vázquez, R. C. Salvarezza, J. M. Vara, A. J. Arvia, 'Scanning-tunneling-microscopy study on the growth mode of vapor-deposited gold films,' *Phys. Rev. A* **45**, 7440–7446 (1992).

[173] C. Herring, 'Effect of change of scale on sintering phenomena,' *J. Appl. Phys.* **21**, 301–303 (1950).

[174] H. J. Herrmann, 'Geometrical cluster growth models and kinetic gelation,' *Phys. Rep.* **136**, 153–227 (1986).

[175] H. J. Herrmann and S. Roux, *Statistical Models for the Fracture of Disordered Media* (North-Holland, Amsterdam, 1990).

[176] H. J. Herrmann and H. E. Stanley, 'The fractal dimension of the minimum path in two- and three-dimensional percolation,' *J. Phys. A* **21**, L829–L833 (1988).

[177] J. C. Heyraud and J. J. Metois, 'Growth slopes of metallic crystals and roughening transition,' J. Cryst. Growth **82**, 269–278 (1987).

[178] R. Hirsch and D. E. Wolf, 'Anisotropy and scaling of Eden clusters in two and three dimensions,' *J. Phys. A* **19**, L251–L256 (1986).

[179] P. C. Hohenberg and B. I. Halperin, 'Theory of dynamic critical phenomena,' *Rev. Mod. Phys.* **49**, 435–479 (1977).

[180] V. K. Horváth, F. Family and T. Vicsek, 'Anomalous noise distribution of the interface in two-phase fluid flow,' *Phys. Rev. Lett.* **67**, 3207–3210 (1991).

[181] V. K. Horváth, F. Family and T. Vicsek, 'Dynamic scaling of the interface in two-phase fluid flows,' *J. Phys. A* **24**, L25–L29 (1991).

[182] A. W. Hunt, C. Orme, D. R. M. Williams, B.G. Orr and L. M. Sander, 'Instabilities in MBE growth,' *Europhys. Lett.* **27**, 611–616 (1994).

[183] D. A. Huse, J. G. Amar and F. Family, 'Relationship between a generalized restricted solid-on-solid growth model and a continuum equation for interface growth,' *Phys. Rev. A* **41**, 7075–7077 (1990).

[184] D. A. Huse and C. L. Henley, 'Pinning and roughening of domain walls in Ising systems due to random impurities,' *Phys. Rev. Lett.* **54**, 2708–2711 (1985).

[185] D. A. Huse, C. L. Henley and D. S. Fisher, 'Huse, Henley and Fisher Respond,' *Phys. Rev. Lett.* **55**, 2924–2924 (1985).

[186] T. Hwa, *Statistical mechanics and dynamics of surfaces and membranes* (Ph.D. thesis, MIT, 1991).

[187] T. Hwa and E. Frey, 'Exact scaling function of interface growth dynamics,' *Phys. Rev. A* **44**, R7873–R7876 (1991).

[188] T. Hwa and M. Kardar, 'Dissipative transport in open systems: An investigation of self-organized criticality,' *Phys. Rev. Lett.* **62**, 1813–1816 (1989).

[189] T. Hwa and M. Kardar, 'Avalanches, hydrodynamics and discharge events in models of sandpiles,' *Phys. Rev. A* **45**, 7002–7021 (1992).

[190] T. Hwa, M. Kardar and M. Paczuski, 'Growth-induced roughening of crystalline facets,' *Phys. Rev. Lett.* **66**, 441–444 (1991).

[191] T. Hwa, M. Kardar and M. Paczuski, 'Reply to "Renormalization of the driven sine–Gordon equation in $2+1$ dimensions,"' *Phys. Rev. Lett.* **72**, 785–785 (1994).

[192] R. Q. Hwang and R. J. Behm, 'Scanning tunneling microscopy studies on the growth and structure of thin metallic films on metal substrates,' *J. Vac. Sci. Technol. B* **10**, 256–261 (1992).

[193] R. Q. Hwang, J. Schroder, G. Gunther and R. J. Behm, 'Fractal growth of two-dimensional islands: Au on Ru(0001),' *Phys. Rev. Lett.* **67**, 3279–3282 (1991).

[194] H. K. Janssen and B. Schmittmann, 'Field theory of long time behavior in driven diffusive systems,' *Z. Phys. B* **63**, 517–520 (1986).

[195] M. H. Jensen and I. Procaccia, 'Unusual exponents in interface roughening: The effect of pinning,' *J. de Physique II* **1**, 1139–1146 (1991).

[196] P. Jensen, A.-L. Barabási, H. Larralde, S. Havlin and H. E. Stanley, 'Controlling nanostructures,' *Nature* **368**, 22–22 (1994).

[197] P. Jensen, A.-L. Barabási, H. Larralde, S. Havlin and H. E. Stanley, 'Model incorporating deposition, diffusion, and aggregation in submonolayer nanostructures,' *Phys. Rev. E* **50**, 618–621 (1994).

[198] P. Jensen, A.-L. Barabási, H. Larralde, S. Havlin and H. E. Stanley, 'Connectivity of diffusing particles continually deposited on a surface: Relation to LECBD experiments' [Proc. of 1993 ETOPIM-3], *Physica A* **207**, 219–227 (1994).

[199] P. Jensen, A.-L. Barabási, H. Larralde, S. Havlin and H. E. Stanley, 'Deposition, diffusion and aggregation of atoms on surfaces: A model for nanostructure growth,' *Phys. Rev. B* **50**, 15316–15329 (1994).

[200] P. Jensen, P. Melinon, A. Hoareau, J. X. Hu, B. Cabaud, M. Treilleux, E. Bernstein and D. Guillot, 'Experimental achievement of 2D percolation

and cluster–cluster aggregation models by cluster deposition,' *Physica A* **185**, 104–110 (1992).

[201] H. Jeong, B. Kahng and D. Kim, 'Dynamics of a Toom interface in three dimensions,' *Phys. Rev. Lett.* **71**, 747–749 (1993).

[202] H. Ji and M. O. Robbins, 'Transition from compact to self-similar growth in disordered systems: Fluid invasion and magnetic domain growth,' *Phys. Rev. A* **44**, 2538–2542 (1991).

[203] H. Ji and M. O. Robbins, 'Percolative, self-affine, and faceted domain growth in random three-dimensional magnets,' *Phys. Rev. B* **46**, 14519–14527 (1992).

[204] Z. Jiang and H. G. E. Hentschel, 'Interfaces driven by quenched random fields,' *Phys. Rev. A* **45**, 4169–4172 (1992).

[205] M. D. Johnson, C. Orme, A. W. Hunt, D. Graff, J. Sudijono, L. M. Sander, and B. G. Orr, 'Stable and unstable growth in molecular beam epitaxy,' *Phys. Rev. Lett.* **72**, 116–119 (1994).

[206] R. Jullien and R. Botet, 'Scaling properties of the surface of the Eden model in $d = 2, 3, 4$,' *J. Phys. A* **18**, 2279–2287 (1985).

[207] R. Jullien and R. Botet, 'Surface thickness in the Eden model,' *Phys. Rev. Lett.* **54**, 2055–2055 (1985).

[208] R. Jullien, J. Kertész, P. Meakin and D. E. Wolf, eds., *Surface Disordering: Growth, Roughening and Phase Transitions* (Nova Science, New York, 1992).

[209] A. Juneja, D. P.Lathrop, K. R. Sreenivasan and G. Stolovitzky, 'Synthetic turbulence' *Phys. Rev. E* **49**, 5179–5194 (1994).

[210] L. P. Kadanoff, S. R. Nagel, L. Wu and S. M. Zhou, 'Scaling and universality in avalanches,' *Phys. Rev. A* **39**, 6524–6537 (1989).

[211] G. L. M. K. S. Kahanda, X.-q. Zou, R. Farell and P.-z. Wong, 'Columnar growth and kinetic roughening in electrochemical deposition,' *Phys. Rev. Lett.* **68**, 3741–3744 (1992).

[212] N. G. van Kampen, *Stochastic Processes in Physics and Chemistry* (North-Holland, Amsterdam, 1981).

[213] M. Kardar, 'Roughening by impurities at finite temperature,' *Phys. Rev. Lett.* **55**, 2923–2923 (1985).

[214] M. Kardar and D. Ertas, 'Nonequilibrium dynamics of fluctuating lines,' in *Scale Invariance, Interfaces and Non-Equilibrium Dynamics*, Proc. 1994 NATO Adv. Study Inst. (Newton Institute, Cambridge [UK], 1994).

[215] M. Kardar and J. O. Indekeu, 'Wetting of fractally rough surfaces,' *Phys. Rev. Lett.* **65**, 662–662 (1990).

[216] M. Kardar and J. O. Indekeu, 'Adsorption and wetting transition on rough substrates,' *Europhys. Lett.* **12**, 161–166 (1990).

[217] M. Kardar, G. Parisi and Y.-C. Zhang, 'Dynamic scaling of growing interfaces,' *Phys. Rev. Lett.* **56**, 889–892 (1986).

[218] M. Kardar and Y.-C. Zhang, 'Scaling of directed polymers in random media,' *Phys. Rev. Lett.* **58**, 2087–2090 (1987).

[219] M. Kardar and Y.-C. Zhang, 'Transfer matrix simulations of 2d-interfaces with three-dimensional random media,' *Europhys. Lett.* **8**, 233–238 (1989).

[220] P. Kablinski, A. Maritan, F. Toigo, J. Koplik and J. R. Banavar, 'Dynamics of rough surfaces with an arbitrary topology,' *Phys. Rev. E* **49** R937–R940 (1994).

[221] R. P. U. Karunasiri, R. Bruinsma and K. Rudnick, 'Thin-film growth and the shadow instability,' *Phys. Rev. Lett.* **62**, 788–791 (1989).

[222] J. Kertész, V. K. Horváth and F. Weber, 'Self-affine rupture lines in paper sheets,' *Fractals* **1**, 67–74 (1993).

[223] J. Kertész and T. Vicsek, 'Diffusion-limited aggregation and regular patterns: Fluctuations versus anisotropy,' *J. Phys. A* **19** L257–L260 (1986).

[224] J. Kertész and T. Vicsek, 'The growth of rough surfaces and interfaces,' in *Fractals in Science*, edited by A. Bunde and S. Havlin (Springer-Verlag, Berlin, 1994).

[225] J. Kertész and D. E. Wolf, 'Noise reduction in Eden models II: Surface structure and intrinsic width,' *J. Phys. A* **21**, 747–761 (1988).

[226] J. Kertész and D. E. Wolf, 'Anomalous roughening in growth processes,' *Phys. Rev. Lett.* **62**, 2571–2574 (1989).

[227] D. A. Kessler, H. Levine and L. M. Sander, 'Molecular-beam epitaxial growth and surface diffusion,' *Phys. Rev. Lett.* **69**, 100–103 (1992).

[228] D. Kessler, H. Levine and Y. Tu, 'Interface fluctuations in random media,' *Phys. Rev. A* **43**, 4551–4554 (1992).

[229] H. Kesten, *Percolation Theory for Mathematics* (Birkhäuser, Boston, 1982).

[230] J. M. Kim and S. Das Sarma, 'Discrete model for conserved growth equations,' *Phys. Rev. Lett.* **72**, 2903–2907 (1994).

[231] J. M. Kim and J. M. Kosterlitz, 'Growth in a restricted solid-on-solid model,' *Phys. Rev. Lett.* **62**, 2289–2292 (1989).

[232] J. M. Kim, M. A. Moore and A. J. Bray, 'Zero-temperature directed polymers in a random potential,' *Phys. Rev. A* **44**, 2345–2351 (1991).

[233] W. Kinzel, 'Directed percolation,' in *Percolation Structures and Processes*, edited by G. Deutscher, R. Zallen and J. Adler (A. Hilger, Bristol, 1983), pp. 425–446.

[234] B. Koiller, H. Ji and M. O. Robbins, 'Effect of disorder and lattice type on domain-wall motion in two dimensions,' *Phys. Rev. B* **46**, 5258–5265 (1992).

[235] B. Koiller, M. O. Robbins, H. Ji and C. S. Nolle, 'Morphology and dynamics of domain-wall motion in disordered two-dimensional magnets,' in *New Trends in Magnetic Materials and their Applications*, edited by J. L. Moran-Lopez and J. M. Sanchez (Plenum, New York, 1993).

[236] E. Kopatzky, S. Gunther, W. Nichtl-Pecher and R. J. Behm, 'Homoepitaxial growth on Ni(100) and its modification by a preadsorbed oxygen adlayer,' *Surf. Sci.* **284**, 154–166 (1993).

[237] J. M. Kosterlitz and D. J. Thouless, 'Ordering, metastability and phase transitions in two-dimensional systems,' *J. Phys. C* **6**, 1181–1203 (1973).

[238] M. Kotrla and A. C. Levi, 'Kinetic six-vertex model as model of bcc crystal growth,' *J. Stat. Phys.* **64**, 579–604 (1991).

[239] M. Kotrla and A. C. Levi, 'Kinetic roughness in the BCSOS model,' *J. Phys. A* **25**, 3121–3132 (1992).

[240] M. Kotrla and A. C. Levi, 'Kinetics of crystal growth near the roughening transition: A Monte Carlo study,' *Surf. Sci.* **317**, 183–193 (1994).
[241] M. Kotrla, A. C. Levi and P. Šmilauer, 'Roughness and nonlinearities in (2+1)-dimensional growth models with diffusion,' *Europhys. Lett.* **20**, 25–30 (1992).
[242] J. Krim, I. Heyvaert, C. Van Haesendock and Y. Bruynseraede, 'Scanning tunneling microscopy observation of self-affine fractal roughness in ion-bombarded film surfaces,' *Phys. Rev. Lett.* **70**, 57–60 (1993).
[243] J. Krim, D. H. Solina and R. Chiarello, 'Nanotribology of a Kr monolayer: A quartz-crystal microbalance study of atomic-scale friction,' *Phys. Rev. Lett.* **66**, 181–184 (1991).
[244] J. Krug, 'Kinetic roughening by exceptional fluctuations,' *J. Phys. I (France)* **1**, 9–12 (1991).
[245] J. Krug, 'Turbulent interfaces,' *Phys. Rev. Lett.* **72**, 2907–2910 (1994).
[246] J. Krug and P. Meakin, 'Microstructure and surface scaling in ballistic deposition at oblique incidence,' *Phys. Rev. A* **40**, 2064–2077 (1989).
[247] J. Krug and P. Meakin, 'Columnar growth in oblique incidence ballistic deposition: Faceting, noise reduction, and mean-field theory,' *Phys. Rev. A* **43**, 900–919 (1991).
[248] J. Krug and P. Meakin, 'Scaling properties of the shadowing model for sputter deposition,' *Phys. Rev. E* **47**, R17–R20 (1993).
[249] J. Krug, P. Meakin and T. Halpin-Healy, 'Amplitude universality for driven interfaces and directed polymers in random media,' *Phys. Rev. A* **45**, 638–653 (1992).
[250] J. Krug, M. Plischke and M. Siegert, 'Surface diffusion currents and the universality classes of growth,' *Phys. Rev. Lett.* **70**, 3271–3274 (1993).
[251] J. Krug and H. Spohn, 'Universal classes for deterministic surface growth,' *Phys. Rev. A* **38**, 4271–4283 (1988).
[252] J. Krug and H. Spohn, 'Anomalous fluctuations in the driven and damped Sine–Gordon chain,' *Europhys. Lett.* **8**, 219–224 (1989).
[253] J. Krug and H. Spohn, 'Mechanism for rough-to-rough transitions in surface growth,' *Phys. Rev. Lett.* **64**, 2232–2332 (1990).
[254] J. Krug and H. Spohn, 'Kinetic roughening of growing surfaces,' in *Solids Far From Equilibrium: Growth, Morphology and Defects*, edited by C. Godrèche (Cambridge University Press, Cambridge, 1991).
[255] R. Kunkel, B. Poelsema, L. K. Verheij and G. Comsa, 'Reentrant layer-by-layer growth during molecular-beam epitaxy of metal-on-metal substrates,' *Phys. Rev. Lett.* **65**, 733–736 (1990).
[256] M. G. Lagally, ed., *Kinetics of Ordering and Growth at Surfaces* (Plenum Press, New York, 1990).
[257] M. G. Lagally, 'Atom motion on surfaces' *Phys. Today* **46**, No. 11, 24-31 (1993).
[258] Z.-W. Lai and S. Das Sarma, 'Kinetic roughening with surface relaxation: Continuum versus atomistic models,' *Phys. Rev. Lett.* **66**, 2348–2351 (1991).
[259] C.-H. Lam and L. M. Sander, 'Surface growth with power-law noise,' *Phys. Rev. Lett.* **69**, 3338–3341 (1992).
[260] C.-H. Lam and L. M. Sander, 'Exact scaling in surface growth with power-law noise,' *Phys. Rev. E* **48**, 979–987 (1993).

[261] C.-H. Lam and L. M. Sander, 'Inverse method for interface problems,' *Phys. Rev. Lett.* **71**, 561–564 (1993).

[262] C.-H. Lam, L. M. Sander and D. E. Wolf, 'Surface growth with temporally correlated noise,' *Phys. Rev. A* **46**, R6128–R6131 (1992).

[263] P. M. Lam and F. Family, 'Dynamics of a height conserved surface growth model with spatially correlated noise,' *Phys. Rev. A* **44**, 4854–4860 (1991).

[264] C. J. Lanczycki and S. Das Sarma, 'Nonequilibrium influence of upward atomic mobility in one-dimensional molecular-beam epitaxy,' *Phys. Rev. E* **50**, 213–223 (1994).

[265] J. Lapujoulade, 'The roughening of metal surfaces,' *Surf. Sci. Rep.* **20**, 191–249 (1994).

[266] K. B. Lauritsen and H. C. Fogedby, 'Critical exponents from power spectra,' *J. Stat. Phys.* **72**, 189–205 (1993).

[267] M. J. Leaseburg and R. B. Pandey, 'Driven front and interface of a fluid-flow model in $2+1$ dimensions,' *Phys. Rev. E* **50**, 3730–3736 (1994).

[268] J. L. Lebowitz, C. Maes and E. R. Speer, 'Statistical mechanics of probabilistic cellular automata,' *J. Stat. Phys.* **59**, 117–170 (1990).

[269] J. Lee, S. Schwarzer, A. Coniglio and H. E. Stanley, 'Localization of growth sites in DLA clusters: Multifractality and multiscaling,' *Phys. Rev. E* **48**, 1305–1315 (1993).

[270] L. Lenormand, in *Fractal Forms*, edited by E. Guyon and H. E. Stanley (Elsevier/North Holland, Amsterdam, 1991).

[271] Y. Lereah, I. Zarudi, E. Grünbaum, G. Deutscher, S. V. Buldyrev and H. E. Stanley, 'Morphology of Ge:Al thin films: Experiments and model,' *Phys. Rev. E* **49**, 649–656 (1993).

[272] H. Leschhorn, 'Interface depinning in a disordered medium: Numerical results,' *Physica A* **195**, 324–335 (1993).

[273] H. Leschhorn, 'Grenzflächen in ungeordneten Medien,' Ph.D. Thesis, Ruhr University, Bochum, 1994.

[274] H. Leschhorn and L.-H. Tang, 'Comment on "Elastic string in a random potential",' *Phys. Rev. Lett.* **70**, 2973–2973 (1993).

[275] H. Leschhorn and L.-H. Tang, 'Avalanches and correlations in driven interface depinning,' *Phys. Rev. E* **49**, 1238–1245 (1994).

[276] D. Liu and M. Plischke, 'Universality in two- and three-dimensional growth and deposition models,' *Phys. Rev. B* **38**, 4781–4787 (1988).

[277] S. K. Ma, *Modern Theory of Critical Phenomena* (Benjamin/Cummings Publishing Company, Reading, 1976).

[278] S. K. Ma and G. F. Mazenko, 'Critical dynamics of ferromagnets in $6-\epsilon$ dimensions: General discussion and detailed calculation,' *Phys. Rev. B* **11**, 4077–4100 (1975).

[279] K. V. McCloud and J. V. Maher, 'Experimental perturbations to Saffman-Taylor flow,' *Phys. Rep.* **xx**, xxx (1995).

[280] H. Makse, A.-L. Barabási and H. E. Stanley, 'Elastic string in an anisotropic random medium near the depinning transition,' preprint, 1994.

[281] H. Makse, S. Havlin, H. E. Stanley and M. Schwartz, 'Novel method for generating long-range correlations,' *Chaos, Solitons, and Fractals* **6**, 295–303 (1995).

[282] H. Makse, S. Havlin, M. Schwartz and H. E. Stanley, 'Generating long-range correlations,' preprint, 1995.

[283] K. J. Måløy, J. Feder and T. Jøssang, 'Viscous fingering fractals in porous media,' *Phys. Rev. Lett.* **55**, 2688–2691 (1985).

[284] K. J. Måløy, A. Hansen, E. L. Hinrichsen and S. Roux, 'Experimental measurements of the roughness of brittle cracks,' *Phys. Rev. Lett.* **68**, 213–215 (1992).

[285] B. B. Mandelbrot, 'A fast fractional Gaussian noise generator,' *Water Resour. Res.* **7**, 543 (1971).

[286] B. B. Mandelbrot, 'Intermittent turbulence in self-similar cascades: Divergence of high moments and dimension of the carrier,' *J. Fluid. Mech.* **62**, 331–358 (1974).

[287] B. B. Mandelbrot, *The Fractal Geometry of Nature* (Freeman, San Francisco, 1982).

[288] B. B. Mandelbrot, 'Self-affine fractals and fractal dimension,' *Physica Scripta* **32**, 257–260 (1985).

[289] B. B. Mandelbrot, 'Self-affine fractal sets, I: The basic fractal dimensions,' in *Fractals in Physics*, edited by L. Pietronero and E. Tosatti (Elsevier Science Publishers, Amsterdam, 1986).

[290] R. N. Mantegna, S. V. Buldyrev, A. L. Goldberger, S. Havlin, C.-K. Peng, M. Simons and H. E. Stanley, 'Linguistic features of noncoding DNA sequences,' *Phys. Rev. Lett.* **73**, 3169–3172 (1994).

[291] A. Margolina and H. E. Warriner, 'Growth in a restricted solid on solid with correlated noise,' *J. Stat. Phys.* **60**, 809–821 (1990).

[292] A. Maritan, F. Toigo, J. Koplik and J. R. Banavar, 'Dynamics of growing interfaces,' *Phys. Rev. Lett.* **69**, 3193–3195 (1992).

[293] N. Martys, M. Cieplak and M. O. Robbins, 'Critical phenomena in fluid invasion of porous media,' *Phys. Rev. Lett.* **66**, 1058–1061 (1991).

[294] N. Martys, M. O. Robbins and M. Cieplak, 'Finite-size scaling studies of fluid invasion in porous media,' in *Extended Abstracts on Scaling in Disordered Materials: Fractal Structure and Dynamics*, edited by J. P. Stokes, M. O. Robbins and T. A. Witten (Materials Research Society, Pittsburgh, 1990), pp. 67–70.

[295] N. Martys, M. O. Robbins and M. Cieplak, 'Scaling relations for interface motion through disordered media: Application to fluid invasion,' *Phys. Rev. B* **44**, 12294–12306 (1991).

[296] M. Matsushita and H. Fujikawa, 'Diffusion-limited growth in bacterial colony formation,' *Physica A* **168**, 498–506 (1990).

[297] M. Matsushita, Y. Hayakawa and Y. Sawada, 'Fractal structure and cluster statistics of zinc-metal trees deposited on a line electrode,' *Phys. Rev. A* **32**, 3814–3820 (1985).

[298] M. Matsushita and S. Ouchi, 'On the self-affinity of various curves,' *Physica D* **38**, 246–251 (1989).

[299] S. Matsuura and S. Miyazima, 'Colony of fungus aspergillus oryzae and self-affine fractal geometry of growth fronts,' *Fractals* **1**, 11–19 (1993).

[300] S. Matsuura and S. Miyazima, 'Formation of ramified colony of fungus aspergillus oryzae on agar media,' *Fractals* **1**, 336–345 (1993).

[301] T. M. Mayer, E. Chason and A. J. Howard, 'Roughening instability and ion-induced viscous relaxation of SiO_2 surfaces,' *J. Appl. Phys.* **76**, 1633–1643 (1994).

[302] P. Meakin, 'Diffusion-controlled cluster formation in 2–6-dimensional space,' *Phys. Rev. A* **26**, 1495–1507 (1983).

[303] P. Meakin, 'Ballistic deposition onto inclined surfaces,' *Phys. Rev. A* **38**, 994–1003 (1988).

[304] P. Meakin, 'The growth of rough surfaces and interfaces,' *Phys. Rep.* **235**, 189–289 (1993).

[305] P. Meakin, A. Coniglio, H. E. Stanley and T. A. Witten, 'Scaling properties for the surfaces of fractal and nonfractal objects: An infinite hierarchy of critical exponents,' *Phys. Rev. A* **34**, 3325–3340 (1986).

[306] P. Meakin and R. Jullien, 'Simple ballistic deposition models for the formation of thin films,' *SPIE* **821**, 45–56 (1987).

[307] P. Meakin and R. Jullien, 'Restructuring effect in the rain model for random deposition,' *J. Physique* **48**, 1651–1662 (1987).

[308] P. Meakin and R. Jullien, 'Spatially correlated ballistic deposition,' *Europhys. Lett.* **9**, 71–76 (1989).

[309] P. Meakin and R. Jullien, 'Spatially correlated ballistic deposition on one- and two-dimensional surfaces,' *Phys. Rev. A* **41**, 983–993 (1990).

[310] P. Meakin, R. Jullien and R. Botet, 'Large-scale numerical investigation of the surface of Eden clusters,' *Europhys. Lett.* **1**, 609–615 (1986).

[311] P. Meakin and J. Krug, 'Scaling structure in simple screening models for columnar growth,' *Phys. Rev. A* **46**, 4654–4660 (1992).

[312] P. Meakin, P. Ramanlal, L. M. Sander and R. C. Ball. 'Ballistic deposition on surfaces,' *Phys. Rev. A* **34**, 5091–5103 (1986).

[313] P. Meakin, H. E. Stanley, A. Coniglio and T. A. Witten, 'Surfaces, interfaces, and screening of fractal structures,' *Phys. Rev. A* **32**, 2364–2369 (1985).

[314] E. Medina, T. Hwa, M. Kardar and Y.-C. Zhang, 'Burgers equation with correlated noise: Renormalization-group analysis and applications to directed polymers and interface growth,' *Phys. Rev. A* **39**, 3053–3075 (1989).

[315] C. Meneveau and K. R. Sreenivasan, 'Simple multifractal cascade model for fully developed turbulence,' *Phys. Rev. Lett.* **59**, 1424–1427 (1987).

[316] R. Messier and J. E. Yehoda, 'Geometry of thin-film morphology,' *J. Appl. Phys.* **58**, 3739–3746 (1985).

[317] H. Metiu, 'Epitaxial growth and the art of computer simulations,' *Science* **255**, 1088–1092 (1992).

[318] T. Michely, M. Hohage, M. Bott and G. Comsa, 'Inversion of growth speed anisotropy in two dimensions,' *Phys. Rev. Lett.* **70**, 3943–3946 (1993).

[319] J. J. de Miguel, A. Sanches, A. Cebollada, J. M. Gallego, J. Ferron and S. Ferrer, 'The surface morphology of a growing crystal studied by thermal energy atom scattering (TEAS),' *Surf. Sci.* **180/190**, 1062–1068 (1987).

[320] L. V. Mikheev, 'Linear mobility at the equilibrium roughening transition: Is the λ term relevant?' *Phys. Rev. Lett.* **71**, 2347–2347 (1993).

[321] D. J. Miller, K. E. Gray, R. T. Kampwirth and J. M. Murduck, 'Studies of growth instabilities and roughening in sputtered NbN films using a multilayer decoration technique,' *Europhys. Lett.* **19**, 27–32 (1992).

[322] V. Y. Milman, N. A. Stelmashenko and R. Blumenfeld, 'Fracture surfaces: A critical review of fractal studies and a novel morphological analysis of scanning tunneling microscopy measurements,' *Prog. Mat. Sci.* **38**, 425–474 (1994).

[323] S. Miyazima and H. E. Stanley, 'Intersection of two fractal objects: Useful method of estimating the fractal dimension,' *Phys. Rev. B* **35**, 8898–8900 (1987).

[324] Y.-W. Mo, J. Kleiner, M. B. Webb and M. G. Lagally, 'Activation energy for surface diffusion of Si on Si(001): A scanning tunneling microscopy study,' *Phys. Rev. Lett.* **66**, 1998–2001 (1991).

[325] Y.-W. Mo, J. Kleiner, M. B. Webb and M. G. Lagally, 'Surface self-diffusion of Si on Si(001),' *Surf. Sci.* **268**, 275–295 (1992).

[326] Y. W. Mo, J. Kleiner, M. B. Webb and M. G. Lagally, 'Reply to "Surface diffusion and island density,"' *Phys. Rev. Lett.* **69**, 986–986 (1992).

[327] E. W. Montroll and G. Weiss, 'Random walks on lattices. II,' *J. Math. Phys.* **6**, 167–181 (1965).

[328] M. A. Moore, T. Blum, J. P. Doherty, J.-P. Bouchaud and P. Claudin, 'Glassy solutions of the Kardar–Parisi–Zhang equation,' preprint, 1994.

[329] K. Moser, *Untersuchung rauher Oberflächen durch Simulation von Langevin-Gleichungen* (Diplomarbeit, Universität zu Köln, 1990).

[330] K. Moser, D. E. Wolf and J. Kertész, 'Numerical solution of the Kardar–Parisi–Zhang equation in one, two and three dimensions,' *Physica A* **178**, 215–226 (1991).

[331] K. Moser and D. E. Wolf, 'Kinetic roughening of vicinal surfaces,' in *Surface Disordering: Growth, Roughening and Phase Transitions*, edited by R. Jullien, J. Kertész, P. Meakin and D. E. Wolf (Nova Science, New York, 1992), pp. 21–30.

[332] B. A. Movchan and A. V. Demchishin, *Phys. Met. Metallorg. USSR* **28**, 83 (1969).

[333] W. W. Mullins, 'Theory of thermal grooving,' *J. Appl. Phys.* **28**, 333–339 (1957).

[334] T. Nagatani, 'Scaling structure in a simple growth model with screening: Forest formation model,' *J. Phys. A* **24**, L449–L454 (1991).

[335] T. Nagatani, J. Lee, and H. E. Stanley, 'Crossover effects in chemical-dissolution phenomena: A renormalization-group study,' *Phys. Rev. A* **45**, 2471–2479 (1992).

[336] O. Narayan and D. S. Fisher, 'Threshold critical dynamics of driven interfaces in random media,' *Phys. Rev. B* **48**, 7030–7042 (1993).

[337] T. Nattermann, 'Ising domain wall in a random pinning potential,' *J. Phys. C* **18**, 6661–6679 (1985).

[338] T. Nattermann and P. Rujan, 'Random field and other systems dominated by disorder fluctuations,' *Int. J. Mod. Phys.* **B3**, 1597–1654 (1989).

[339] T. Nattermann, S. Stepanow, L.-H. Tang and H. Leschhorn, 'Dynamics of interface depinning in a disordered medium,' *J. Physique II* **2**, 1483–1488 (1992).

[340] T. Nattermann and L.-H. Tang, 'Kinetic surface roughening: I. The Kardar–Parisi–Zhang equation in the weak coupling regime,' *Phys. Rev. A* **45**, 7156–7161 (1992).

[341] T. Nattermann and J. Villain, 'Random field Ising systems: A survey of current theoretical views,' *Phase Transitions* **11**, 5–51 (1988).

[342] J. H. Neave, P. J. Dobson, B. A. Joyce and J. Zhang, 'Reflection high-energy electron diffraction oscillations from vicinal surfaces: A new approach to surface diffusion measurements,' *Appl. Phys. Lett.* **47**, 100–102 (1985).

[343] M. Nelkin, 'Universality and scaling in fully developed turbulence,' *Adv. Phys.* **43**, 143–181 (1994).

[344] D. Nelson, T. Piran and S. Weinberg, eds., *Statistical Mechanics of Membranes and Surfaces* (World Scientific, Singapore, 1989).

[345] L. Niemeyer, L. Pietronero and H. J. Wiesmann, 'Fractal dimension of dielectric breakdown,' *Phys. Rev. Lett.* **52**, 1033–1036 (1984).

[346] J. Nittmann, G. Daccord and H. E. Stanley, 'Fractal growth of viscous fingers: A quantitative characterization of a fluid instability phenomenon,' *Nature* **314**, 141–144 (1985).

[347] J. Nittmann and H. E. Stanley, 'Tip splitting without interfacial tension and dendritic growth patterns arising from molecular anisotropy,' *Nature* **321**, 663–668 (1986).

[348] J. Nittmann and H. E. Stanley, 'Non-deterministic approach to anisotropic growth patterns with continuously tunable morphology: The fractal properties of some real snowflakes,' *J. Phys. A* **20**, L1185–L1191 (1987).

[349] J. Nittmann and H. E. Stanley, 'Role of fluctuations in viscous fingering and dendritic crystal growth: A noise-driven model with non-periodic sidebranching and no threshold for onset,' *J. Phys. A* **20**, L981–L986 (1987).

[350] C. S. Nolle, B. Koiller, N. Martys and M. O. Robbins, 'Morphology and dynamics of interfaces in random two-dimensional media,' *Phys. Rev. Lett.* **71**, 2074–2077 (1993).

[351] C. S. Nolle, B. Koiller, N. Martys and M. O. Robbins, 'Effect of quenched disorder on moving interfaces in two dimensions,' *Physica A* **205**, 342–354 (1994).

[352] P. Nozieres, 'Shape and growth of crystals,' in *Solids Far From Equilibrium: Growth, Morphology and Defects*, edited by C. Godrèche (Cambridge University Press, Cambridge, 1991).

[353] P. Nozieres and F. Gallet, 'The roughening transition of crystal surfaces. I. Static and dynamic renormalization theory, crystal shape and facet growth,' *J. Physique* **48**, 353–367 (1987).

[354] Z. Olami, I. Procaccia and R. Zeitak, 'Theory of self-organized interface depinning,' *Phys. Rev. E* **49**, 1232–1237 (1994).

[355] B. G. Orr, D. Kessler, C. W. Snyder and L. Sander, 'A model for strain-induced roughening and coherent island growth,' *Europhys. Lett.* **19**, 33–38 (1992).

[356] P. Ossadnik, 'Multiscaling analysis of large-scale off-lattice DLA,' *Physica A* **176**, 454–462 (1991).

[357] S. Pal and D. P. Landau, 'Monte Carlo simulation and dynamic scaling of surfaces in MBE growth,' *Phys. Rev. B* **49**, 10597–10606 (1994).

[358] V. Panella and J. Krim, 'Adsorption isotherm study of the fractal scaling behavior of vapor-deposited silver films,' *Phys. Rev. E* **49**, 4179–4183 (1994)

[359] G. Parisi, 'On surface growth in random media,' *Europhys. Lett.* **17**, 673–678 (1992).

[360] H. Park, A. Provata and S. Redner, 'Interface growth with competing surface currents,' *J. Phys. A* **24**, L1391–L1397 (1991).

[361] H.-O. Peitgen, H. Jürgens and D. Saupe, *Chaos and Fractals: New Frontiers of Science* (Springer-Verlag, New York, 1992).

[362] Y. P. Pellegrini and R. Jullien, 'Roughening transition and percolation in random ballistic deposition,' *Phys. Rev. Lett.* **64**, 1745–1748 (1990).

[363] Y. P. Pellegrini and R. Jullien, 'Kinetic roughening phase transition in surface growth: Numerical study and mean-field approach,' *Phys. Rev. A* **43**, 920–929 (1991).

[364] C.-K. Peng, S. Havlin, M. Schwartz and H. E. Stanley, 'Directed polymer and ballistic-deposition growth with correlated noise,' *Phys. Rev. A* **44**, R2239–R2242 (1991).

[365] C. K. Peng, S. Buldyrev, A. Goldberger, S. Havlin, F. Sciortino, M. Simons and H. E. Stanley, 'Long-range correlations in nucleotide sequences,' *Nature* **356**, 168–171 (1992).

[366] C. K. Peng, S. V. Buldyrev, S. Havlin, M. Simons, H. E. Stanley and A. L. Goldberger, 'Mosaic organization of DNA nucleotides,' *Phys. Rev. E* **49**, 1685–1689 (1994).

[367] P. Pfeifer and D. Avnir, 'Chemistry in noninteger dimensions between two and three. I. Fractal theory of heterogeneous surfaces,' *J. Chem. Phys.* **79**, 3558–3565 (1983).

[368] P. Pfeifer, M. W. Cole and J. Krim, 'Reply to "Wetting of fractally rough surfaces,"' *Phys. Rev. Lett.* **65**, 663–663 (1990).

[369] P. Pfeifer, Y. J. Wu, M. W. Cole and J. Krim, 'Multilayer adsorption on a fractally-rough surface,' *Phys. Rev. Lett.* **62**, 1997–2000 (1989).

[370] Y. H. Phang, D. E. Savage, R. Kariotis and M. G. Lagally, 'X-ray diffraction measurements of partially correlated interfacial roughness in multilayers,' *J. Appl. Phys.* **74**, 3181–3188 (1993).

[371] A. Pimpinelli, J. Villain and D. E. Wolf, 'Surface diffusion and island density,' *Phys. Rev. Lett.* **69**, 985–985 (1992).

[372] M. Plischke and Z. Rácz, 'Dynamic scaling on the surface structure of Eden clusters,' *Phys. Rev. A* **32**, 3825–3828 (1985).

[373] M. Plischke, Z. Rácz and D. Liu, 'Time-reversal invariance and universality of two-dimensional growth models,' *Phys. Rev. B* **35**, 3485–3495 (1987).

[374] M. Plischke, J. D. Shore, M. Schroeder, M. Siegert and D. E. Wolf, 'Comment on "Solid-on-solid rules and models for nonequilibrium growth in 2+1 dimensions",' *Phys. Rev. Lett.* **71**, 2509–2509 (1993).

[375] S. Prakash, S. Havlin, M. Schwartz and H. E. Stanley, 'Structural and dynamical properties of long-range correlated percolation,' *Phys. Rev. A* **46**, R1724–R1727 (1992).

[376] Z. Rácz, M. Siegert, D. Liu and M. Plischke, 'Scaling properties of driven interfaces: Symmetries, conservation laws, and the role of constraints,' *Phys. Rev. A* **43**, 5275–5283 (1991).

[377] Z. Rácz and T. Vicsek, 'Diffusion controlled deposition: Cluster statistics and scaling,' *Phys. Rev. Lett.* **51**, 2382–2385 (1983).

[378] C. Ratsch, A. Zangwill, P. Šmilauer and D. D. Vvedensky, 'Saturation and scaling of epitaxial island densities,' *Phys. Rev. Lett.* **72**, 3194–3197 (1994).

[379] M. O. Robbins, M. Cieplak, H. Ji, B. Koiller and N. Martys, 'Growth in systems with quenched disorder' in *Growth Patterns in Physical Sciences and Biology*, edited by J. M. Garcia-Ruiz, E. Louis, L. Sander and P. Meakin (Plenum, New York, 1993), pp. 65–75.

[380] H. Roder, E. Hahn, H. Brune, J.-P. Bucher and K. Kern, 'Building one- and two-dimensional nanostructures by diffusion-controlled aggregation at surfaces,' *Nature* **366**, 141–143 (1993).

[381] C. Roland and H. Guo, 'Interface Growth with a Shadow Instability,' *Phys. Rev. Lett.* **66**, 2104–2107 (1991).

[382] M. Rost and H. Spohn, 'Renormalization of the driven Sine-Gordon equation in 2+1 dimensions,' *Phys. Rev. E* **49**, 3709–3716 (1994).

[383] R. A. Roy and R. Messier, 'Preparation-physical structure relations in SiC sputtered films,' *J. Vac. Sci. Technol. A* **2**, 312–315 (1984).

[384] M. A. Rubio, C. A. Edwards, A. Dougherty and J. P. Gollub, 'Self-affine fractal interfaces from immiscible displacement in porous media,' *Phys. Rev. Lett.* **63**, 1685–1688 (1989).

[385] T. Salditt, T. H. Metzger and J. Peisl, 'Kinetic roughness of amorphous multilayers studied by diffuse X-ray scattering,' *Phys. Rev. Lett.* **73**, 2228–2231 (1994).

[386] T. Salditt, T. H. Metzger and J. Peisl, 'Characterisation of roughness correlations in W/Si multilayers by diffuse X-ray scattering,' *J. Physique III* **4**, C9–C171 (1994).

[387] T. Salditt, T. H. Metzger, J. Peisl and X. Jiang, 'Diffuse X-ray Scattering of Amorphous Multilayers,' *J. Physique III* **4**, 1573–1580 (1994).

[388] R. C. Salvarezza, L. Vázquez, P. Herrasti, P. Ocón, J. M. Vara and A. J. Arvia, 'Self-affine fractal vapour-deposited gold surfaces characterization by scanning-tunneling microscopy,' *Europhys. Lett.* **20**, 727–732 (1992).

[389] L. M. Sander and H. Yan, 'Temporal characteristics in nonequilibrium surface-growth models,' *Phys. Rev. A* **44**, 4885–4892 (1991).

[390] B. Schmittmann and R. K. P. Zia, 'Statistical mechanics of driven diffusive systems,' in *Phase Transitions and Critical Phenomena, Vol. 17*, edited by C. Domb and J. Lebowitz (Academic Press, London, 1994).

[391] M. Schroeder, M. Siegert, D. E. Wolf, J. D. Shore and M. Plischke, 'Scaling of growing surfaces with large local slopes,' *Europhys. Lett.* **24**, 563–568 (1993).

[392] M. Schwartz and S. F. Edwards, 'Nonlinear deposition: A new approach,' *Europhys. Lett.* **20**, 301–305 (1992).

[393] R. L. Schwoebel, 'Step motion on crystal surfaces II,' *J. Appl. Phys.* **40**, 614–619 (1968).
[394] R. L. Schwoebel and E. J. Shipsey, 'Step motion on crystal surfaces,' *J. Appl. Phys.* **37**, 3682–3686 (1966).
[395] R. B. Selinger, J. Nittmann and H. E. Stanley, 'Inhomogeneous diffusion-limited aggregation,' *Phys. Rev. A* **40**, 2590–2601 (1989).
[396] J. A. Shapiro. 'Organization of developing *Escherichia coli* colonies viewed by scanning electron microscopy,' *J. Bacteriol.* **169**, 142–156 (1987).
[397] M. Siegert and M. Plischke, 'Instability in surface growth with diffusion,' *Phys. Rev. Lett.* **68**, 2035–2038 (1992).
[398] M. Siegert and M. Plischke, 'Slope selection and coarsening in molecular beam epitaxy,' *Phys. Rev. Lett.* **73**, 1517–1520 (1994).
[399] P. Sigmund, 'Theory of sputtering: Sputtering yield of amorphous and polycrystalline targets,' *Phys. Rev.* **184**, 383–416 (1969).
[400] P. Sigmund, 'A mechanism of surface micro-roughening by ion bombardment,' *J. Matls. Sci.* **8**, 1545–1553 (1973).
[401] S. K. Sinha, E. B. Sirota, S. Garoff and H. E. Stanley, 'X-ray scattering from rough surfaces,' *Phys. Rev. B* **38**, 2297–2311 (1988).
[402] P. Šmilauer and M. Kotrla. 'Crossover effects in the Wolf-Villain model of epitaxial growth in $1+1$ and $2+1$ dimensions,' *Phys. Rev. B* **49**, 5769–5772 (1994).
[403] P. Šmilauer and M. Kotrla, 'Kinetic roughening in growth models with diffusion in higher dimensions,' *Europhys. Lett.* **27**, 261–266 (1994).
[404] P. Šmilauer and D. D. Vvedensky, 'Step edge barriers on GaAs(001),' *Phys. Rev. B* **48**, 17603–17606 (1993).
[405] P. Šmilauer, M. R. Wilby and D. D. Vvedensky, 'Reentrant layer-by-layer growth: A numerical study,' *Phys. Rev. B* **47**, 4119–4122 (1993).
[406] P. Šmilauer, M. R. Wilby and D. D. Vvedensky, 'Morphology of singular and vicinal metal surfaces sputtered at different temperatures,' *Surf. Sci. Lett.* **291**, L733–L738 (1993).
[407] G. W. Smith, A. J. Pidduck, C. R. Whitehouse, J. L. Glasper, A. M. Keir, and C. Pickering, 'Surface topography changes during the growth of GaAs by molecular beam epitaxy,' *Appl. Phys. Lett.* **59**, 3282–3284 (1991).
[408] M. von Smoluchowski, 'Versuch einer mathematischen Theorie der Koagulationskinetik kolloider Lösungen,' *Z. Phys. Chem.* **92**, 129 (1918).
[409] K. Sneppen, 'Self-organized pinning and interface growth in a random medium,' *Phys. Rev. Lett.* **69**, 3539–3542 (1992).
[410] K. Sneppen and M. H. Jensen, 'Reply to "Self-organized interface depinning",' *Phys. Rev. Lett.* **70**, 3833–3833 (1993).
[411] K. Sneppen and M. H. Jensen, 'Colored activity in self-organized critical interface dynamics,' *Phys. Rev. Lett.* **71**, 101–104 (1993).
[412] D. Spasojevic and P. Alstrøm. 'Interfacial growth in inhomogeneous media,' *Physica A* **201**, 482–495 (1993).
[413] J. L. Spouge, 'Exact solutions for a diffusion-reaction process in one dimension,' *Phys. Rev. Lett.* **60**, 871–874 (1988).
[414] H. E. Stanley, *Introduction to Phase Transitions and Critical Phenomena* (Oxford University Press, New York, 1971).

[415] H. E. Stanley, 'Role of Fluctuations in Fluid Mechanics and Dendritic Solidification,' *Phil. Mag. B* **56**, 665–686 (1987).

[416] H. E. Stanley, 'Fractal landscapes in physics and biology' [Thirtieth Saha Memorial Lecture, Calcutta], *Physica A* **186**, 1–32 (1992).

[417] H. E. Stanley, A. Bunde, S. Havlin, J. Lee, E. Roman and S. Schwarzer, 'Dynamic mechanisms of disorderly growth: Recent approaches to understanding diffusion limited aggregation,' *Physica A* **168**, 23–48 (1990).

[418] H. E. Stanley, A. Coniglio, S. Havlin, J. Lee, S. Schwarzer and M. Wolf, 'Diffusion limited aggregation: A paradigm of disorderly cluster growth,' *Physica A* **205**, 254–271 (1994).

[419] H. E. Stanley and P. Meakin, 'Multifractal phenomena in physics and chemistry,' *Nature* **335**, 405–409 (1988).

[420] H. E. Stanley and N. Ostrowsky, eds., *On Growth and Form: Fractal and Non-Fractal Patterns in Physics* (Martinus Nijhoff Publishers, Dordrecht, 1986).

[421] H. E. Stanley and N. Ostrowsky, eds., *Random Fluctuations and Pattern Growth: Experiments and Models* (Kluwer Academic Publishers, Dordrecht, 1988).

[422] H. E. Stanley and N. Ostrowsky, eds., *Correlations and Connectivity: Geometric Aspects of Physics, Chemistry and Biology* (Kluwer Academic Publishers, Dordrecht, 1990).

[423] H. E. Stanley, P. J. Reynolds, S. Redner and F. Family, 'Position-space renormalization group for models of linear polymers, branched polymers, and gels,' in *Real-Space Renormalization*, edited by T. W. Burkhardt and J. M. J. van Leeuwen (Springer-Verlag, Berlin, 1982), pp. 171–208.

[424] D. Stauffer and A. Aharony. *Introduction to Percolation Theory, 2nd Edition* (Taylor & Francis, London, 1992).

[425] D. Stauffer and H. E. Stanley, *From Newton to Mandelbrot: A Primer in Theoretical Physics* (Springer-Verlag, Berlin, 1990).

[426] D. G. Stearns, 'X-ray scattering from interfacial roughness in multilayer structures,' *J. Appl. Phys.* **71**, 4286–4298 (1992).

[427] J. P. Stokes, A. P. Kushnick and M. O. Robbins, 'Interface dynamics in porous media: A random-field description,' *Phys. Rev. Lett.* **60**, 1386–1389 (1988).

[428] S. Stoyanov, 'Layer growth of epitaxial films and superlattices,' *Surf. Sci.* **199**, 226–242 (1988).

[429] S. Stoyanov and D. Kaschiev, 'Thin film nucleation and growth theories: A confrontation with experiment,' *Current Topics in Material Sciences, Vol.* 7, edited by E. Kaldis (North-Holland, Amsterdam, 1981), pp. 69–141.

[430] J. A. Stroscio, D. T. Pierce and R. A. Dragoset, 'Homoepitaxial growth of iron and a real space view of reflection high energy electron diffraction,' *Phys. Rev. Lett.* **70**, 3615–3618 (1993).

[431] B. Suki, A.-L. Barabási, Z. Hantos, F. Peták and H. E. Stanley, 'Avalanches and power law behaviour in lung inflation,' *Nature* **368**, 615–618 (1994).

[432] T. Sun, H. Guo and M. Grant, 'Dynamics of driven interfaces with a conservation law,' *Phys. Rev. A* **40**, 6763–6766 (1989).

[433] T. Sun and M. Plischke, 'Field-theory renormalization approach to the Kardar–Parisi–Zhang equation,' *Phys. Rev. E* **49**, 5046–5057 (1994).

[434] D. N. Sutherland, 'Comment on Vold's simulation of floc formation,' *J. Colloid Interface Sci.* **22**, 300–302 (1966).

[435] J. Szép, J. Cserti and J. Kertész, 'Monte Carlo approach to dendritic growth,' *J. Phys. A* **18**, L413–L418 (1985).

[436] H. Takayasu, *Fractals in the Physical Sciences* (Manchester University Press, Manchester, 1990).

[437] P. Tamborenea and S. Das Sarma, 'Surface-diffusion-driven kinetic growth on one-dimensional substrates,' *Phys. Rev. E* **48**, 2575–2594 (1993).

[438] C. Tang, 'Diffusion-limited aggregation and the Saffman–Taylor problem,' *Phys. Rev. A* **31**, 1977–1979 (1985).

[439] C. Tang, S. Alexander and R. Bruinsma, 'Scaling for the growth of amorphous films,' *Phys. Rev. Lett.* **64**, 772–775 (1990).

[440] L.-H. Tang, 'Island formation in submonolayer epitaxy,' *J. Physique I* **3**, 935–950 (1993).

[441] L.-H. Tang, B. M. Forrest and D. E. Wolf, 'Kinetic surface roughening. II. Hypercube-stacking models,' *Phys. Rev. A* **45**, 7162–7179 (1992).

[442] L.-H. Tang, M. Kardar and D. Dhar, 'Driven depinning in anisotropic media,' *Phys. Rev. Lett.* **74**, xxx (1995).

[443] L.-H. Tang and H. Leschhorn, 'Pinning by directed percolation,' *Phys. Rev. A* **45**, R8309–R8312 (1992).

[444] L.-H. Tang and H. Leschhorn, 'Self-organized interface depinning,' *Phys. Rev. Lett.* **70**, 3832–3832 (1993).

[445] L.-H. Tang and T. Nattermann, 'Kinetic roughening in molecular-beam epitaxy' *Phys. Rev. Lett.* **66**, 2899–2902 (1991).

[446] C. Thompson, C. Palasantzas, Y. P. Feng, S. K. Sinha and J. Krim, 'X-ray reflectivity study of the growth kinetics of vapor-deposited silver films,' *Phys. Rev. B* **49**, 4902–4907 (1994).

[447] W. M. Tong, E. J. Snyder, R. S. Williams, A. Yanase, Y. Segawa and M. S. Anderson, 'Atomic force microscope studies of CuCl island formation on $CaF_2(111)$ substrates,' *Surf. Sci. Lett.* **277**, L63–L69 (1992).

[448] W. M. Tong and R. S. Williams, 'Kinetics of surface growth: Phenomenology, scaling and mechanisms of smoothening and roughening,' *Ann. Rev. Phys. Chem.* **45**, 401–438 (1994).

[449] W. M. Tong, R. S. Williams, A. Yanase, Y. Segawa and M. S. Anderson, 'Atomic force microscope studies of growth kinetics: Scaling in the heteroepitaxy of CuCl on $CaF_2(111)$,' *Phys. Rev. Lett.* **72**, 3374–3377 (1994).

[450] A. L. Toom, 'Stable and attractive trajectories in multicomponent systems,' in *Multicomponent Random Systems*, edited by R. L. Dobrushin and Ya. G. Sinai (Marcel Dekker, New York, 1980).

[451] T. T. Tsong, 'Experimental studies of the behaviour of single adsorbed atoms on solid surfaces,' *Rep. Prog. Phys.* **51**, 759–832 (1988).

[452] Y. Tu, 'Instability in a continuum kinetic-growth model with surface relaxation,' *Phys. Rev. A* **46**, R729–R732 (1992).

[453] Y. Tu, 'Absence of finite upper critical dimension in the spherical KPZ model,' *Phys. Rev. Lett.* **73**, 3109–3112 (1994).

[454] P. A. Tydeman and A. M. Hiron, 'The tensile rupture of paper,' *B.P. & B.I.R.A. Bulletin* **35**, 9–21 (1964).

[455] J. A. Venables, G. D. T. Spiller and M. Hanbrucken, 'Nucleation and growth of thin films,' *Rep. Prog. Phys.* **47**, 300–459 (1984).

[456] T. Vicsek, *Fractal Growth Phenomena, 2nd Edition* (World Scientific, Singapore, 1992).

[457] T. Vicsek, 'Kinetic roughening with multiplicative noise,' in *Surface Disordering: Growth, Roughening and Phase Transitions*, edited by R. Jullien, J. Kertész, P. Meakin and D. E. Wolf, (Nova Science, New York, 1992), pp. 155–162.

[458] T. Vicsek and A.-L. Barabási, 'Multi-affine model for the velocity distribution in fully turbulent flows,' *J. Phys. A* **24**, L845–L851 (1991).

[459] T. Vicsek, M. Cserzö and V. K. Horváth. 'Self-affine growth of bacterial colonies,' *Physica A* **167**, 315–321 (1990).

[460] T. Vicsek, A. Czirók, O. Shochet and E. Ben-Jacob. 'Self-affine Roughening of Bacteria Colony Surfaces,' in *Proc. NATO Advanced Research Workshop on Spatio-Temporal Patterns in Nonequilibrium Complex Systems* (1993).

[461] T. Vicsek, E. Somfai and M. Vicsek, 'Kinetic roughening with multiplicative noise,' *J. Phys. A* **25**, L763–L768 (1992).

[462] J. Villain, 'Nonequilibrium "critical" exponents in the random-field Ising model,' *Phys. Rev. Lett.* **52**, 1543–1546 (1984).

[463] J. Villain, 'Continuum models of crystal growth from atomic beams with and without desorption,' *J. Phys. I* **1**, 19–42 (1991).

[464] J. Villain, A. Pimpinelli, L. Tang and D. Wolf, 'Terrace sizes in molecular beam epitaxy,' *J. Phys. I France* **2**, 2107–2121 (1992).

[465] J. Villain, A. Pimpinelli and D. Wolf, 'Layer by layer growth in molecular beam epitaxy,' *Comm. Cond. Mat. Phys.* **16**, 1–18 (1992).

[466] M. J. Vold, 'A numerical approach to the problem of sediment volume,' *J. Coll. Sci.* **14**, 168–174 (1959).

[467] M. J. Vold. 'Sediment volume and structure in dispersions of anisotropic particles,' *J. Phys. Chem.* **63**, 1608–1612 (1959).

[468] R. F. Voss, 'Random fractals: Characterisation and measurement,' in *Scaling Phenomena in Disordered Systems*, edited by R. Pynn and A. Skjeltorp (Plenum Press, New York, 1985).

[469] R. F. Voss, in *Fundamental Algorithms in Computer Graphics*, edited by R. Earnshaw (Springer-Verlag, Berlin, 1985), pp. 805–835.

[470] D. D. Vvedensky, A. Zangwill, C. N. Luse and M. R. Wilby, 'Stochastic equations of motion for epitaxial growth,' *Phys. Rev. E* **48**, 852–861 (1993).

[471] S. C. Wang and G. Ehrlich, 'Atom incorporation at surface clusters: An atomic view,' *Phys. Rev. Lett.* **67**, 2509–2512 (1991).

[472] S. C. Wang and G. Ehrlich, 'Adatom motion to lattice steps: A direct view,' *Phys. Rev. Lett.* **70**, 41–44 (1993).

[473] S. C. Wang and G. Ehrlich, 'Atom condensation at lattice steps and clusters,' *Phys. Rev. Lett.* **71**, 4174–4177 (1993).

[474] W. Weber and B. Lengeler, 'Diffuse scattering of hard X-rays from rough surfaces,' *Phys. Rev. B* **46**, 7953–7956 (1992).

[475] J. D. Weeks and G. H. Gilmer, 'Dynamics of crystal growth,' *Adv. Chem. Phys.* **40**, 157–228 (1979).

[476] J. D. Weeks, G. H. Gilmer and K. Jackson, 'Analytical theory of crystal growth,' *J. Chem. Phys.* **65**, 712–720 (1976).

[477] M. R. Wilby, D. D. Vvedensky and A. Zangwill, 'Scaling in a solid-on-solid model of epitaxial growth,' *Phys. Rev. B* **46**, 12896–12898 (1992).

[478] E. D. Williams and N. C. Bartelt, 'Thermodynamics of surface morphology,' *Science* **251**, 393–400 (1991).

[479] K. G. Wilson and J. Kogut, 'The renormalization group and the ϵ expansion,' *Phys. Rep.* **12**, 75–200 (1974).

[480] T. A. Witten and L. M. Sander, 'Diffusion-limited aggregation,' *Phys. Rev. Lett.* **47**, 1400–1403 (1981).

[481] T. A. Witten and L. M. Sander, 'Diffusion-limited aggregation: A kinetic critical phenomenon,' *Phys. Rev. B* **27**, 5686–5697 (1983).

[482] D. E. Wolf, 'Kinetic roughening of vicinal surfaces,' *Phys. Rev. Lett.* **67**, 1783–1786 (1991).

[483] D. E. Wolf, 'Kinetic roughening,' *IFF-Bulletin* **39**, 4–33 (1991).

[484] D. E. Wolf, 'Computer simulation of molecular beam epitaxy,' in *Scale Invariance, Interfaces and Non-Equilibrium Dynamics*, Proc. 1994 NATO Adv. Study Inst. (Newton Institute, Cambridge [UK], 1994).

[485] D. E. Wolf and J. Kertész, 'Noise reduction in Eden models I,' *J. Phys. A* **20**, L257–L261 (1987).

[486] D. E. Wolf and J. Kertész, 'Surface width exponents for three- and four-dimensional Eden growth,' *Europhys. Lett.* **4**, 651–656 (1987).

[487] D. E. Wolf and L.-H. Tang, 'Inhomogeneous growth processes,' *Phys. Rev. Lett.* **65**, 1591–1594 (1990).

[488] D. E. Wolf and J. Villain, 'Growth with surface diffusion,' *Europhys. Lett.* **13**, 389–394 (1990).

[489] P. E. Wolf, F. Gallet, S. Balibar, E. Rolley and P. Nozieres, 'Crystal growth and crystal curvature near roughening transition in hcp4He,' *J. Physique* **46**, 1987–2007 (1985).

[490] P.-z. Wong, 'Flow in porous media: Permeability and displacement patterns,' *Mat. Res. Soc. Bulletin* **19**, 32–38 (1994).

[491] P.-z. Wong and A. J. Bray, 'Scattering by rough surfaces,' *Phys. Rev. B* **37**, 7751–7758 (1988).

[492] D. P. Woodruff, *The Solid-Liquid Interface* (Cambridge University Press, London, 1973).

[493] H. Yan. 'Kinetic growth with surface diffusion: The scaling aspect,' *Phys. Rev. Lett.* **68**, 3048–3051 (1992).

[494] H. Yan, D. Kessler and L. M. Sander, 'Roughening phase transition in surface growth,' *Phys. Rev. Lett.* **64**, 926–929 (1990).

[495] H.-N. Yang, T.-M. Lu and G.-C. Wang, 'High-resolution low-energy electron diffraction study of Pb(110) surface roughening transition,' *Phys. Rev. Lett.* **63**, 1621–1624 (1989).

[496] H.-N. Yang, T.-M. Lu and G.-C. Wang, 'Time-invariant structure factor in an epitaxial growth front,' *Phys. Rev. Lett.* **68**, 2612–2615 (1992).

[497] H.-N. Yang, T.-M. Lu and G.-C. Wang, 'Diffraction from a surface growth front,' *Phys. Rev. B* **47**, 3911–3922 (1993).

[498] H.-N. Yang, G.-C. Wang and T.-M. Lu. *Diffraction from Rough Surfaces and Dynamic Growth Fronts* (World Scientific, Singapore, 1994).

[499] J. H. Yao, C. Roland and H. Guo, 'Interfacial dynamics with long-range screening,' *Phys. Rev. A* **45**, 3903–3912 (1992).

[500] H. You, R. P. Chiarello, H. K. Kim and K. G. Vandervoort, 'X-ray reflectivity and scanning-tunneling-microscope study of kinetic roughening of sputter-deposited gold films during growth,' *Phys. Rev. Lett.* **70**, 2900–2903 (1993).

[501] J. G. Zabolitzky and D. Stauffer, 'Simulation of large Eden clusters,' *Phys. Rev.* **34**, 1523–1530 (1986).

[502] A. Zangwill, *Physics at Surfaces* (Cambridge University Press, Cambridge, 1988).

[503] P. Zeppenfeld, K. Kern, R. David and G. Comsa, 'No thermal roughening on Cu(110) up to 900 K,' *Phys. Rev. Lett.* **62**, 63–66 (1989).

[504] J. Zhang, Y.-C. Zhang, P. Alstrøm and M. T. Levinsen, 'Modeling forest fire by a paper-burning experiment, a realization of the interface growth mechanism,' *Physica A* **189**, 383–389 (1992).

[505] Y.-C. Zhang, 'Replica scaling analysis of interfaces in random media,' *Phys. Rev. A* **42**, 4897–4900 (1990).

[506] Y.-C. Zhang, 'Growth anomaly and its implications,' *Physica A* **170**, 1–13 (1990).

[507] Y.-C. Zhang, 'Non-universal roughening of kinetic self-affine interfaces,' *J. Physique* **51**, 2129–2134 (1990).

[508] J.-K. Zuo, J. F. Wendelken, H. Durr and C.-L. Liu, 'Growth and coalescence in submonolayer homoepitaxy on Cu(100) studied with high-resolution low-energy electron diffraction,' *Phys. Rev. Lett.* **72**, 3064–3067 (1994).

Index